Views of Nature and Dualism

Thomas John Hastings • Knut-Willy Sæther
Editors

Views of Nature and Dualism

Rethinking Philosophical, Theological, and Religious Assumptions in the Anthropocene

palgrave
macmillan

Editors
Thomas John Hastings
International Bulletin of Mission Research, Princeton Theological Seminary, Princeton, NJ, USA

Knut-Willy Sæther
Department of Religious Studies
Volda University College
Volda, Norway

ISBN 978-3-031-42901-9 ISBN 978-3-031-42902-6 (eBook)
https://doi.org/10.1007/978-3-031-42902-6

This Palgrave Macmillan imprint is published by the registered company Springer Nature Switzerland AG.
The registered company address is: Gewerbestrasse 11, 6330 Cham, Switzerland

Paper in this product is recyclable.

Preface and Acknowledgements

In our previous book, *The Grace of Being Fallible in Philosophy, Theology and Religion* (2020), our aim was to explore how fallibilism as an epistemological position and orientation to truth claims might help fund more spacious approaches in philosophy, theology, and religion. Our current volume is a follow-up to the first volume. Again, the topic was developed over time by Tom and Knut-Willy's continuing conversations. In March 2022, we invited a small group of European and American scholars to a research symposium, entitled "Views of Nature and Dualism." The symposium was hosted by OMSC at Princeton Theological Seminary and the international research group "Nature, Aesthetics, and Ethics," based at Volda University College and led by Knut-Willy. Our gathering in Princeton was the first step to the contributions in this volume. We wish to express our thanks to Volda University College and OMSC@PTS for supporting our meetings and the process that made this book possible.

Princeton, NJ, USA — Thomas John Hastings
Volda, Norway — Knut-Willy Sæther

Contents

Notes on Contributors

Lisa E. Dahill, Miriam Therese Winter Professor of Transformative Leadership and Spirituality, Hartford International University for Religion & Peace.

Joel D. Daniels, Protestant Interfaith Chaplain at the Claremont Colleges and Research Fellow at Claremont Graduate University.

Caleb Gordon, Associate Tutor, Lindisfarne College of Theology, UK.

Mark Graves, Research Associate Professor, School of Psychology, Fuller Theological Seminary, Pasadena, California, USA.

Thomas John Hastings, Former Executive Director of Overseas Ministries Study Center, Princeton Theological Seminary, New Jersey, USA. Editor of *International Bulletin of Mission Research.*

Lina Langby, PhD Candidate, Department of Theology, Uppsala University, Sweden.

Jonas Gamborg Lillebø, Associate Professor, Department of Religious Studies, Volda University College, Norway.

Hannah Malcolm, Curate in the diocese of Newcastle and PhD Candidate in the Department of Theology and Religion, Durham University, UK.

Elaine Padilla, Professor of Philosophy and Religion, LatinX/Latin American Studies, University of La Verne, California, USA.

Knut-Willy Sæther, Professor, Department of Religious Studies, Volda University College, Norway.

Victor Thasiah, Professor of Religion, California Lutheran University, California, USA.

Dag Erik Wold, PhD Candidate and Assistant Professor, Department of Physical Education, Volda University College, Norway.

CHAPTER 1

Attending to the Human-Nature Relationship: Approaches, Contexts, and Challenges

Thomas John Hastings and Knut-Willy Sæther

In the face of the anthropogenic threats to the singular planetary habitat we share with other human beings and non-human species, humanities scholars feel a renewed sense of urgency (1) to acknowledge the ways our species has funded particular histories of environmental exploitation, alienation, and collapse, (2) to unpack inherited assumptions that impact our views of nature and interspecies relations, and (3) to suggest ways of thinking and acting that seek to repair the damage and promote mutual flourishing for all of earth's inhabitants. This volume brings together

T. J. Hastings (✉)
International Bulletin of Mission Research, Princeton Theological Seminary, Princeton, NJ, USA
e-mail: thomas.hastings@ptsem.edu

K.-W. Sæther
Department of Religious Studies, Volda University College, Volda, Norway
e-mail: knut-willy.saether@hivolda.no

T. J. Hastings, K.-W. Sæther (eds.), *Views of Nature and Dualism*,
https://doi.org/10.1007/978-3-031-42902-6_1

scholars in philosophy, theology, and religion who take up this urgent ethical task from a broad range of perspectives and locations.

Given this overreaching perspective, we direct our attention to views of nature and dualism, and in particular the relationship between human beings and non-human nature. Undoubtedly, the title of the book brings together two highly contested notions: "nature" and "dualism." Our task is not to provide an exhaustive exploration of these notions as they surface in different academic fields or lived contexts, but to shed light on how human encounters with nature may be more adequately elaborated in the Anthropocene. And the "views" in the title was born of our invitation to a diverse group of colleagues to attend to this encounter in the hope that paying closer attention may be both self-involving and self-transcending. The underlying context for our volume is the current debate on environmental issues, where we recognize a gap between the need to emphasize a renewed relationship with nature while, at the same time, we struggle with feelings of alienation in our encounters with nature.

To shed new light on views of nature and dualism, we acknowledge the importance of giving voice to more theoretically grounded research as well as to studies addressing concrete contexts. There are a number of important questions related to "views of nature and dualism," and we find research that creatively combines both fruitful and desirable. In this introductory chapter, we will address our epistemological point of departure, followed by an identification and problematization of various notions of nature and dualism. In light of the environmental challenge, we argue that views of nature and dualism need to be addressed in specific contexts. Our elaboration of this problematic is followed by an exploration of dualism in religion and a problematization of human-nature dualism.

Epistemological Point of Departure

In our previous volume, *The Grace of Being Fallible in Philosophy, Theology, and Religion* (2020), we explored the many faces of fallibilism. There we acknowledged that our search for knowledge is characterized by "verisimilitude" and found the notion of constructive critical realism a fruitful epistemological point of departure.[1] That book addressed particular

[1] Thomas John Hastings and Knut-Willy Sæther, *The Grace to be Fallible in Philosophy, Theology, and Religion* (New York: Palgrave Macmillan, 2020), 4.

epistemological questions and topics related to fallibilism. Our current volume has the same epistemological basis as presented in the introductory chapter of *The Grace of Being Fallible*. However, we are now turning our attention to topics of how to understand nature per se and the human-nature relationship, hence in one sense, we are moving from epistemology in the direction of ontology. In this respect, we find it helpful to express a relationship between the two volumes in terms of *epistemology mirroring ontology*. It is important to note that ontology exercises marginal control over epistemology. Such a formulation is not contradictory but expresses a core element of a constructive critical realism.

Our encounter with nature and how we articulate knowledge and understanding of nature, in which we ourselves are instantiated, calls for cautious navigation and awareness of the limits of our explorations. At the same time, a fallibilist approach, which allows that we may have gotten it wrong, does not mean succumbing to relativism. In our current situation of global environmental threats, relativism and fatalism are paths we wish to avoid. Instead, we suggest ways of understanding nature and, in particular, the human encounter with nature, which can contribute to repairing the damage and promoting the mutual flourishing of all of earth's inhabitants. Hence, the unavoidably intertwined relationship between humans and nature needs to be reconsidered and examined with care. This is our main task in this volume.

A fallible consciousness as a condition for our explorations does not mean withdrawing from challenges that are urgent in our time, but it is acknowledged as a necessary point of departure for the contributors to this volume. In chapter 3, Lina Langby expresses it this way: "Human beings are fallible by nature, and no matter what worldview we believe in or try to act by, we cannot guarantee ethically good behavior on our part." Here, Langby touches on the need for critical self-reflection. In our attempts to attend to reality as a whole in terms of views of nature, and thus to gain new insights with relevance for understanding the relationship between humans and nature, there is no guarantee, on the one hand, that such insights will necessarily issue in right action. On the other hand, we believe that a deeper understanding of the complexities of our world might help us in doing what is right while being aware that knowing the right does not necessarily lead to right actions. This tension or challenge is touched on in Sæther's chapter, where he explores how perspectives from ecospirituality may be helpful.

The Ambiguity of Nature and Dualism

Notions of nature and dualism are ambiguous. We are not going to elaborate these in depth here but only emphasize a few points relevant to our task. We tend to hold an approach to and an understanding of nature as the non-human nature "surrounding" us. This might seem at first glance to be a given understanding and one that accepts some sort of gap or even a duality between humans and nature. But this is not the case as we see it. Rather, it is exactly this topic that we attempt to address and problematize. Nature does not exist at some meta level. Nature is concrete, observable, and explorable, and we ourselves inhabit this world and are surrounded by and interwoven with nature. In addition, informed by different religions and, for us, in particular the Christian milieu, humans also transcend nature, which implies notions of spirituality, i.e., having faith in something more than nature.

At first glance, our use of dualism in this context might be taken as pejorative, which brings us to the notion of dualism in our title. Our task is not to dispose of dualism but to try to unpack this notion in our discussions of views of nature. We might find other approaches to nature more relevant, however, this discussion needs to carefully consider the notions involved. Neither will we settle on a particular definition of dualism, but we will leave room for different approaches and problematizations.

Another clarification is whether the singular "dualism" or plural "dualisms" is most appropriate for our explorations. Although we use dualism to describe a common characteristic, for example, in what Hastings indicates by the ubiquitous metaphor of the line, we are aware that dualism can take different shapes, hence a more adequate term might be "dualisms." By addressing these notions in the plural, we indicate a variety and multiplicity of meanings. In a similar way, cognate or contrasting notions such as monism and holism might also have to be considered. In some of the essays, these notions are problematized in more depth, since an important intention of the volume is to address how to approach and identify what we are dealing with when highlighting views of nature.

By addressing dualism in relation to views of nature, we acknowledge that dualism, in different ways and understandings, is a notion that inevitably arises and challenges us in our approach to nature. This is also the case in the current situation with environmental issues. Some argue that dualism, and in particular "substance dualism" per the legacy of René Descartes, is the root of almost all problems related to the human detachment from nature that has led to our current environmental crisis. In such cases, dualism is used pejoratively to indicate almost everything that has gone wrong in the human-nature relationship. In addition, related terms like chasm, abyss, separation, and split are used in the same negative way, leaving the impression that we have to leave dualism behind and point in a different direction in order to move forward. We think such descriptions and judgments are oversimplified.

We can identify notions of dualism in a wide range of contexts. The contexts indicate what kind of duality is present. For example, we might speak of dualism in terms of transcendence-immanence, reason-senses, nature-culture, intrinsic human dualism (mind-body) or human-nature. These different dualisms may or may not be related. That is, we might argue for a duality in terms of transcendence and immanence, yet not advocate a human-nature dualism. One of these approaches to dualism above—nature-culture—is closely related to human-nature emphasized in this volume. Finnish environmentalist Yrjö Haila describes the Western view of humanity's place in nature as dominated by a dualistic opposition between nature and culture.[2] Further, Haila argues that dualism tends to be reproduced in certain ways: "I think the dualism tends to be reproduced because of two main reasons. First, it arises in human interactions with the world, which tend to turn into subject-object relationships. Second, these specific subject-object relationships are generalized to a totalizing distinction between 'us' and 'the environment.'"[3] Based on this observation, Haila then describes this kind of dualism as a dead end, since there is no way of creating alternative views without lapsing into new modes of dualisms. At least, Haila says, we might "construct context-specific alternatives demonstrating that culture and nature belong together."[4]

Views of Nature and Dualism: Contexts and Space/Place

Our approaches to and understandings of nature are always contextual. We explore nature in space and time. We approach nature and our surroundings within a given geographical context. Such a contextual

[2] Yrjö Haila, "Beyond the Nature-Culture Dualism," *Biology and Philosophy* 15 (2000): 155–175.

[3] Haila, "Beyond the Nature-Culture Dualism," 155.

[4] Haila, "Beyond the Nature-Culture Dualism," 156.

consciousness is also echoed in the spatial turn in religion. The renewed interest in space or place was first addressed in geography, but in recent years it has appeared in a variety of fields—also within the humanities.[5] Hence, in our task of dealing with views of nature and dualism, notions of place and space are important. A wide range of topics can be treated within this spatial turn. We will mention just a few.

The spatial turn is not only about geographical or physical space. Our understanding of the human-nature relationship in terms of space/place is also constructed or created. In our view, the reciprocity between humans and nature is important. Peter Nynäs puts it this way: "There is a significant reciprocity in the correspondence between the human self and place. Places also interpret us. They express and articulate us and are thus constitutive of who we are and become."[6] Such a view challenges the constructivist perspective and interrogates the traditional dichotomy between subject and object, an issue highly relevant for our understanding of the human-nature relationship.

The spatial turn in religion might not be considered as something new, as it has been present in the phenomenology of religion's emphasis on holy places, as in Mircea Eliade.[7] However, a heightened emphasis has been evident since the 2000s. In theology, Sigurd Bergmann is a leading voice addressing this turn as follows: "Theology's reflections about space and place provide a deep challenge and an urgent necessity for theology to become aware of its embeddedness in the existential spatiality of life."[8] For Bergmann, a contextual consciousness is needed to address views of nature with particular reference to environmental issues. Such a grounded approach is resonant, for example, with this volume's contributions by Elaine Padilla and Dag Erik Wold. In philosophy, we find a slightly different emphasis on the spatial turn. Jeff Malpas's *Place and Experience: A*

[5] See Barney Warf and Santa Arias (eds.), *The Spatial Turn: Interdisciplinary Perspectives* (London and New York: Routledge, 2008), and Marijn Nieuwenhuis and David Crouch, *The Question of Space: Interrogating the Spatial Turn between Disciplines* (Lanham, MD: Rowman & Littlefield International, 2017).

[6] Peter Nynäs, "From Sacred Place to an Existential Dimension of Mobility," in *The Ethics of Mobilities: Rethinking Place, Exclusion, Freedom and Environment*, eds. Sigurd Bergmann and Tore Sager (Aldershot: Ashgate, 2008), 157–176.

[7] Kim Knott, *The Location of Religion: A Spatial Analysis* (London/New York: Routledge, 2014).

[8] Sigurd Bergmann, "Theology in its Spatial Turn: Space, Place and Built Environments Challenging and Changing the Images of God," *Religion Compass*, 1(3), (May 2007), 353.

Philosophical Topography is one example, Arne Næss addresses it in his ecophilosophy, and we even find the notion of "geophilosophy" developed, for example, in Mark Bonta and John Protevi.[9]

Emphasizing space and place makes us aware of our own contextual point of departure. Our cultural backgrounds and understandings of nature influence our approaches to nature and what we identify as the most pressing challenges in our time. While this is not a problem per se, it is a fact of which we need to be aware. However, having a contextual awareness of space and place does not contradict philosophical or theological "meta-reflections" on nature, as views of nature and dualism may be explored both as "top-down" and "bottom-up."

Dualisms and Religions

In religion, dualism is mainly associated with belief in two opposing gods or powers, where notions of transcendence-immanence are present. In some forms of Christianity, we find a fundamental dualism, which is also related to other dual phenomena of reality, such as God-nature, God-human, Spirit-matter, God-evil, and heaven-earth. The transcendence-immanence duality expresses a divided reality, and through the centuries, this duality has been problematized by theologians in different ways. What is the relationship between God and nature or the worldly and the heavenly? Are there some connections? If so, what kind of "connections" are we talking about? In Christian theology, the fundamental story of Jesus problematizes this core challenge. The incarnation, or in John's terms, the divine logos made flesh, the cross and resurrection, and the two natures of Christ are key topoi where the duality of heaven and earth interconnect.

Dualism in terms of dividing and splitting resonates in the depth of the Christian worldview with notions of good and evil. Biblical notions of evil are often correlated with a breach, alienation, or loss of relationality that surfaces in divisions between people (community), people and the natural world (creation), and people and God (ultimacy). Hence, while evil moves in the direction of separating, splitting up, tearing down, or putting into

[9] See Jeff Malpas, *Place and Experience: A Philosophical Topography* (London and New York: Routledge, 2008); Arne Næss, *Ecology of Wisdom* (London: Penguin Books, 2008), and Mark Bonta and John Protevi, *Deleuze and Geophilosophy* (Edinburgh: Edinburgh University Press, 2004).

conflict, the divine will seeks to unify, bring together, and heal. Paul's eschatological vision of "God in all" (1 Corinthians 15:28) brings this uniting intention to a crescendo. We cannot unpack other interpretations from the biblical accounts or the history of theology here, but there are various interesting ambiguities and contestations regarding dualism in the Christian tradition. We might mention here, for example, Thomas F. Torrance's "Barthian" criticism of dualism in the Latin church fathers.[10]

Further, religious dualism is strongly related to more specific understandings of particular gods or powers. In Christianity, for example, monotheism has been advocated traditionally in contrast to views such as polytheism or pantheism. However, these notions are not unambiguous and can be understood in different ways. Hence, attempts to describe God and God's relation to the world in religions such as Christianity is an ongoing discussion.

We should also note in passing the presence of strong dualisms in early Christianity found, for example, in various early iterations of Gnosticism. In the contestation of "orthodox" doctrine, extreme views of dualism were often deemed heretical. Gnosticism, a general notion that took different directions, emphasized a rupture between the spiritual and the material worlds, wherein the spirit's deliverance from matter was the main goal.

Speaking of dualism in the context of religion can lead to another form of oversimplification. The "dual" might not be the only way of parsing "contrasting" phenomena. We might very well think in terms of a different number of "poly-isms," involving more than two contrasting phenomena. In any case, religious dualism (or "poly-isms"), in terms of God/gods' relation to the world, expresses some degree of separation. These separations can take many forms, such as between the divine and the human, the divine and nature, etc.

One clear critique of the "separation of reality" has been offered by Charles S. Peirce. He blames William of Ockham for putting the unity of reality into question. According to Pierce, Ockham's nominalism contributed to a departmentalization of reality, which has led to a disunity in our understanding of reality.[11] Later philosophers and theologians have been

[10] Thomas F. Torrance, "Karl Barth and the Latin Heresy," *Scottish Journal of Theology* 39 no. 4 (1986), 461–482.

[11] Charles S. Peirce, *Philosophical Writings of Peirce* (ed. Justus Buchler) (Mineola: Dover Publications, 1955), 248.

inspired by Peirce's critique and used this as a backdrop for developing their own approach for understanding reality. Peirce's solution for thinking reality as a whole is developed in his theory of signs. We find another notable critique of dualism in Baruch Spinoza. According to later philosophers, such as Arne Næss, Spinoza's elaboration of *natura naturans* and *natura naturata* gives resources for a holistic understanding of reality.[12]

Whether scholars are inspired by Peirce, Spinoza, or other thinkers, we find most of the criticism of dualism focusing on Rene Descartes. Specifically, his substance dualism is seen as the main problem. Even if criticisms of Descartes might be more or less univocal, attempts to think of reality as interwoven take a very different shape in later philosophical or theological approaches. In any case, as both Langby and Sæther touch upon in this volume, dualism needs clarification depending on the context in which it is used. So far, we have addressed some important topics on dualism related to religion and theology.

The Human-Nature Dualism

As described above, dualism has traditionally been identified by contrasting terms such as immanent-transcendent, God-world, Spirit-matter, etc. The duality of human-nature is slightly different. Such a dualism, described by Langby as an "in-worldly" dualism, can be considered as something other than "traditional" dualism. Langby and Sæther problematize this in different contexts and suggest that it might be more adequate to describe this duality as contrasts, opposites, differences, or distinctions. Lillebø explores in depth some philosophical and historical sources for problematizing distinctions. Anyway, dualism is also an established term for describing a view of the human-nature relationship. Such an in-worldly dualism is also elaborated by Wold in the specific context of Norway's mountaineering philosophers.

During the last decades' discussions on environmental issues, one of the most highlighted topics has been how we can preserve nature or save what is left of nature, often in terms of wilderness. In the current situation on climate and environmental issues, one often-highlighted topic is our need to preserve nature. In recent years, in the context of climate change and being aware of the damage humans have done to earth, we do not only talk about preserving but also restoring the earth. One proposed way

[12] Næss, *Ecology of Wisdom*, 230ff.

out of the destructive human-nature relationship is to "set nature free" from humans. Nature then ought to be preserved so we can protect its "original" qualities, i.e., more or less untouched by humans. In this connection, the notion of wilderness has been emphasized. However, one interesting reflection with such an agenda is the intrinsic dualism present in such a view. It is a paradox that our understanding of wilderness actually embodies a dualistic vision wherein the human is considered outside of this wilderness, as pointed out by William Cronon in "The Trouble with Wilderness, or, Getting Back to the Wrong Nature": "To the extent that we celebrate wilderness as the measure with which we judge civilization, we reproduce the dualism that sets humanity and nature at opposite poles."[13] In addition, Cronon explores what he describes as the myth of the wilderness as virgin uninhabited land. Such a "wilderness imaginary" has its roots in European romantic writers and was picked up later by settlers in North America, and he argues that notions of wilderness as autonomous and free spaces beyond all human interaction are more likely an invention or fiction.[14]

Might we find a renewed approach to the human-nature relationship that accommodates some aspects of both dualism and monism or holism without lapsing exclusively into one or the other? We think this is both feasible and necessary. Indeed, as Hastings suggests, our propensity for drawing lines of distinction *and* circles of connection is rooted in biology and socialization. In his illuminating introduction to *Envisioning Nature, Science and Religion*, James D. Proctor describes some selected approaches to nature, or what he describes as *visions*, as "evolutionary nature," "emergent nature," "malleable nature," "nature as sacred," and "nature as culture."[15] Intrinsic to these visions are potential dualistic challenges that also show how our views of nature are closely tied with dualism. Inspired by Proctor's notion of visions, our volume addresses visions for how to approach and rethink views of nature and dualism, which may inspire more positive responses to the current environmental challenges.

With these clarifications and problematizations in mind, we will now introduce the essays in the book that explore views of nature and dualism

[13] William Cronon, "The Trouble with Wilderness: Or, Getting Back to the Wrong Nature," *Environmental History*, 1 No 1(1996): 17.

[14] Cronon, "The Trouble with Wilderness: Or, Getting Back to the Wrong Nature," 15.

[15] James D. Proctor, "Visions of Nature, Science, and Religion," in *Envisioning Nature, Science and Religion*, ed. James D. Proctor (Conshohocken, PA: Templeton Press, 2009), 3–35.

from different angles and contexts. Some are more theoretical while others address specific contexts.

Scholars of ecospirituality argue for a novel and profound understanding of the human-nature relationship, which is relevant for the current environmental situation. In Chapter 2, *Knut-Willy Sæther* identifies and analyzes topics within ecospirituality that appear to be crucial for understanding this relationship. He takes up notions of seeing nature/awareness, place/space, belonging/feeling at home, and aesthetic experiences, which provide insight for leaving dualism behind. However, recognizing that ecospirituality uses dualism in a vague and pejorative way, he problematizes notions of dualism, monism, and holism, and argues for a "dynamic relational holism" as a more suitable way to characterize the human-nature relationship in ecospirituality.

In Chapter 3, *Lina Langby* argues that a panentheistic conception of God can contribute better to our environmental concerns than classical theism, because nature is more inherently valuable in panentheism than in classical theism. Furthermore, pantheism could contribute to an even deeper environmental ethics than classical theism and panentheism since, in pantheism, all aspects of the world are seen as sacred and divine. Nevertheless, the benefits of the ontological God-world monism essential to pantheism are less significant than one might think. Both panentheism and pantheism emphasize the divinity of the physical, bodily, and natural world, making the relative benefits of pantheism insignificant compared to panentheism.

In Chapter 4, *Elaine Padilla* leads us into the depths of nature's unconscious with her Jungian-based study of artist Myrna Baez's *El Mangle*. The trees, whose entangled roots are immersed in Caribbean waters, exemplify Jung's Philosophical Tree and a religious longing for a mystical harmony that has the potential to become actual through processes similar to entering a mother's womb. Padilla makes use of the Caribbean philosophy of Edouard Glissant and Corrington's ordinal psychoanalysis on melancholy to bring out the themes of the unconscious desire for mutually beneficial interconnectivity (the trees), melancholic expressions of nature, and the maternal as the womblike depth of nature that nurtures human awakening and creativity.

In Chapter 5, *Dag Erik Wold* explores how the Norwegian pedagogical tradition of *friluftsliv* offers resources for building a more nature-friendly society. Emphasizing the human encounter with nature, he introduces the concepts of "non-conventional goals" and "ferd," which are identified with mountaineering philosophers in Norway whose long-distance journeys provide insight into the *friluftsliv* pedagogy and lifestyle. In this context, notions of "free nature," duration, and reflection are crucial. While admitting that the *friluftsliv* project has been criticized for being elitist or naïve, Wold argues that it provides an important pedagogical impetus for a deeper love of nature.

Dualism divides something into two contrasted aspects. In Chapter 6, *Mark Graves* focuses on the division itself and the places and ways in which that division occurs in similar ways across different types of dualism. Drawing upon a scientific and religious examination of nature, of human existence, and of human experience of those realities, Graves uses a systems approach to organize five common types of division. The relative emphases placed on the different divisions by various dualisms and monisms classify these philosophical and theological positions for systematic study.

Considering growing accounts of grief, anger, and anxiety in response to climate change and ecological collapse, in Chapter 7 *Hannah Malcolm* offers a theological reading of how these accounts might be interpreted in light of modernity's assumed nature-culture dualism. Can they be trusted as 'true' if also culturally conditioned? She proposes that this cultural conditioning is not necessarily a barrier to revelation of truth about 'nature' and begins by critically examining modernity's reading of nature-culture dualism and considering its impact on climate change discourse. She then turns to a semiotic reading of encounters between creatures to clarify the relationship between nature/culture and human feeling.

Chapter 8 explores a non-dualistic reading of Dietrich Bonhoeffer grounded in Bonhoeffer's own rejection of God-world dualism in his *Ethics*, his insistence there that God and world are one reality in Jesus Christ, not two. *Lisa Dahill* takes this conviction as a methodological lens for interpreting Bonhoeffer himself, surfacing pieces of his thinking that contribute to a larger non-dualistic vision of God and world in Jesus Christ and critiquing aspects that fall into the dualism he has rejected. In this way, she shows how reading according to this criterion can generate an ecologically fruitful theology of God-world reality in Jesus Christ.

Pentecostal spirituality emphasizes experiential emotionalism, as individuals and communities holistically experience the Holy Spirit, creating a non-dual reality where the material and supernatural overlap and interact. In Chapter 9, *Joel D. Daniels* argues that Pentecostal non-dualism offers potential ecological solutions, returning to human emotion as the key to decision making that leads to change. He begins with a brief outline of Pentecostal cosmology and then moves to emotions. Pentecostals claim that it is the Spirit moving in and through the world, inviting people to experience and participate in creation, providing potential resources for addressing our ecological reality.

In Chapter 10, *Jonas Gamborg Lillebø* highlights how "formal distinction" is a core concept for understanding "univocity of being" in the thinking of John Duns Scotus and Baruch Spinoza. For Scotus, the formal distinction is explicitly a part of his argument against other thinkers of his time, such as Henry of Ghent or Thomas Aquinas. For Spinoza, formal distinction is more implicit in his critique of René Descartes. In both Scotus and Descartes, a discussion of distinctions constitutes a part of their reasoning. However, even though the formal distinction is pivotal for seeing being as univocal, the two thinkers have very different versions of univocity. For logical and theological reasons, Scotus neutralizes the univocity of being, while Spinoza affirms it.

In Chapter 11, *Caleb Gordon* compares the materialist ontologies of David Bentley Hart and Gilles Deleuze to propose an aesthetic approach to environmental ethics. Environmental aesthetics is often associated with a principle of 'disinterestedness' when evaluating relationships between subjects and their environments; disinterestedness requires that an individual's aesthetic responses are separate from their desires. The materialist ontologies he explores, however, do not support firm boundaries between individuals and their environments. They are thus open to the possibility of aesthetic approaches to environmental ethics which do not require disinterestedness; the interests and desires of a subject are necessary considerations for an ethical evaluation of their relationship to their environment.

Zen Master Thich Nhat Hanh's environmental storytelling offers a vision for environmental stewardship that contributes to a much-needed transition from ecological alienation to ecological alignment. In Chapter 12, *Victor Thasiah's* use of "ecological alienation" refers to the fundamental disconnect between state, corporate, and consumer behavior on the one hand, and directly and indirectly related negative environmental consequences on the other. His term "ecological alignment" refers to a certain affection for and attachment to the natural world, which involves participation in both nature-based and culture-based collective efforts toward the mutual flourishing of ecosystems, communities, and ourselves.

Acknowledging the biological and cultural bases for making distinctions, as symbolized by the line, and for making connections, as symbolized by the circle, and touching on ethical and moral problems attending particular uses of lines and circles, in Chapter 13 *Thomas John Hastings* draws attention to the ways academia has prioritized intellect over intuition. In the face of the shared threat to our planetary habitat, the author calls for an embodied, holistic epistemology wherein lines of distinction are softened, and circles of connection are left open. Kagawa Toyohiko (1888–1960) and Nalini Jayasuriya (1927–2014) are introduced as two exemplars of a more hopeful holistic epistemology.

Bibliography

Bergmann, Sigurd. 2007. Theology in its Spatial Turn: Space, Place and Built Environments Challenging and Changing the Images of God. *Religion Compass* 1 (3): 353–379.

Bonta, Mark, and John Protevi. 2004. *Deleuze and Geophilosophy*. Edinburgh: Edinburgh University Press.

Cronon, William. 1996. The trouble with wilderness: or, getting back to the wrong nature. *Environmental History* 1 (1): 7–28.

Haila, Yrjö. 2000. Beyond the Nature-Culture Dualism. *Biology and Philosophy* 15: 155–175.

Hastings, Thomas John, and Knut-Willy Sæther. 2020. A fallibilist approach in the age of COVID-19 and climate change. In *The grace of being fallible in philosophy, theology, and religion*, ed. Thomas John Hastings and Knut-Willy Sæther, 1–11. New York: Palgrave Macmillan.

Knott, Kim. 2014. *The Location of Religion: A Spatial Analysis*. London and New York: Routledge.

Nieuwenhuis, Marijn, and David Crouch. 2017. *The Question of Space: Interrogating the Spatial Turn between Disciplines*. Lanham, MD: Rowman & Littlefield International.

Nynäs, Peter. 2008. From Sacred Place to an Existential Dimension of Mobility. In *The Ethics of Mobilities: Rethinking Place, Exclusion, Freedom and Environment*, ed. Sigurd Bergmann and Tore Sager, 157–176. Aldershot: Ashgate.

Næss, Arne. 2008. *Ecology of Wisdom*. London: Penguin Books.

Peirce, Charles S. 1955. *Philosophical Writings of Peirce (edited by Justus Buchler)*. Mineola: Dover Publications.

Proctor, James D. 2009. Visions of Nature, Science, and Religion. In *Envisioning Nature, Science and Religion*, ed. James D. Proctor, 3–35. Conshohocken, PA: Templeton Press.

Torrance, Thomas F. 1986. Karl Barth and the Latin Heresy. *Scottish Journal of Theology* 39 (4): 461–482.

Warf, Barney, and Santa Arias. 2008. *The Spatial Turn: Interdisciplinary Perspectives*. London and New York: Routledge.

CHAPTER 2

Seeing Nature as a Whole: Ecospirituality and the Human-Nature Relationship

Knut-Willy Sæther

> The view of the world provided by Google Maps is not enough. We need to be moved, to experience wonder, to feel ourselves a part of this cosmos, of this Earth, a part of nature, as self-conscious beings with an awareness of our ancestors and our deepest reality. These sorts of emotional experiences are what changes our lives.[1]

Introduction

In the current situation of climate change and discourses on sustainability and environmental issues, many academic voices have been important. One such voice is ecospirituality. We find it as a part of other streams of thought within the humanities, such as ecophilosophy and ecotheology. All three can only be understood adequately against the historical

[1] Leonardo Boff, *Toward an Eco-Spirituality* (New York: The Crossroad Publishing Company 2014), 46.

K.-W. Sæther (✉)
Department of Religious Studies, Volda University College, Volda, Norway
e-mail: knut-willy.saether@hivolda.no

T. J. Hastings, K.-W. Sæther (eds.), *Views of Nature and Dualism*,
https://doi.org/10.1007/978-3-031-42902-6_2

background of environmental challenges beginning in the 1960s and 1970s. Ecospirituality draws on resources from the first two but attempts to bring novel perspectives to the current debates. Like the more general notion of spirituality, ecospirituality is multifaceted, encompassing many approaches and positions. It covers perspectives from traditional religions, non-religious spiritualities, and environmental sciences, as well as religious movements inspired by new religiosity such as New Age.[2]

All three streams of thought attempt to explain *why* our environmental crisis has reached its current tipping point and to suggest *how* we might solve—or at least alter—this challenging global situation. One of the main answers to the "how" question is related to our need to rethink our understanding of nature, and, in particular, the human-nature relationship.[3] For ecospirituality, such a rethinking is unavoidably intertwined with action.[4] Hence, ecospiritual scholars tend to be closely affiliated with different types of activism, such as environmentalism.

I find insights from ecospirituality helpful for developing a more profound understanding of the human-nature relationship, which is relevant to the current situation. My task is, therefore, to analyze how ecospirituality understands the human-nature relationship. One spin-off of the analysis will be to shed light on ecospirituality's criticism of dualistic approaches to nature.

In addressing "the human-nature relationship," a few points need to be clarified. First, we do not necessarily find one specific, clearly defined understanding of this core relationship in ecospirituality. Second, the use of term "nature" often overlaps with "cosmos" or "reality."[5] Hence, when

[2] See Philip Sheldrake, *Spirituality: A Brief History* (Hoboken: Wiley-Blackwell, 2013). Ivo Jirásek, "Verticality as Non-Religious Spirituality," *Implicit Religion* 16:2 (2013): 191–201.

[3] By using the term "nature" in this context, I have in mind nature as humans' surroundings and in particular their relationship with the living nature. However, I am not only limited to this, as the term is often understood more widely, corresponding to, for example, Earth and Cosmos. See Haydn Washington, *A Sense of Wonder Towards Nature: Healing the Planet through Belonging* (London and New York: Routledge, 2019), 1–5. For the term "the human-nature relationship," see Valentine Seymour, "The Human–Nature Relationship and Its Impact on Health: A Critical Review," *Frontiers in Public Health*, 4:260. doi: 10.3389/fpubh.2016.00260.

[4] Boff, *Toward an Eco-Spirituality*.

[5] See, for example, Alejandro García-Rivera, *Garden of God: A Theological Cosmology* (Minneapolis: Fortress Press, 2009), 29ff.

studying this topic in ecospirituality, we can also talk about the relationship between humans and the cosmos. Despite this ambiguity, some common characteristics may be highlighted when identifying ecospirituality. Third, in exploring ecospirituality, I draw mainly but not exclusively on resources from scholars affiliated with Christianity. Scholars from various Christian traditions have led the way in addressing the human-nature relationship. Fourth, it is somewhat difficult to characterize scholars as ecospiritual scholars, as this would indicate the existence of a defined scholarly field, which is not the case. Hence, my contribution is first and foremost topic-driven, and my discussion partners are scholars who highlight what may be described as ecospiritual concerns.

My task is not to analyze a specific ecospirituality, either within or outside of the Christian tradition, but to identify common topics that provide a clearer understanding of the human-nature relationship. Thus, I leave out a wide range of topics identified as "Christian" topics related to nature, such as the theology of creation, natural theology, etc. Nevertheless, one reason for drawing on resources from scholars within the Christian tradition is that they attempt to address ecospirituality as a broader concept, thereby expanding somewhat limited views of Christian ecospirituality.

The topics I identify as crucial for understanding the human-nature relationship in ecospirituality may be listed as follows: seeing nature/awareness, place, and space, belonging and feeling at home in nature, and, finally, the role of aesthetic experiences in nature. Of course, since the notion is fluid, these topics are not limited to ecospirituality, per se, as we find similar issues addressed in other scholarly contexts.

By exploring these topics, I find resources for attending to the human-nature relationship in the contexts of dualism, monism, and holism. Certainly, notions of dualism, monism, and holism are highly ambiguous, and my task is not to problematize these ideas thoroughly by reference to different fields, such as philosophy and natural science. However, they will be clarified and problematized in greater depth to provide an overall picture of how to understand the human-nature relationship within ecospirituality.

My analysis draws on the work of scholars who explicitly use the term "ecospirituality" in their writings, such as Leonardo Boff, Charles Cummings, and Harald Olsen.[6] In addition, I also find resources from

[6] See Boff, *Toward an Eco-Spirituality*; Charles Cummings, *Eco-Spirituality: Toward a Reverend Life* (Mahwah: Paulist Press, 1991), and Harald Olsen, "Økospiritualitet—for det hele menneske," in *Økoteologi: Kontekstuelle perspektiver på miljø og teologi*, eds. Bård Mæland and Tom Sverre Tomren (Trondheim: Tapir Akademisk Forlag, 2007), 251–263.

scholars who traditionally have been identified as ecotheologians, such as Sigurd Bergmann and Roald Kristiansen, as well as others who may—due to their thematic perspectives—be considered ecospiritual scholars, such as Alejandro García-Rivera. These scholars bring insights into the human-nature relationship that, in my opinion, extend traditional ecotheology, and these insights help give ecospirituality its distinctive character.

First, I will map out the landscape of ecospirituality as a way of identifying its primary agenda. This chapter lays a necessary foundation for my analysis of the chosen topics, leading to a discussion of the human-nature relationship in the context of dualism, monism, and holism.

Mapping the Landscape of Ecospirituality: Visions and Actions

The notion of ecospirituality is a complex one. Hence, "spirituality" and its prefix "eco" need to be clarified. Spirituality spans a wide range of meanings. The origin of using the term is within a Christian context, but it has developed to include discussions in other religions. We now find it applied within various world views and religions and even outside traditional religious contexts. The latter has even given rise to so-called "non-religious" spirituality.[7] Also, in some traditions, we find spirituality closely related to mysticism.[8]

According to Walter Principe, we can identify at least three main understandings of spirituality: (1) as a phenomenon taking place at a real or existential level; (2) as a teaching about lived reality as constructed by spiritual leaders; and (3) as a growing academic field that studies spirituality as a second-order phenomenon.[9] My approach to spirituality is to take all three approaches into account when exploring the narrower concept of ecospirituality, while remaining aware of the elusiveness and subtleness of spirituality.

Adding "eco" attempts to address a specific context for spirituality: the ecological one. Ecospiritual perspectives may take different forms;

[7] Jirásek, "Verticality as Non-Religious Spirituality," 191–201.

[8] See Kathleen Fischer, "Christian Spirituality in a Time of Ecological Awareness," *Theology Today* 67 (2010): 142ff.

[9] Walter Principe, "Toward Defining Spirituality," *Studies in Religion* 12 (1983): 127–141, and Lucy Bregman, *The Ecology of Spirituality: Meanings, Virtues, and Practice in a Post-Religious Age* (Waco: Baylor University Press, 2013), 10–16.

however, we may mention some common streams of thought. One the one hand, without making overly substantive clarifications of what it might cover, ecospirituality at the very least advocates a revised or even new approach to nature, compared with, for example, ecophilosophy and ecotheology.[10] On the other hand, there is no clear demarcation line between the latter two and the former. Some scholars who labeled themselves as ecotheologians in the 1960s and 1970s have stressed some weaknesses with the original ecotheology and have in recent years highlighted how to move forward in response to the current environmental crisis. Hence, it also makes sense to describe ecospirituality as a given perspective, which contributes to its distinctiveness.

In addition to its close relations with ecophilosophy and ecotheology, ecospirituality is also partly influenced by process philosophy (Alfred North Whitehead), process theology (Charles Hartshorne), the theology of ecology (John B. Cobb), and environmental aesthetics (Allen Carlson).[11] How these fields have influenced ecospirituality varies, but they share an understanding of reality as dynamic (evolutionary nature), which for ecospirituality is related to some sort of spiritual understanding of nature.[12] Further, these fields have influenced how ecospirituality emphasizes experiences in nature, especially experiences of beauty, the sublime, and wonder.

A common point of departure for ecospirituality is to highlight a skeptical attitude toward dualistic views of the human-nature relationship. Dualism, pejoratively used, has led to a dangerous separation between humans and nature, where an anthropocentric approach to nature has been predominant. Briefly described, this has led to a climate crisis, which now challenges us to rethink this relationship and how we interact with nature. Kathleen Fischer puts it this way: "The ecological crisis has revealed

[10] Olsen, "Økospiritualitet—for det hele menneske," 259, and Knut-Willy Sæther, "Naturen som sted: En eksplorativ studie av. økofilosofi, økoteologi og kristen økospiritualitet," in *Rom og Sted: Religionsfaglige og interdisiplinære bidrag*, eds. Knut-Willy Sæther and Anders Aschim (Oslo: Cappelen Damm Akademisk, 2020), 175ff.

[11] See Alfred N. Whitehead, *Process and Reality* (New York: The Free Press, 1978); John B. Cobb, *Is It Too Late? A Theology of Ecology* (Minneapolis: Fortress Press, 1972), and Allen Carlson, *Aesthetics and the Environment: The Appreciation of Nature, Art and Architecture* (London and New York: Routledge, 2000).

[12] The notion of spirituality can be understood in different ways, including within ecospirituality. For Rachel Wheeler, a working definition of ecospirituality describes how one relates to the sacred within the context of our natural, global, and even cosmic ecosystems. See Rachel Wheeler, *Eco-Spirituality: An Introduction* (Minneapolis: Fortress Press, 2022). I will return to the "spiritual component" in ecospirituality later in my analysis.

a problematic aspect of how we see the human person, shining the spotlight on how anthropocentric, or human centered, theology has become."[13] Ecospirituality tends to use notions of anthropocentrism and dualism in a pejorative way as legacies we need to abandon.

According to ecospirituality, the grand narrative of the human-nature relationship is a story of how we—through modernity—lost our connection to nature and how we now need to reestablish a more positive relationship with nature.[14] Cummings argues that through the enlightenment, with its scientific understanding of nature, the holy dimension of nature was lost, and in particular the sense of nature's awesomeness.[15] He adds that the holy dimension became compartmentalized as belonging solely to the Church, i.e., within a narrowed religious sphere. For ecospirituality, the story of modernity and industrialization collides with the ecological story that has arisen in recent decades. Back in the 1960s, the problems of industrial pollution and environmental issues were already being highlighted. This was actually the context for the growth of both ecotheology and ecophilosophy, the latter, for example, in the tradition of the Norwegian philosopher Arne Næss and his deep ecology.

Olsen questions the actual effect of these voices. Clearly, there was nothing wrong with the arguments, but no substantive changes took place. According to Olsen, this was because the discourse gave weight to rational arguments. That is, we thought that *knowing* the right things made us prepared to *do* the right things. But this turned out not to be true. What was lacking was a sensibility that paid sufficient attention to the role of emotions and the aesthetic dimension (i.e., aesthetic experiences in nature). In a similar way, Leonardo Boff argues for a new and broader vision for the earth.[16] Through modernity, our encounter with the earth has failed because we developed a certain type of utilitarian and analytical-instrumental reason at the expense of "…the reason of the senses, the reason of the heart, our emotional and spiritual intelligence."[17] For Boff, these "lost" reasons are about ethical perception without which life loses its meaning and luster. Boff's perspective echoes in García-Rivera's writings, where he addresses our need to explore what "moves the heart."[18]

[13] Fischer, "Christian Spirituality in a Time of Ecological Awareness," 170.
[14] Boff, *Toward an Eco-Spirituality*, 39.
[15] Cummings, *Eco-Spirituality*, 5.
[16] Boff, *Toward an Eco-Spirituality*, 37ff.
[17] Boff, *Toward an Eco-Spirituality*, 34.
[18] García-Rivera, *The Garden of God*, x.

For him, emotions play a major role in how to proceed from our current environmental situation. As in Boff, García-Rivera expresses a vision for the earth toward the "Garden of God."[19] This backdrop and context for ecospirituality, as exemplified by some selected scholars, illustrates its overall concern and direction, which lays a foundation for the next section.

Main Topics in Ecospirituality

Seeing Nature and Awareness

I find the notion of "seeing nature" as a first step, or thematic entrance, for unpacking ecospirituality's understanding of the human-nature relationship. From this starting point, other topics can be elaborated. How we are related to and interact with nature is conditioned by how we see nature. This "seeing" is about how we see nature in a particular way. By highlighting the role of seeing nature, ecospirituality stresses important role of perception. According to Olsen, we need to see the details and relations in nature in all their variety, e.g., landscapes, geology, vegetation, and animals.[20] Fischer says that seeing nature in its physical appearance is the first step in our encounter with nature.[21]

Seeing nature makes room for imagination.[22] In one sense, this way of seeing goes deeper in terms of experiencing meaning in nature. Olsen argues that imagination expresses itself in metaphorical language, a move that allows nature an opportunity to regain its symbolic function. Imagining nature, which for Olsen is about a certain sensitivity, can open up the opportunity for insight where the "imaginary" nature is real, but also goes behind what is immediately sensible. Hence, as I read Olsen, seeing nature is both about seeing the real physical nature and seeing nature with an internal, deeper eye. Together, these two ways of seeing endows nature with a renewed view, deeper sensitivity, and richer language. Seeing nature with such an internal, deeper eye is experienced through practice. It leads to growth in awareness, acceptance, and appreciation for others and our earthly home, and such an awareness may be

[19] García-Rivera, *The Garden of God*, vii–xii.

[20] Olsen, "Økospiritualitet—for det hele menneske," 258.

[21] Fischer, "Christian Spirituality in a Time of Ecological Awareness," 178.

[22] See Fischer, "Christian Spirituality in a Time of Ecological Awareness," 178, and Olsen, "Økospiritualitet—for det hele menneske," 258.

described as spiritual awareness.[23] Further, seeing nature is about *seeing nature as a whole*. This "wholeness" points toward an overall understanding of nature and more comprehensive notion of reality. According to Olsen, to see nature as a whole is to recognize how nature is deeply interconnected and how humans can develop skills to see nature as a whole.

However, seeing nature in a specific way involves even more than seeing nature in terms of physical nature, imaginative nature, or as a whole. In ecospirituality, I find the notion of *awareness* crucial. First, we find a broad approach to awareness that emphasizes an awareness of our surroundings. Bergmann and Eaton show that the origin of "awareness" from Greek expresses ongoing reflection on ourselves in the midst of disturbing worldly attractions. As an ongoing reflection, awareness seems to have an element of cognition. However, the other aspect of awareness is its sensory role, as in wakefulness and attentive observation. Informed by modern phenomenology, Bergmann and Eaton say awareness has an integrative function "… among what takes place and what we perceive and then how we reflect upon and interpret experiences."[24] For them, this creates a zigzag pattern where senses and cognition work together in a complex way and not in a straightforward process directly from the senses to cognitive reflection. This interplay brings forth questions of values and ethics: "What we become aware of influences the kinds of questions we ask, how and what we reflect upon, and ultimately how we answer our queries."[25] For Bergmann and Eaton, such a broad understanding of awareness has clear relevance for our global environmental challenges.

In addition to this first approach to awareness, I find two more specific kinds of awareness described by Bergmann and Eaton: awareness of the vulnerability and fragility of life, and awareness of the evolutionary power and potential of life processes in earth's dynamics. They argue that when these two senses are combined, awareness becomes a deep value at work in several endeavors, such as environmental science and social/religious

[23] Rachel Wheeler, *Eco-Spirituality: An Introduction* (Minneapolis: Fortress Press, 2022), 3.

[24] Sigurd Bergmann and Heather Eaton, "Awareness Matters: Introductory Remarks About the Interwoven Gifts of Life and Belief," in *Ecological Awareness: Exploring Religion, Ethics and Aesthetics*, eds. Sigurd Bergmann and Heater Eaton (Münster: LiT Verlag, 2011), 3.

[25] Bergmann and Eaton, "Awareness Matters: Introductory Remarks About the Interwoven Gifts of Life and Belief," 3.

movements. Both of these types of awareness may be described as being life orientated.

Awareness also has an intrinsic component of *time*, in terms of duration. Being aware is about being aware over time. With Bergmann and Eaton, we may speak about becoming aware and staying aware. Thus, awareness is not only about the here and now, but takes place by paying attention to, and dwelling on our surroundings, both in terms of the proximity of nature and the cosmos as a whole.[26] Further, awareness does not "kick in" accidentally and occasionally but is a skill to be nurtured and developed. Therefore, paying attention and exhibiting an openness to our surroundings is crucial, i.e., *how* we pay attention and open up to seeing nature, both externally and internally. Openness activates all our senses and "influences the kinds of questions we ask, how and what we reflect upon, and ultimately how we answer our queries."[27] Hence, for Bergmann and Eaton, awareness has an intrinsic potential for action. Being alert to what is happening within our surroundings forms our thinking and practical problem-solving. This is also a key point for Lucy Bregman, when she argues that spirituality, in general, is a turn from simply believing toward doing something.[28]

Given ecospirituality's environmental concern, we may summarize the various perspectives of awareness, as follows. It is about being aware of the changes we see here and now. It is about being aware over time of what happens around us and how this affects our thinking. And, finally, it is about how our awareness might lead to action for changing the environmental situation. Thus, awareness has a direction, it is a skill, and it leads to action. In ecospirituality, all of these aspects of awareness are crucial, for without them our connection to nature is lost, which in turn creates a state of alienation. More broadly, seeing nature through an ecospiritual lens is about *how* we see nature, *how* we pay attention, and *how* we are aware of our surroundings. This leads me to the next topic on place and space, which is about the context for seeing nature and being aware.

[26] Bregman, *The Ecology of Spirituality: Meanings, Virtues, and Practice in a Post-Religious Age*, 15–16.

[27] Bergmann and Eaton, "Awareness Matters: Introductory Remarks About the Interwoven Gifts of Life and Belief," 3.

[28] Bregman, *The Ecology of Spirituality: Meanings, Virtues, and Practice in a Post-Religious Age*, 29.

Place and Space

The role of place and space in ecospirituality is emphasized in recent writings.[29] One reason for this is "the spatial turn" that has taken place in many academic disciplines. However, place, and even more space, are highly complex notions.[30] I will not discuss the nuances between place and space as this is done elsewhere.[31] However, I will highlight some perspectives for understanding the human-nature relationship.

In general, the spatial turn is characterized by accentuating a given perspective for gaining new insight and giving voice to spatial practices.[32] According to Peter Nynäs, notions of place and space are developed in the following three dimensions: physical-geographical, sense-emotional, and sociocultural.[33] All three dimensions are evident in ecospirituality and help shed light on *where* awareness takes place and develops.

Insights from ecotheology have shaped key topics in ecospiritual thinking on place and space. Kristiansen, for example, emphasizes that ecotheology is first and foremost about developing a consciousness of locality. For him, locality is about belonging to a place. This not only encompasses our relationship with other humans at a given place, but includes a relationship with the whole local environment: trees, mountains, animals, and so on.[34] This consciousness of locality fosters a comprehensive understanding of reality where humans and the surrounding landscape are

[29] Sæther, "Naturen som sted: En eksplorative studie av. økofilosofi, økoteologi og kristen økospiritualitet," 167–187.

[30] Noel Castree, "Place: Connections and Boundaries in an Interdependent World," in *Key Concepts in Geography*, eds. Nicholas Clifford, Sarah Holloway, Stephen. P. Rice, and Gill Valentine (London: Sage, 2009), 156.

[31] See, for example, Knut-Willy Sæther and Anders Aschim (eds.), *Rom og Sted: Religionsfaglige og interdisiplinære bidrag* (Oslo: Cappelen Damm Akademisk, 2020).

[32] Henri Lefebvre, *The Production of Space* (Hoboken: Blackwell, 1991 [1974]), and Peter Nynäs, "From Sacred Place to an Existential Dimension of Mobility," in *The Ethics of Mobilities: Rethinking Place, Exclusion, Freedom and Environment*, eds. Sigurd Bergmann and Tore Sager (London and New York: Routledge, 2008), 157–176.

[33] Peter Nynäs, "From Sacred Place to an Existential Dimension of Mobility," in *The Ethics of Mobilities: Rethinking Place, Exclusion, Freedom and Environment*, eds. Sigurd Bergmann and Tore Sager (London and New York: Routledge, 2008), 158.

[34] Roald Kristiansen, "Økoteologi anno 2007," in *Økoteologi: Kontekstuelle perspektiver på miljø og teologi*, eds. Bård Mæland and Tom Sverre Tomren (Trondheim: Tapir Akademisk Forlag, 2007), 22–24. See also David Abram, *The Spell of the Sensuous: Perception and Language in a More-Than-Human World* (Vintage, 1997).

intertwined in a binding relation.[35] For Kristiansen, such an approach is a first step toward developing a theological covenant where our commitments to our surroundings are highlighted.

Kristiansen's take on locality and place is echoed in more recent ecospiritual thinking.[36] Place is not only a location, in terms of placement, where humans live. Humans are an integrated part of a place, or, more adequately, a space, in the sense that we may talk about an intertwined external and internal space.[37] Bergmann puts it this way: "our surrounding world and our inner world represents a common continuum."[38] Hence, Bergmann argues against a strong dichotomy between the subject and its surroundings. Referring to French poet Noël Arnaud's expression, "I am the space where I am," Bergmann provides an insight into how to understand place and space. If I am the space where I am, then I am not an independent, constructed self, but always a part of my surroundings and part of a deeper reality that grounds, anchors, and embraces the living, he says. Such a relationality is explored by Bergmann along the lines of inhabitation, atmosphere, and "beheimatung" ("making oneself at home").

Within the framework of Christianity, Bergmann says inhabitation is an ongoing dynamic process where God goes into and beyond the world and transfigures it from within.[39] This, for Bergmann, is about the Spirit's presence in nature and is an important contribution to constructing pneumatology along ecological grounds. In a similar way, Rachel Wheeler understands the Spirit as the enlivening agent, energy, or depth dimension throughout and within creation.[40] In more specifically Christian terms, García-Rivera says that God sent the Holy Spirit to the world to renew the gift of our humanity so that we may heal the trauma ever present in the universe.[41] For García-Rivera, this is about restoring the Garden of God.

[35] Kristiansen, "Økoteologi anno 2007," 22.

[36] See, for example, García-Rivera, *Garden of God*, 44ff.

[37] García-Rivera, *Garden of God*, 14–15.

[38] Sigurd Bergmann, "... the Space where I am": Decolonising, Re-sacralising and Transfiguring Landscapes through the Aest/ethic Lens," in *Aesth/Ethics in Environmental Change: Hiking Through the Arts, Ecology, Religion and Ethics of the Environment*, eds. Sigurd Bergmann, Blindow Irmgard, and Konrad Ott (Münster: LiT Verlag, 2013), 39.

[39] Sigurd Bergmann, "Aware of the Spirit: In the Lens of a Trinitarian Aesth/Ethics of Lived Space," in *Ecological Awareness: Exploring Religion, Ethics and Aesthetics*, eds. Sigurd Bergmann and Heather Eaton (Münster: LiT Verlag, 2011), 32.

[40] Wheeler, *Eco-Spirituality: An Introduction*, 7.

[41] García-Rivera, *Garden of God*, 22.

Atmosphere, or the plural atmospheres, helps us to understand the notion of *relations* in a different way than the traditional subject-object scheme. Atmosphere is what connects the interior and the external (the bodily and spiritual) surroundings. For Bergmann, the Cartesian dichotomy and the distinction between subject and object in identity philosophy have been a hinderance in achieving a pneumatology of inhabitation. The notion of atmosphere can dissolve such a dualism, he says. According to Bergmann, atmosphere is characterized as having a dynamic character: "Atmospheres emerge, they can endure, and they can disappear in the spaces between something and us."[42] Atmosphere enables us to understand the world as dynamic and interwoven where a dualistic demarcation line between subject and object is absent. Further, atmosphere is not only what emerges and exists in the interspace between the internal and external, such as in the human encounter with nature, but also what takes place in our "inner bodily-soul-spiritual being," says Bergmann. At first glance, the notion of atmosphere may seem somewhat elusive or diffuse, but it helps articulate how reality is intertwined and can be correlated with the spiritual aspect in ecospirituality. For Bergmann, atmosphere is a fruitful notion for what he describes as an aesth/ethical theology of the Holy Spirit in space and place.

These notions of inhabitation and atmosphere resonate to some degree with Teilhard de Chardin's *The Divine Milieu*, which has also been a partial influence on García-Rivera.[43] On the one hand, Teilhard argues that humans' spiritual life is a participation in the destiny of the world, or, as described by Philip Hefner, humans are building the earth.[44] On the other hand, informed by the natural sciences and philosophical aesthetics after Teilhard, García-Rivera relates the Holy Spirit to the intrinsic beauty in nature with his notion of the "beauty of living, dynamic forms."[45] For García-Rivera, beauty and the Holy Spirit are tied together, in the sense that beauty is understood as the most visible sign of the work of the Holy

[42] Bergmann, "Aware of the Spirit: In the Lens of a Trinitarian Aesth/Ethics of Lived Space," 33.

[43] Pierre Teilhard de Chardin, *The Divine Milieu* (New York: Harper and Row Publishers, 1960).

[44] Philip Hefner, "Teilhard's Spiritual Vision for the Mystical Milieu," in *The Legacy of Pierre Teilhard de Chardin*, ed. James Salmon, SJ, and John Farina (Mahwah: Paulist Press, 2011), 92.

[45] García-Rivera, *Garden of God*, 98ff.

Spirit.[46] Hence, for both García-Rivera and Bergmann, aesthetics seems to be of major importance in how to understand reality as interwoven.

Even though I am not primarily focusing on the Christian understanding of the Holy Spirit, Christian ecospirituality gives, as shown, more specific direction to how to understand the "spiritual" component. Boff, also within a Christian tradition, uses more extensive terms for describing "the spiritual" in ecospirituality, such as the Holy, Spiritus Creator, and the sacred.[47]

Based on my analysis of inhabitation and atmosphere, where the Spirit is crucial, it seems permissible to talk about a *spatial-spiritual awareness* in the human-nature relationship. Nevertheless, we still need to ask what kind of experiences in nature might support such an understanding of space and place. Bergmann puts it this way: "How do we conceive the space in which we move? Is it a mere container or a living all-embracing space? Is it an absolute entity or do we perceive it as a complex relational space?"[48] Bergmann points here toward his third construct, "beheimatung," which I will develop along the lines of "belonging" and "feeling at home."

Belonging and Feeling at Home in Nature

Awareness of one's surroundings is not solely about being aware in an intellectual way. It also entails emotions playing an important role in how awareness unfolds in place and time. In ecospirituality, emotions may be explored in different ways, and I will emphasize the notions of "belonging" and "feeling at home." These two are unquestionably about "emotions" that arise in our encounter with nature. For ecospirituality, these perspectives also draw on resources from both ecophilosophy and ecotheology.

Both belonging and feeling at home cover different perspectives, and in my context, it is about belonging to and feeling at home *in nature*. We find related terms, such as the feeling of a place (Arne Næss), biophilia

[46] García-Rivera, *Garden of God*, 51–52.

[47] Boff, *Toward an Eco-Spirituality*, 42ff.

[48] Sigurd Bergmann, "The Beauty of Speed or the Discovery of Slowliness," in *The Ethics of Mobilities: Rethinking Place, Exclusion, Freedom and Environment*, eds. Sigurd Bergmann and Tore Sager (London and New York: Routledge, 2008), 22.

(Edward O. Wilson), and topophilia (Roald Kristiansen).[49] Some of these terms are more comprehensive, others are narrower: biophilia toward bios as home (the living), cosmos toward the whole universe as home. In environmental science, expressions like "earth as our home" are used.

Despite the different aspects of "home," they all mediate a common message closely related to place and space: Humans are part of a larger whole and emotionally interwoven with this, whether it is expressed as bios, nature, earth, or cosmos. For ecospirituality, the origin of the prefix "eco" and the term "ecology" supports such an approach. "Eco-" takes its root from the Greek *oikos*, expressing both one's house and one's surroundings.[50] Hence, the notion of home can be identified at several levels. First, "eco" can be understood as having an interior dimension, a "house," where we live together in a community. Second, it can be identified as having an exterior dimension. A household is always situated in a particular place, a bioregion. A bioregion is a geographically defined area with some specific common characteristics. They are not defined as national boundaries but as physical geographical features (e.g., geology, geomorphology, climate, flora, and fauna). A more or less defined region creates an *oikos*. In the words of Cummings, ecospirituality fosters a certain bioregional loyalty: "the feeling of being at home within one's bioregion, being in communion with one's proximate environment."[51]

The notions of belonging and feeling at home are about recognizing nature as our home in the sense of being a fundamental place for humans. But it also encompasses a more active component, according to Bergmann. He uses the German term "Beheimatung," understood as making oneself at home, for describing "home" as having a formative quality in addition to naming a given location. The environmental relevance is clear: How are we going to make our earth, which we are now destroying, our home (again)? In a similar way, García-Rivera questions whether the cosmos is our home by asking, in the current situation of environmental challenges, whether we are at home in the cosmos. For García-Rivera, the way to move forward is to rediscover our home in the cosmos.[52] For Boff, the active dimension is expressed even more clearly: Our task is to build a

[49] Arne Næss, *Ecology of Wisdom* (London: Penguin Books, 2008), 45ff; Edward Wilson, *Biophilia: The Human Bond with Other Species* (Cambridge: Harvard University Press, 1984), and Kristiansen, "Økoteologi anno 2007."

[50] Cummings, *Eco-Spirituality*, 65.

[51] Cummings, *Eco-Spirituality*, 66.

[52] García-Rivera, *The Garden of God*, xii.

home for all people.[53] This active dimension is needed, he says, because we do not feel at home in the current situation. The lack of feeling at home, or at least having limited experiences of feeling at home may be described in negative terms as a feeling of alienation.

The Role of Aesthetic Experiences in Nature

I find the emotions of belonging and feeling at home closely related to aesthetic experiences. These emotions can be understood as being evoked by aesthetic experiences in nature.[54] The role of aesthetic experiences appears to be important for the distinctiveness of ecospirituality, however, this aspect is still awaiting more extensive articulation.[55] The backdrop for this revitalization of aesthetics is the previously mentioned rational and dualistic approach to nature through modernity. Fischer puts it this way: "A dichotomous understanding of the human person has dissociated reason not only from emotion but also from imagination"[56] Again, we notice that the main concern is leaving behind what has been going wrong in modernity. However, we can certainly problematize ecospirituality's narrative of modernity as stressing rationality at the expense of emotions and aesthetic experiences. Did modernity actually lose its connection to nature? Was there really a decline of emotion, mystery, and wonder (from Descartes, through Newton, to Darwin), leaving only a mechanical picture of nature? A short reminder from the conclusion of Darwin's *On the Origin of Species* may correct, or at least modify, such a sweeping anti-modernity narrative:

> There is grandeur in this view of life, with its several powers, having been originally breathed into a few forms or into one; and that, whilst this planet has gone cycling on according to the fixed law of gravity, from so simple a

[53] Boff, *Toward an Eco-Spirituality*, 32.

[54] Helga Synnevåg Løvoll, Knut-Willy Sæther, and Mark Graves, "Feeling at Home in the Wilderness: Environmental Conditions, Well-Being and Aesthetic Experience," *Frontiers in Psychology*, 11: 402, doi:10.3389/fpsyg.2020.00402.

[55] On this topic, I find that ecospirituality draws on resources from environmental aesthetics.

[56] Fischer, "Christian Spirituality in a Time of Ecological Awareness," 178.

beginning endless forms most beautiful and most wonderful have been, and are being, evolved.[57]

Hence, I think a more nuanced view of what has been going on in modernity is needed. The notions of modernity, enlightenment, and rationality can easily be used in a pejorative way for negatively making statements about the understanding of the human-nature relationship. That said, I think ecospiritual scholars have a point when they more clearly address the role of emotions and aesthetic experiences in nature. Our encounter is not only about understanding nature as our home, but also about feeling and experiencing nature as our home.

Another take on the apparent gap between rationality and emotions is to argue for how they are intertwined. I cannot explore this ongoing discussion in depth here but will just sketch one possible trajectory for thinking these two notions together. In his elaboration of ecophilosophy, Næss draws on Spinoza's notions of *amor intellectualis* and *ratio*.[58] For Spinoza, Næss says, *amor intellectualis* unites emotions and reason in a large whole. Further, ratio is the inner voice of reason and points out a direction in achieving the goals of freedom and being in itself. Hence, ratio is about a deeper rationality (where both cognition and emotions find their place), which has a *telos* with ethical implications.[59]

Aesthetic-emotional encounters with nature are subtle and complex. Thomas Berry says our encounter with nature includes the following broad range of elements: "The natural world is the life-giving nourishment of our physical, emotional, aesthetic, moral, and religious existence."[60] Even when we focus solely on aesthetic experience, we still must struggle with a highly fluid notion. Nevertheless, some salient points may be highlighted. Recent research has stressed three aesthetic experiences in

[57] Charles Darwin, *On the Origin of Species by Means of Natural Selection, or the Preservation of Favoured Races in the Struggle for life* (London: John Murray, 1859), 490.

[58] Næss, *Ecology of Wisdom*, 235.

[59] This is further elaborated in Knut-Willy Sæther, "Rationality in Play? A Philosophical Journey in the Current Landscape of Facts and Truth," in *Navigating Post-Truth and Alternative Facts: Religion and Science as Political Theology*, ed. Jennifer Baldwin (Lanham: Lexington Books, 2018), 63–79.

[60] Thomas Berry, "Economics: Its Effect on the Life Systems of the World," in *Thomas Berry and the New Cosmology*, eds. Anne Lonergan and Caroline Richards (Mystic: Twenty-Third Publications, 1990), 18.

nature: beauty, the sublime, and wonder.[61] In particular, awe is prominent in nature-based experiences of the sublime and wonder.[62] Jeffrey Sobosan puts it this way: "What [the stars] did stir in me (...) was wonder, and I have turned this wonder loose in my imagination many times as I have looked into the clear night sky. And each time I have been given joy."[63] Robert Fuller describes wonder as having an intensified and transformative character, expanding the range of our needs, and fostering empathy and compassion.[64] Whether we talk about experiences of beauty, the sublime, or wonder in nature, they may all be understood broadly as aesthetic experiences, and these expansive experiences foster something in us that may be interpreted as self-transcendent.[65]

At first glance, aesthetic experiences such as beauty and wonder are recognized as positive phenomena. By contrast, experience of the sublime might evoke an ambiguity. We are attracted to the sublime yet also feel a need to withdraw. The sublime simultaneously fascinates and frightens us. Experiences of the sublime have echoes in mysticism and deep experiences of the holy described by Rudolf Otto's classic term *tremendum et fascinans*.[66] Indeed, recent research suggests that these kinds of experiences may yield a sense of self-transcendence, foster meaning, or make us feel a part of the larger whole.[67]

However, in order to address the negative emotions that arise in the face of climate change, we need to nuance the received tradition of

[61] Mark Graves, Helga Løvoll, and Knut-Willy Sæther, "Friluftsliv: Aesthetic and Psychological Experience of Wilderness Adventure," in *Issues in Science and Theology: Nature and Beyond*, eds. Michael Fuller, Dirk Evers, Anne Runehov, Knut-Willy Sæther, and Bernard Michollet (Dordrecht/Heidelberg/New York/London: Springer, 2020), 207–220.

[62] Robert Fuller, "From Biology to Spirituality: The Emotional Dynamics of Wonder," in *Practices of Wonder: Cross Disciplinary Perspectives*, ed. Sophia Vasalou (New York: Pickwick Publications, 2012), and Knut-Willy Sæther, *Naturens skjønnhet* (Oslo: Cappelen Damm Akademisk, 2017).

[63] Jeffrey G. Sobosan, *Romancing the Universe: Theology, Science, and Cosmology* (Grand Rapids: Wm. B. Eerdmans Publishing Co., 1999), 1.

[64] Fuller, "From Biology to Spirituality: The Emotional Dynamics of Wonder," 85.

[65] Helga Synnevåg Løvoll and Knut-Willy Sæther, "Awe Experiences, the Sublime, and Spiritual Well-being in Arctic Wilderness," *Frontiers in Psychology*, 13:973922, doi: 10.3389/fpsyg.2022.973922.

[66] Rudolf Otto, *The Idea of the Holy* (Oxford: Oxford University Press, 1923).

[67] Mark Graves, Helga Løvoll, and Knut-Willy Sæther, "Friluftsliv: Aesthetic and Psychological Experience of Wilderness Adventure," in *Issues in Science and Theology: Nature and Beyond*, eds. Michael Fuller, Dirk Evers, Anne Runehov, Knut-Willy Sæther, and Bernard Michollet (Dordrecht/Heidelberg/New York/London: Springer, 2020), 207–220.

ascribing positive emotional and aesthetic effects to our encounters with nature. Indeed, experiences of negative emotions such as fear, anger, guilt, and shame have been largely responsible for launching movements of environmental advocacy. Such engagements provoke an awareness that something is terribly wrong and open up the possibility of experiences of ugliness, disorder, lack of connection, grief (the term "ecogrief" is even used), anxiety, and alienation. As one striking example, García-Rivera describes his "vision of hell" when working on a missile project as an engineer.[68] There is an undeniable tension between feeling at home in nature while experiencing alienation. Such an emotional state corresponds to what Glenn Albrecht has described as solastalgia, "the homesickness you have when you are still at home."[69] Our environmental home is changing radically, and this creates anxiety and fear. At the same time, aesthetic experiences of beauty, wonder, and even the sublime lead in a more positive direction with implications for ethics.

Bergmann and Sobosan link aesthetic experiences with ethical reflection. For Sobosan, gazing at the stars evokes several intertwined feelings:

> The experience is both aesthetic and moral. It is aesthetic because the panorama of shapes and colors my eyes take in produces an experience of beauty; it is moral because knowledge of the sheer size and age of what I am seeing produces an experience of humility.[70]

Similarly, Bergmann correlates aesthetics and ethics in his notion of aesth/ethics. According to Bergmann, aesthetic experiences are about perception, and if ethics is defined as a discursive reflection on moral problems, we cannot exclude people's mental capacities by separating perception from ethics. In relation to moral problems, perception is necessarily prior to reflection and pursuing solutions.[71] Perceiving ourselves and others within a common environment tends to encourage moral agency and reflection. I find here traits for further development in how to understand the human-nature relationship.

[68] García-Rivera, *The Garden of God*, viii.

[69] Glenn Albrecht, "Solastalgia: The Distress Caused by Environmental Change," *Australasian Psychiatry*, 15 (2007): 95.

[70] Sobosan, *Romancing the Universe: Theology, Science, and Cosmology*, 1.

[71] Sigurd Bergmann, "Atmospheres of Synergy: Towards an Eco-theological aesth/ethics of Space," *Ecotheology* 11(2006): 326–356.

First, aesthetic experiences in nature and the evaluation of these experiences entail value judgments. Even before we evaluate, we are dealing with values by giving priority to exposing ourselves to experiences of nature. That is, aesthetic experiences presuppose value judgments by ascribing value to nature. Second, given their intrinsic value, aesthetic experiences may engender moral agency. However, this needs to be problematized, because is it the case that aesthetic experiences in nature necessarily result in moral agency and action? From our discussion of Olsen's work above, we established that knowing what is right does not necessarily lead to right action. Indeed, knowledge may not issue in any action. The interconnectedness between aesthetics, ethics, and moral action is undoubtedly complex. Boff makes a helpful comment by pointing out that our experiences of our surroundings might move us toward concrete action that is earth-grounded, as expressed by his "four Rs": reduce, reuse, recycle, and reforest.[72] Hence, we see a possible trajectory from aesthetic experiences to ethics and action, which will be essential if ecospirituality is to avoid being solely a rational endeavor.

Toward a Dynamic Relational Holism

Through my analysis so far, I have unpacked what gives ecospirituality its distinctiveness for understanding the human-nature relationship. Given its strong criticism of dualism, we need to examine what we find as ecospirituality's own alternative. In this final section, I will address this topic in greater depth in light of dualism, monism, and holism. By integrating the topics analyzed above, I will argue that the human-nature relationship in ecospirituality may be described as a *dynamic relational holism*.

As we have seen, ecospirituality's criticism of dualism is expressed by countering Cartesian dualism or dualism in general. The notion of anthropocentrism is also part of that criticism.[73] The "splitting of spheres" taking place in modernity is described in dual terms, such as subject/object,

[72] Boff, *Eco-Spirituality*, 48.

[73] As mentioned earlier, the notion of anthropocentrism is also used in a pejorative way in ecospirituality. However, the term "anthropocentrism" needs to be problematized, as done by Washington in *A Sense of Wonder Towards Nature: Healing the Planet Through Belonging*, 55–60. See also Mary Evelyn Tucker and John Grim, "The Movement of Religion and Ecology: Emerging Field and Dynamic Force," in *Routledge Handbook of Religion and Ecology*, eds. Willie Jenkins, Mary Evelyn Tucker, and John Grim (London and New York: Routledge, 2017), 4–5.

human/nature, spirit/matter, and brain/mind. However, we need to give dualism some more content than the pejorative use we find in ecospirituality. Dualism has a variety of meanings in the history of thought and in current philosophical discussions. In philosophy of mind, which to a large extent has problematized this topic, dualism refers mainly to the radical difference between mind and matter. Further, dualism is not only about "equal" phenomena, as it makes sense to describe dual categories as involving different kinds of hierarchy, such as human beings as superior to nature, man superior to woman, and so on.[74]

My interest here is in what has been described as substance dualism, inherited from Descartes.[75] Such a dualism may, to some degree, be associated with metaphysical dualism. The term "substance" expresses a dualism about different entities, e.g., mind-matter, or in more classical terms soul-body. In such a dualism, the individual entities are substances of contrary natures, although causally, they might influence each other. Based on the analysis above, I find it evident that ecospirituality pushes back on substance dualism and argues for a more integrated understanding of reality, including the human-nature relationship. In particular, ecospirituality suggests how the interior and the exterior of humans and nature are deeply connected and how the notion of Spirit may be developed out of underlying relationality.

At the same time, one might ask whether a substance dualism by necessity leads to a view where humans are separated from nature and further results in alienation from nature and a lack of environmental responsibility. One could explore how some sort of dualistic understanding of reality may very well evoke a concern for our environment. However, this line of thought is not of interest in ecospirituality, which does not dissolve the distinction between humans and nature, but presupposes dualities, or at least distinctions, when talking about the human encounter with nature. This is evident in the role of place and space, how to see nature, and how to be aware of our surroundings. Hence, it seems that the problem of dualism for ecospiritual scholars is actually about dualities understood as substantially separated and separating phenomena, and not about the distinctiveness of parts.

[74] See Heather Eaton, "Christianity, Ecofeminism, and Transformation," in *Religion and Ecology*, ed. John Hart (Oxford: Wiley Blackwell, 2017), 263.

[75] Fischer, "Christian Spirituality in a Time of Ecological Awareness," 176.

Ecospirituality's rejection of substance dualism then brings us to what may be considered the opposite notion: monism.[76] As with dualism, monism can take different shapes, such as substance monism. In one sense, substance monism may be associated with physicalism, which states that the only existing substance is physical. In that case, we are dealing with some sort of reductionist materialistic monism. Clearly, ecospirituality does not follow this line of thought. However, if substance monism does not necessarily lead to reductionism, do we then find a nonreductionist monism? In general, substance monism suggests that reality is substantially one and the same thing and is not dividable. Hence, there are no substantial differences or distinctions per se, but only one substance. Such a view is not evident from my exploration of the material above, thus monism seems to be inadequate for describing ecospirituality inasmuch as it argues for differences or distinctions of parts. On the other hand, ecospirituality does argues for seeing and experiencing reality as a whole or at least striving to develop such a way of seeing. However, this is not a monistic understanding of the human-nature relationship, since seeing nature as a whole and how parts are related does not dissolve distinctions.

Hence, there is an important difference between substance monism and arguing that reality is a whole. The latter presupposes the existence of dualities or entities in terms of different phenomena, e.g., nature and human, which are seen *as related*. Ecospirituality's emphasis on seeing nature and being aware of our surroundings presupposes distinctions. When Bergmann argues against a strong dichotomy between a subject and its surroundings, he presupposes distinctions: a subject (a human who sees) and an object (nature being seen). Otherwise, the notion of surrounding does not make sense.

What, then, is the alternative to dualism and monism? For Fischer, a feasible approach is to consider holistic models of reality.[77] As with dualism and monism, holism is an ambiguous term.[78] Nevertheless, I find two elements in holism appropriate for ecospirituality's description of the human-

[76] Howard Robinson, "Dualism," in *The Stanford Encyclopedia of Philosophy* (Fall 2020 Edition), ed. Edward N. Zalta, https://plato.stanford.edu/archives/fall2020/entries/dualism/

[77] Fischer, "Christian Spirituality in a Time of Ecological Awareness," 177.

[78] Such as ontological holism and property holism. See Richard Healey and Henrique Gomes, "Holism and Nonseparability in Physics," *The Stanford Encyclopedia of Philosophy* (Spring 2022 Edition), Edward N. Zalta (ed.), https://plato.stanford.edu/archives/spr2022/entries/physics-holism/

nature relationship and view of reality as a whole. First, in order to understand the human-nature relationship properly, we need to it (and reality) as a whole. However, such a holism, described as metaphysical, covers more than just seeing reality as a whole. It also claims that the particularities are related in this wholeness. Second, metaphysical holism argues that the whole is greater than the sum of its parts.[79] Such a holism resonates with the notion of emergence.[80] However, compared to holistic understandings, emergent nature embraces a more dynamic view, which also includes the evolutionary story of nature while being more comprehensive than this story. As described by James D. Proctor, an emergent nature is "… in many respects an even more far-sweeping vision than evolutionary nature."[81]

Therefore, an ecospiritual approach to nature encompasses a reflection of how parts are connected while not being reducible from the whole to the parts. Metaphysical holism supports an understanding of humans and nature as intertwined, yet not substantially the same. Wheeler articulates this clearly when she describes how ecospirituality differs from ecomysticism. For her, ecospirituality is "holding in tension the interconnectedness and relationality of all creation, without the collapse of individual parts to accomplish union."[82]

Further, holism is not only about seeing and recognizing that humans and nature are intertwined, but also that they are dependent on each other.[83] More specifically for the human-nature relationship, a holistic approach in this sense may be illustrated on two "levels." First, the human person is an intimate interconnection (a whole) where no aspect of it may live independently of the whole (contrary to mind/body dualism). Second, the whole cosmos is an integrated and interdependent system.

We need to take the holistic approach one step further by asking what binds together the human-nature relationship and reality as a whole.

[79] In particular, this point is stressed in environmental philosophy. See, for example, Michael P. Nelson, "Teaching Holism in Environmental Ethics." *Environmental Ethics* 32 (1): 33–49.

[80] For a problematization of emergence, see Gregory Peterson, "Who needs emergence?" In *Envisioning Nature, Science and Religion*, ed. James D. Proctor (Conshohocken: Templeton Press), 3–35.

[81] James D. Proctor, "Visions of Nature, Science and Religion." In *Envisioning Nature, Science and Religion*, ed. James D. Proctor (Conshohocken: Templeton Press, 2009), 10.

[82] Wheeler, *Eco-Spirituality: An Introduction*, 6.

[83] Abram, *The Spell of the Sensuous: Perception and Language in a More-Than-Human World.*

Substance dualism might also very well argue for relations between the parts. Hence, this is not what distinguishes holism from dualism. Rather, it is *how* parts are related in a larger whole and the degree to which they are intertwined. Two points seem important. First, ecospirituality emphasizes place and space. Humans, as well as nature that surrounds us, are always located. The notions of parts and whole have a concrete context. As Bergmann argues, I am always a part of my surroundings and part of a deeper reality that grounds, anchors, and embraces the living. Therefore, the notions of place and space provide a spatial context of how to understand relationality.

My elaboration of ecospirituality's relational understanding of nature supports such a holistic approach. Relations presuppose differences, or at least distinctions. If not, there are no relations. By arguing for a relational reality in such a way, we can talk about how the relations of parts exhibit holism. Hence, ecospirituality maintains the uniqueness of parts, such as human beings.[84] However, how parts are related, e.g., humans and nature, needs to be more elaborated in ecospirituality in order to give holism a more substantial content.

When ecospirituality argues against dualism it is also as a criticism of separation of the "spiritual" and the "material" (e.g., spirit/matter). How to overcome this dualism toward a more integrated understanding may be developed along the lines of inhabitation, atmosphere, and Spirit. These notions are useful first steps for further explorations of what relationality might cover in this context.[85] In addition, the notions of feeling at home and belonging, and their close connection to aesthetic experiences, give voice to a more profound understanding of "spiritual" and "material." For ecospirituality, the spiritual dimension is not limited to humans, but binds together the whole of reality, and the "material" is present in humans as well as in reality as a whole. Hence the duality of spirit/matter follows different "dual lines" than the distinction between humans and nature.

Inspired by Paul Teller, I suggest that the understanding of ecospirituality may be developed along the lines of *relational holism*. Teller writes

[84] Wheeler, *Eco-Spirituality: An Introduction*, 6.

[85] Overcoming dualism in how to understand the human nature (e.g., matter/mind and body/soul) can be explored in different ways in light of the topics analyzed above, such as the problematic divisions between inner and outer that ecospirituality argues against. See Fischer, "Christian Spirituality in a Time of Ecological Awareness," 176.

within the context of physics and offers a stinging criticism of physicalism. According to Teller:

> By *relational holism* I will mean the claim that objects that in at least some circumstances we can identify as separate individuals have *inherent relations*, that is, relations that do not supervene on the nonrelational properties of the distinct individuals.[86]

Without unpacking Teller's relational holism in physics here, I am using his notion as an attempt to characterize the human-nature relationship in ecospirituality. In addition, and based on my analysis above, we need to incorporate a *dynamic* understanding. The insight from evolution theory and, more extensively, the notion of emergence resonates with ecospirituality's "spiritual component." Another topic from my analysis that supports a relational holism, which includes a dynamic component, is emotional and aesthetic experiences in nature. Two points may be highlighted. First, my elaboration of ecospirituality emphasizes that aesthetic experiences in nature extend humans' daily experiences. They are self-transcendent and bring forth insight into how humans are related to something larger than themselves. One point here is to argue for the transformative character these experiences generate in our encounters with nature, in terms of "feeling a part of a larger whole." Second, the ancient definition of what is considered aesthetically beautiful is unity in variety.[87] Such an expression pinpoints what takes place in the aesthetic experience. We experience both the parts (variety) and at the same time how these parts are related (unity). Hence, aesthetic experiences provide insight into how to understand reality as a whole while leaving room for differences. Through such experiences, a direction is disclosed for how we may deepen our interact with nature. In line with Bergmann and Sobosan, aesthetic experiences are considered value judgments, which act as a motivation for action.

The topics I have identified, seeing nature/awareness, place and space, belonging and feeling at home in nature, and the role of aesthetic experiences in nature provide resources to characterize ecospirituality as advocating a dynamic relational holism. My contribution in this article is a first

[86] Paul Teller, "Relational Holism and Quantum Mechanics," *The British Journal for the Philosophy of Science* 37(1) (1986): 73.

[87] García-Rivera, *The Community of the Beautiful: A Theological Aesthetics* (Liturgical Press, 1999), 72.

step toward how this can be developed further and needs to be critically elaborated in the ongoing discussion on how to understand the human-nature relationship in ecospirituality.

Bibliography

Abram, David. 1997. *The Spell of the Sensuous: Perception and Language in a More-Than-Human World.* Vintage.

Albrecht, Glenn. 2007. Solastalgia: The Distress Caused by Environmental Change. *Australasian Psychiatry* 15: 95–98.

Bergmann, Sigurd. 2006. Atmospheres of Synergy: Towards an Eco-theological Aesth/ethics of Space. *Ecotheology* 11: 326–356.

———. 2008. The Beauty of Speed or the Discovery of Slowliness. In *The Ethics of Mobilities: Rethinking Place, Exclusion, Freedom and Environment*, ed. Sigurd Bergmann and Tore Sager, 13–24. London and New York: Routledge.

———. 2011. Aware of the Spirit: In the Lens of a Trinitarian Aesth/ethics of Lived Space. In *Ecological Awareness: Exploring Religion, Ethics and Aesthetics*, ed. Sigurd Bergmann and Heather Eaton, 23–39. Münster: LiT Verlag.

———. 2013. "'… the Space where I am'": Decolonising, Re-sacralising and Transfiguring Landscapes through the Aest/ethic Lens. In *Aesth/Ethics in Environmental Change: Hiking Through the Arts, Ecology, Religion and Ethics of the Environment*, ed. Sigurd Bergmann, Irmgard Blindow, and Konrad Ott, 39–69. Münster: LiT Verlag.

Bergmann, Sigurd, Irmgard Blindow, and Ott Konrad. 2013. Aesth/Ethics in Environmental Change: Hiking Through the Arts, Ecology. In *Religion and ethics of the environment.* Münster: LiT Verlag.

Bergmann, Sigurd, and Heather Eaton. 2011. Awareness Matters: Introductory Remarks About the Interwoven Gifts of Life and Belief. In *Ecological Awareness: Exploring Religion, Ethics and Aesthetics*, ed. Sigurd Bergmann and Heather Eaton, 1–6. Münster: LiT Verlag.

Berry, Thomas. 1990. Economics: Its Effect on the Life Systems of the World. In *Thomas Berry and the New Cosmology*, ed. Anne Lonergan and Caroline Richards, 5–26. Mystic: Twenty-Third Publications.

Boff, Leonardo. 2014. *Toward an Eco-Spirituality.* New York: The Crossroad Publishing Company.

Bregman, Lucy. 2000. *The Ecology of Spirituality: Meanings, Virtues, and Practice in a Post-Religious Age. Waco: Baylor University Press, 2014. Carlson, Allen. Aesthetics and the Environment: The Appreciation of Nature, Art and Architecture.* London and New York: Routledge.

Castree, Noel. 2009. Place: Connections and Boundaries in an Interdependent World. In *Key Concepts in Geography*, ed. Nicholas Clifford, Sarah Holloway, Stephen P. Rice, and Gill Valentine, 153–172. London: Sage.

Cobb, John B. 1972. *Is It Too Late? A Theology of Ecology*. Minneapolis: Fortress Press.

Cummings, Charles. 1991. *Eco-Spirituality: Toward a Reverend Life*. Mahwah: Paulist Press.

Darwin, Charles. 1859. *On the Origin of Species by Means of Natural Selection, or the Preservation of Favoured Races in the Struggle for Life*. London: John Murray.

Eaton, Heather. 2017. Christianity, Ecofeminism, and Transformation. In *Religion and Ecology*, ed. John Hart, 256–272. Oxford: Wiley Blackwell.

Fischer, Kathleen. 2010. Christian Spirituality in a Time of Ecological Awareness. *Theology Today* 67: 169–181.

Fuller, Robert. 2012. From Biology to Spirituality: The Emotional Dynamics of Wonder. In *Practices of Wonder: Cross Disciplinary Perspectives*, ed. Sophia Vasalou, 64–87. Eugene: Pickwick Publications.

García-Rivera, Alejandro. 2009. *The Garden of God: A Theological Cosmology*. Minneapolis: Fortress Press.

Graves, Mark, Helga Synnevåg Løvoll, and Knut-Willy Sæther. 2020. Friluftsliv: Aesthetic and Psychological Experience of Wilderness Adventure. In *Issues in Science and Theology: Nature and Beyond*, ed. Michael Fuller, Dirk Evers, Anne Runehov, Knut-Willy Sæther, and Bernard Michollet, 207–220. Dordrecht/Heidelberg/New York/London: Springer.

Healey, Richard, and Henrique Gomes. 2022. Holism and Nonseparability in Physics. In *The Stanford Encyclopedia of Philosophy*, ed. Edward N. Zalta. https://plato.stanford.edu/archives/spr2022/entries/physics-holism/.

Hefner, Philip. 2011. Teilhard's Spiritual Vision for the Mystical Milieu. In *The Legacy of Pierre Teilhard de Chardin*, ed. S.J. James Salmon and John Farina, 79–94. Mahwah: Paulist Press.

Jirásek, Ivo. 2013. Verticality as Non-Religious Spirituality. *Implicit Religion* 16 (2): 191–201.

Kristiansen, Roald. 2007. Økoteologi anno 2007. In *Økoteologi: Kontekstuelle perspektiver på miljø og teologi*, ed. Bård Mæland and Tom Sverre Tomren, 13–27. Trondheim: Tapir Akademisk Forlag.

Lefebvre, H. 1974. *The production of space*. Vol. 1991. Oxford: Blackwell.

Løvoll, Helga Synnevåg, Knut-Willy Sæther, and Mark Graves. 2020. Feeling at Home in the Wilderness: Environmental Conditions, Well-Being and Aesthetic Experience. *Frontiers in Psychology* 11: 402. https://doi.org/10.3389/fpsyg.2020.00402.

Løvoll, Helga Synnevåg, and Knut-Willy Sæther. 2022. Awe experiences, the sublime, and spiritual well-being in artic wilderness. *Frontiers in Psychology* 13: 973922. https://doi.org/10.3389/fpsyg.2022.973922.

Nelson, Michael P. 2010. Teaching Holism in Environmental Ethics. *Environmental Ethics* 32 (1): 33–49.

Nynäs, Peter. 2008. From Sacred Place to an Existential Dimension of Mobility. In *The Ethics of Mobilities: Rethinking Place, Exclusion, Freedom and Environment*, ed. Sigurd Bergmann and Tore Sager, 157–176. London and New York: Routledge.

Næss, Arne. 2008. *Ecology of Wisdom*. London: Penguin Books.

Olsen, Harald. 2007. Økospiritualitet–for det hele menneske. In *Økoteologi: Kontekstuelle perspektiver på miljø og teologi*, ed. Bård Mæland and Tom Sverre Tomren, 251–263. Trondheim: Tapir Akademisk Forlag.

Otto, Rudolf. 1923. *The Idea of the Holy*. Oxford: Oxford University Press.

Principe, Walter. 1983. Toward Defining Spirituality. *Studies in Religion* 12: 127–141.

Proctor, James D. 2009. Visions of Nature, Science and Religion. In *Envisioning Nature, Science and Religion*, ed. James D. Proctor, 3–35. Conshohocken: Templeton Press.

Seymour, Valentine. 2016. The human–nature relationship and its impact on health: A critical review. *Frontiers in Public Health* 4: 260. https://doi.org/10.3389/fpubh.2016.00260.

Sheldrake, Philip. 2013. *Spirituality: A Brief History*. Hoboken: Wiley-Blackwell.

Sobosan, Jeffrey G. 1999. *Romancing the Universe: Theology, Science, and Cosmology*. Grand Rapids: Wm. B. Eerdmans Publishing Co.

Sæther, Knut-Willy. 2017. *Naturens skjønnhet: En studie av forholdet mellom estetikk, teologi og naturvitenskap*. Oslo: Cappelen Damm Akademisk.

———. 2018. Rationality in Play? A Philosophical Journey in the Current Landscape of Facts and Truth. In *Navigating Post-Truth and Alternative Facts: Religion and Science as Political Theology*, ed. Jennifer Baldwin, 63–79. Lanham: Lexington Books.

———. 2020. Naturen som sted: En eksplorative studie av økofilosofi, økoteologi og kristen økospiritualitet. In *Rom og Sted: Religionsfaglige og interdisiplinære bidrag*, ed. Knut-Willy Sæther and Anders Aschim, 167–187. Oslo: Cappelen Damm Akademisk.

Sæther, Knut-Willy, and Anders Aschim. 2020. *Rom og Sted: Religionsfaglige og interdisiplinære bidrag*. Oslo: Cappelen Damm Akademisk.

de Chardin, Teilhard. 1960. Pierre. In *The Divine Milieu*. New York: Harper and Row Publishers.

Teller, Paul. 1986. Relational Holism and Quantum Mechanics. *The British Journal for the Philosophy of Science* 37 (1): 71–81.

Tucker, Mary Evelyn, and John Grim. 2017. The Movement of Religion and Ecology: Emerging Field and Dynamic Force. In *Routledge Handbook of Religion and Ecology*, ed. Willie Jenkins, Mary Evelyn Ticker, and John Grim, 3–12. London and New York: Routledge.

Washington, Haydn. 2019. *A Sense of Wonder Towards Nature: Healing the Planet through Belonging*. London and New York: Routledge.
Wheeler, Rachel. 2022. *Eco-Spirituality: An Introduction*. Minneapolis: Fortress Press.
Whitehead, Alfred N. 1978. *Process and Reality*. New York: The Free Press.
Wilson, Edward. 1984. *Biophilia: The human bond with other species*. Cambridge: Harvard University Press.

CHAPTER 3

The Role of Panentheism and Pantheism for Environmental Well-Being

Lina Langby

Introduction

In this chapter, I explore pragmatic arguments favoring panentheism and pantheism, focusing on environmental well-being. In short, pantheism is a conception of God/the divine stating that God is identical to the world. Panentheism is a conception of God/the divine saying that the world is part of, or in, God, while God exceeds the world ontologically. At the heart of the debate about environmental well-being are ***ontological God-world dualism*** and ***in-worldly dualisms***, where ontological God-world dualism is believed to cause harmful in-worldly hierarchical dualisms.

In the philosophical literature and debate about different conceptions of God/the divine, we find various reasons to prefer or reject them. Some arguments are theological, others focus on internal coherence, and yet another revolves around correspondence to reality as described by the

L. Langby (✉)
Department of Theology, Uppsala University, Uppsala, Sweden
e-mail: lina.langby@teol.uu.se

T. J. Hastings, K.-W. Sæther (eds.), *Views of Nature and Dualism*,
https://doi.org/10.1007/978-3-031-42902-6_3

modern natural sciences.[1] When it comes to arguments about the harms or benefits of a particular worldview, the reasons offered are almost exclusively pragmatic. A pragmatic argument in favor of a worldview claims that the worldview in question is *better* than its alternatives for different reasons. In the discussion about environmental well-being, such pragmatic arguments are significant.

Panentheists and pantheists often argue that their worldview and conception of God entails a more ethical approach to and relationship with nature than other conceptions of God. Although there certainly are epistemic truth claims in favor of pantheism or panentheism, or at least claims that a specific worldview is *the most accurate* way to conceptualize the God-world relationship, such truth claims are of less significance when it comes to questions of harm and benefit. In other words, it does not necessarily matter whether the God-world conception in question corresponds to reality or not because the *effects* of our worldviews are real enough. When we act in the world, we affect it, and some actions cause more significant harm than others.

Ontological God-World Dualism

Before turning to the role of panentheism and pantheism for environmental well-being, I wish to clarify more precisely what classical theism, panentheism, and pantheism are and how they relate to ontological God-world dualism.

Classical Theism

Classical theism, or traditional theism as it is sometimes called, is a family of doctrines claiming God to be transcendent, omnipresent, eternal, self-sufficient, ontologically independent, necessary, unaffected by time or contingent events, and the creator and sustainer of everything that exists. The traditional omni-attributes of omnibenevolence (perfect goodness),

[1] See, for example, John Bishop and Ken Perszyk, "Concepts of God and Problems of Evil," in *Alternative Concepts of God: Essays on the Metaphysics of the Divine*, eds. Andrei A. Buckareff and Yujin Nagasawa (Oxford: Oxford University Press, 2016), 106–27, https://doi.org/10.1093/acprof:oso/9780198722250.001.0001; Philip Clayton, *God and Contemporary Science*, Edinburgh Studies in Constructive Theology (Edinburgh: Edinburgh Univ. Press, 1997); Philip Clayton and Arthur Peacocke, *in Whom We Live and Move and Have Our Being: Panentheistic Reflections on God's Presence in a Scientific World* (Grand Rapids, Mich.: William B. Eerdmans Publishing Company, 2004).

omnipotence (perfect power), and omniscience (perfect knowledge) are generally not disputed among classical theists.[2] God, according to classical theism, existed before God created the world and will continue to exist eternally even if the world seizes to exist. The God of classical theism is a pure transcendent spirit—ontologically distinct from the physical world but immanent in it due to divine omnipresence. Classical theists distinguish between God as Creator and the world as the creation. They are two distinct categories. The doctrine of *creatio ex nihilo*—creation out of nothing or no pre-existing thing—thus follows naturally.[3] The world exists because of the Creator's sustaining, voluntary activity. In other words, God does not essentially need the world, but the world essentially needs God to exist. Even if there are classical theists who reject *creatio ex nihilo*, they must accept it as *possible* given the classical theist's belief in a wholly transcendent, ontologically independent, omnipotent, Creator-God. The existence of God is independent of the existence or non-existence of the physical world, which entails that reality as such is constituted by both matter (the world) and spirit (God). Because of this, we see that classical theism entails *ontological God-world dualism*.

I take classical theism to entail *at least* the following claims: (1) God is the creator and sustainer of the world, (2) ontologically, God exists independently of the world, (3) God is a transcendent and non-physical spirit, (4) God is eternal, necessary, and perfect Being.

Panentheism

Etymologically panentheism and pantheism both stem from Greek. *Pan* (πάν) means "all" or "everything," *theos* (θεος) means "God," and *en* (ἐν) means "in." We can thus put together *pan-theism* and *pan-en-theism*: Everything-God (pantheism) and everything-in-God (panentheism).[4]

[2] Thomas Williams, "Introduction to Classical Theism," in *Models of God and Alternative Ultimate Realities*, ed. Jeanine Diller and Asa Kasher (Dordrecht: Springer, 2013); Katherin Rogers, *Perfect Being Theology*, Reason and Religion (Edinburgh: Edinburgh University Press, 2000).

[3] Thomas Aquinas, *Summa Theologiae* I, Q.45, article 5.

[4] I use capital G when discussing classical theism, panentheism, and pantheism. The reason for this is not to imply, for example, a Christian understanding, nor that God must be a person or person-like. I use capital G to signify the ultimacy of the God-conception. In classical theism, panentheism, and pantheism God/the divine is regarded as the ultimate One. In other words, the use of capital G signifies that it is not Thor, Odin, or Zeus we are talking about. Thor, Odin, and Zeus would all be described with the lowercase g. They are gods—not God.

Karl Christian Friedrich Krause, a German idealist philosopher (1781–1832), is often presented as the one who first named the doctrine that everything is in God *panentheism*.[5] The history of panentheism can be traced back at least to Neo-Platonism, and panentheistic thoughts are found in many religions. However, Krause was the first to use the label "panentheism." Panentheism has often been regarded as the middle ground between classical theism and pantheism, and it is sometimes also accompanied by process theology.[6]

According to panentheism, God is both immanent and transcendent in/to the universe.[7] This is the case also with classical theism. However, where the classical theistic God is immanent and omnipresent, God is still an ontologically independent and separate "category" or "entity": God is pure spirit. In panentheism, the world is a part of God. God is both spirit and matter and the world and God affects each other in a mutual relationship.[8]

I claim it is appropriate to ascribe ontological God-world dualism to panentheism because the defining feature of panentheism is the belief that God is the world (the world is a part of God) but that God nevertheless is something *more* than the world. God transcends the part of Godself which is the world. This is the most significant difference between panentheism and pantheism because, in pantheism, God does not transcend the world in any way—God *is* the world and nothing more.

[5] Michael W. Brierley, "Naming a Quiet Revolution: The Panentheistic Turn in Modern Theology," in *In Whom We Live and Move and Have Our Being: Panentheistic Reflections on God's Presence in a Scientific World*, eds. Arthur Peacocke and Philip Clayton (Grand Rapids, Mich.: Eerdmans, 2004), 2. For an additional view, see John Culp, "Panentheism," in *The Stanford Encyclopedia of Philosophy*, ed. Edward N. Zalta (Metaphysics Research Lab, Stanford University, 2017), https://plato.stanford.edu/archives/sum2017/entries/panentheism/

[6] I take process theology, at least of the Whiteheadian tradition, to be a version of panentheism. However, not all versions of panentheism adhere to the core doctrines of process theology.

[7] Mikael Stenmark, "Panentheism and Its Neighbors." *International Journal for Philosophy of Religion* 85, no. 1 (2019): 26–27, https://doi.org/10.1007/s11153-018-9687-9

[8] Niels Henrik Gregersen, "Three Varieties of Panentheism," in *In Whom We Live and Move and Have Our Being: Panentheistic Reflections on God's Presence in a Scientific World*, eds. Philip Clayton and Arthur Peacocke (Grand Rapids, Mich.: William B. Eerdmans Publishing Company, 2004), 22–23.

I take panentheism to entail *at least* the following claims: (1) God is the primary cause and ground for the world's existence, (2) the world is an ontological part of God, (3) the world is part of God, but God is more than the world, (4) God and the world affect each other so that there is a feedback effect between God and the world, and (5) God is perfect in goodness, knowledge, and at least as powerful as one needs to be in order to cause or sustain the world.

Pantheism

The English term pantheism was conjoined by the Irishman John Toland in 1705, although Joseph Ralphson used the Latin term *pantheismus* in 1697 in *De Spatio Reali seu In fi nito*.[9] All pantheists agree that God and the world are identical *somehow* and that the God-world is an ontologically monistic unity.[10] How this monistic unity is to be understood is not agreed upon. Still, regardless of exactly how to understand the identity claim all pantheists reject ontological God-world dualism.[11] Thus, in pantheism, we have ontological God-world monism.

Pantheism is not the same as atheism. Nor is it the same as naturalism.[12] A pantheist need not identify God with a purely materialistic and reductive universe. The pantheism of Baruch Spinoza (1632–1677), the holistic pantheism of Grace Jantzen (1948–2006), and the panpsychist pantheism

[9] Andrei A. Buckareff, *Pantheism*, 1st ed. (Cambridge: Cambridge University Press, 2022), 2, https://doi.org/10.1017/9781108558266

[10] Luca Valera and Gabriel Vidal, "Pantheism, Panentheism, and Ecosophy: Getting Back to Spinoza?," *Zygon®* 0, no. 0 (2022): 3, https://doi.org/10.1111/zygo.12800. For an overview of different ways to conceptualize the pantheistic God-world identity, see William Mander, "Pantheism," in *The Stanford Encyclopedia of Philosophy*, ed. Edward N. Zalta, Spring 2020 (Metaphysics Research Lab, Stanford University, 2020), https://plato.stanford.edu/archives/spr2020/entries/pantheism/

[11] Michael P. Levine, *Pantheism: A Non-Theistic Concept of Deity* (London, United States: Taylor & Francis Group, 1994), 2, http://ebookcentral.proquest.com/lib/uu/detail.action?docID=169344; Timothy Sprigge, "Pantheism," *The Monist* 80, no. 2 (1997): 197; Mary-Jane Rubenstein, *Pantheologies: Gods, Worlds, Monsters* (New York: Columbia University Press, 2021), 25.

[12] Compare with Levine, *Pantheism*, 2; Martin O. Yalcin, "American Naturalism on Pantheism," *American Journal of Theology & Philosophy* 32, no. 2 (2011): 156–179; Brian Leftow, "Naturalistic Pantheism," in *Alternative Concepts of God: Essays on the Metaphysics of the Divine*, eds. Andrei Buckareff and Yujin Nagasawa, 1st ed. (Oxford: Oxford University Press, 2016), 66.

of Timothy Sprigge (1932–2007) are examples of pantheistic positions that adhere to ontological God-world monism without reducing the God-world in terms of materialism or physicalism. Pantheism can have a naturalistic component in that pantheism need not go beyond the "natural." However, the broader metaphysical picture needs to include something which makes the world into a divine unity, for example, mentality, intentionality, or teleology.

According to Spinoza, the world is perfect since God is perfect and this has nothing to do with whether we like the world or not.[13] A pantheistic ethics is non-anthropocentric and grounded in the divine unity.[14] A pantheistic ethics therefore claims that "Everything that is part of the divine Unity (as everything is) is also part of the moral community."[15] A naturalist on the other hand cannot ascribe inherent value to anything. According to a naturalist, value must always be relative to us and our desires. It is *we* who postulate intention in nature, but a naturalist cannot think that *nature* itself is intentional, valuable, moral, or that it has a telos. A pantheist thus claims something beyond what naturalists do. Pantheists claim that something unites all nature/the universe, namely an inherently and supremely valuable telos: the divine unity. The divine is somehow (perhaps by consciousness) present in all the parts,[16] or, as in other versions of pantheism, there are no parts at all but only the One indivisible God-world reality.[17] However, a naturalist cannot claim that nature as a whole is a divine unity.[18]

I take pantheism to entail *at least* the following claims: (1) The Universe/nature is an all-encompassing divine monistic Unity (God), (2) this divine Unity (God) is all there is, and God is nothing more than this, (3) the all-encompassing Universe/nature/Unity (God) is supremely good, valuable, or sacred.

[13] Benedict de Spinoza, *The Ethics (Ethica Ordine Geometrico Demonstrata)*, trans. R.H.M. Elwes (Mineola, New York: Dover Publications, 2018), 34. Part II, Definition VI.

[14] Michael P. Levine, "Pantheism, Ethics and Ecology," *Environmental Values* 3, no. 2 (1994): 122–131.

[15] Levine, "Pantheism, Ethics and Ecology," 132.

[16] Sprigge, "Pantheism," 192.

[17] Arvind Sharma, *The Philosophy of Religion and Advaita Vedānta: A Comparative Study in Religion and Reason*, Hermeneutics (University Park, Pa: Pennsylvania State Univ. Press, 1995), 1–3.

[18] Yalcin, "American Naturalism on Pantheism," 172.

In-Worldly Dualisms

I have mentioned what I intend by ontological God-world dualism and explained how it relates to classical theism and panentheism. Pantheism on the other hand ascribes to ontological God-world monism. Before turning to the arguments of environmental well-being, it is time to look at the relationship between ontological God-world dualism and what I call in-worldly dualisms.

When examining arguments about environmental well-being offered by panentheists and pantheists, the notion of in-worldly dualisms will occur. For example, the traditional dualisms of nature-God, transcendent-immanent, man-woman, and human-animal are mentioned. If the term dualism entails two non-reducible independent categories, the notion of in-worldly dualisms applied to man-woman, or nature-human, may be misleading. Strictly speaking, these *contrasts* or *opposites* may not be dualisms in the sense that they are non-reducible distinct categories such as mind-body often are said to be. Nevertheless, I shall use the term in-worldly dualisms about distinctions such as human-nature, human-animal, man-woman, reason-emotion, spirit-body, and so on because this label captures what panentheists and pantheists often regard as the problem with classical dualistic theism.

As I see it, the problem panentheists and pantheists sometimes seek to address is when ontological God-world dualism spills over into the world. The problem is when belief in a wholly transcendent and independent God—supreme in value—affects the way we conceptualize and categorize in-worldly entities and phenomena.

Claiming that there is something problematic about conceptualizing the world in terms of in-worldly dualisms and contraries does not necessarily entail rejecting that the world actually consists of a plurality of beings and entities (although some versions of pantheism actually reject the reality of anything but the Brahman).[19] Instead, panentheists and pantheists generally accept the reality of all sorts of in-worldly pluralism such as different genders, animals, and plants. The problem is not the plurality of beings. The problem is when we conceptualize the plurality of beings in valued and hierarchal ways, causing harmful consequences.

[19] R. Puligandla, "God and Ultimate Reality: An Analytical Interpretation of Śaṇkara's Philosophy," in *Models of God and Alternative Realities*, ed. Jeanine Diller and Asa Kasher (Dordrecht: Springer, 2013), 615–23; Purushottama Bilimoria and Ellen Stansell, "Suturing the Body Corporate (Divine and Human) in the Brahmanic Traditions," *Sophia* 49, no. 2 (June 2010): 237–259, https://doi.org/10.1007/s11841-010-0183-7

Feminist philosophers and theologians such as Grace Jantzen, Pamela Sue Anderson, Beverly Clack, and Nancy Frankenberry have pointed out how much in Western culture and history to a very large extent have valued the masculine higher than the feminine.[20] A significant reason for this is the historical—*man-made*—connection and linkage between the transcendent/spiritual and the male/masculine. Simultaneously, the female and feminine have been historically connected to the bodily and the sinful.[21] The Western culture and its classical dualistic theology are heavily influenced by the philosophy of ancient Greece and the Platonic notions of an ideal, unspoiled world. God/the good is put on the top of the hierarchy, women/the female at the bottom, and men in between. This hierarchical and patriarchal worldview is seen in Aristotle's (in)famous table of dualisms in his *Metaphysics*, which has influenced the Western conceptualization of reality.[22] These dualisms or contraries put men, reason, light, spirit, good, and order on one side and women, chaos, body, darkness, and evil on the other side.

Caroline Criado Perez has recently shown how valued in-worldly hierarchies still affect our daily lives. Cars, medical care, and indoor room temperature are constructed for the male human body as if a "human being" is a man and a woman only a deviant form of man.[23]

[20] Grace Jantzen, *Becoming Divine: Towards a Feminist Philosophy of Religion* (Bloomington, Ind: Indiana University Press, 1999); Pamela Sue Anderson, "Feminist Challenges to Conceptions of God: Exploring Divine Ideals," *Philosophia* 35, no. 3–4 (2007): 361–370, https://doi.org/10.1007/s11406-007-9083-7; Beverley Clack, "Feminist Approaches to Religion," in *The Routledge Handbook of Contemporary Philosophy of Religion*, ed. Graham Oppy (London: Routledge Taylor and Francis Group, 2017); Nancy Frankenberry, "Feminist Philosophy of Religion," in *The Stanford Encyclopedia of Philosophy*, ed. Edward N. Zalta, Summer 2018 (Metaphysics Research Lab, Stanford University, 2018), https://plato.stanford.edu/archives/sum2018/entries/feminist-religion/

[21] Grace M. Jantzen, "Feminism and Pantheism," *The Monist* 80, no. 2 (1997): 266–285.

[22] See Mary Daly, *The Church and the Second Sex* (Boston, Mass.: Beacon, 1999).

[23] Beverley Clack, *Feminism, Religion and Practical Reason*, ed. Yujin Nagasawa, 1st ed., Cambridge Elements: Elements in the Philosophy of Religion (Cambridge: Cambridge University Press, 2021), 19, https://doi.org/10.1017/9781108859653. In *Invisible Women*, Caroline Criado Perez (2019) provides data showing how society is constructed for men to a large extent. For example, women have a naturally lower body temperature than men, thus making a system based on the temperature of a middle-aged man of 70 kg significantly colder to the majority of women who, naturally, are not themselves middle-aged men of 70 kg. In the case of health care, it is a known fact among those working with a cardiac arrest that women rate their experience of chest pain lower than men even if they, in fact, have a more serious cardiac condition. This makes it difficult to assess which patient to prioritize.

A theology and a worldview where, for example, men and human beings (especially human men) are valued as standing closer to God than the rest of creation will affect how we structure society and how we interact with each other and the rest of creation. It will affect our shared reality and the possibility of promoting human and environmental flourishing.

In-worldly dualisms or contraries are not necessarily harmful. The potential harm that ontological God-world dualism brings is the risk of it spilling over into the world, thus creating harmful value hierarchies that hinder human and environmental flourishing. This is the risk and potential problem with ontological God-world dualism.

Before turning to the different reasons in favor of panentheism and pantheism, it is essential to acknowledge that neither classical theism, panentheism, nor pantheism *necessarily* has harmful consequences for the environment. Even if there are reasons to prefer one conception of God over the others, this does not entail that one cannot be deeply engaged in environmental concerns regardless of which conception of God one adheres to. The objective here is not to argue that, for example, classical theism has inherently negative implications for the environment. The aim is to examine whether a religious person who prioritizes environmental well-being has pragmatic reasons to be a panentheist or a pantheist rather than a classical theist.

Inclusive Metaphors and Environmental Theology

Sallie McFague (1933–2019) devotes her book *Models of God: Theology for an Ecological Nuclear Age* (1987) to the question of how we today best ought to talk about God. As she puts it, a theology for our time must speak of God with inclusive metaphors to create a theology and an environment where we appreciate our role and place in nature in an inclusive way, and we need metaphors of God as an actively present part of nature. According to her, panentheism entails such inclusive metaphors. What drives McFague to develop her ecological theology is the belief that life—both human and animal lives, and nature as such—is in grave danger if we continue to make the traditional sharp distinction between divine/human, human/nature, and spirit/body.

Classical theism and traditional Western theology have used metaphors of God that contribute to an unhealthy understanding of our own place in creation. Such harmful metaphors include masculine and hierarchal ones that contribute to a worldview in which human being—especially Man—is regarded as the all-powerful ruler of the earth. The conception of God as an omnipotent and essentially independent spirit that does not essentially need the physical, *sinful,* body/earth has contributed to a worldview in which humans—especially Men—believe that they can use and treat nature only as a means to their end. When God is conceptualized as a pure transcendent spirit, many have drawn the conclusion that the physical and embodied must be the opposite of God: sinful, weak, and not inherently valuable. This, essentially, is the problem with the traditional conception of God found in classical theism. The ontological God-world dualism of classical theism spills over into in-worldly dualism and makes the in-worldly distinctions into *valued* ones—where some aspects of the world become regarded as divine (particularly the masculine, power, knowledge, reason, and individuality) while other aspects have become regarded as non-divine (the female, weakness, feeling, embodiedness, and relations).

According to McFague, the problem of traditional, often Christian, metaphors, and descriptions of God are the dualisms and the hierarchies implied in the models of God as *king, Lord, ruler, master, judge*. These analogies or metaphors are bad because they affect our understanding of God and ourselves as masters of the universe where the earth is something possessed—a mere object for our (and God's) enjoyment. Without a mentality that invokes a sense of holiness and respect for the environment, the risk is grave that we end up with an instrumental view of nature. Such a mentality can ultimately result in disrespect for nature, pollution, and animal cruelty. And is this not what the Sixth Assessment Report of the Intergovernmental Panel on Climate Change bears witness to? According to the report, there is no longer any doubt that human beings are part of the cause of our current climate crisis.[24]

If God truly is the world plus something more, as panentheism says, we must change the traditional conception of God because it does not indicate that God also *is* the world. Even if this panentheistic belief is false, it

[24] Veronika Eyring et al., "Human Influence on the Climate System," in *Climate Change 2021: The Physical Science Basis. Contribution of Working Group I to the Sixth Assessment Report of the Intergovernmental Panel on Climate Change*, ed. Valérie Masson-Delmotte et al. (Cambridge: Cambridge University Press, 2021).

would contribute to us *acting* as if it is true—and that is what makes the difference.

Thus, what McFague and other panentheists want is a change in how we conceptualize the God-world relationship. We cannot continue to conceptualize ourselves, God, and nature in the traditional dualistic way. We must change our attitudes and conceptions so that we see ourselves, God, and nature as necessarily bound together as part of a shared world.[25] If we want a world of environmental well-being and flourishing, we should not regard nature as something dead, as only a disposable means to achieve our own desires. This means that we should change the way we conceptualize our relationship with nature and God. We need more inclusive metaphors that speak of relationality instead of independence and hierarchal ruling.

The arguments in favor of a panentheistic conception of God offered here are pragmatic or *value oriented*. McFague offers no metaphysical arguments and takes good care *not* to do so since her concern lies not in epistemology and truth claims but in pragmatism and how we ought to think and speak about God *for our sake* and *nature's sake*. Her point is that a panentheistic model, emphasizing motherhood, natality, and the bodily nature of the divine, is a *better* model of God than a classical theistic model because the conceptions, metaphors, and models in panentheism contribute to a more inclusive and environmentally prosperous worldview. We ought to think of God with panentheistic metaphors.

The traditional conceptions of God as *absolute, complete, perfect, transcendent,* and *omnipotent* are problematic metaphors in our time for similar reasons.[26] McFague proposes that we use panentheistic conceptualizations of God, and she examines three metaphors that are supposed to have the desired inclusive effect.[27] She argues in favor of conceptualizing God as *mother, lover*, and *friend*. Unfortunately, the metaphors and analogies in classical theism (which McFague calls triumphalist models) have resulted in dualistic thinking where God is the separate transcendent mighty Lord. The classical view of God as the ruling king who conquers death (Jesus's resurrection) and eventually destroys evil (as in

[25] Sallie McFague, *Models of God: Theology for an Ecological, Nuclear Age* (London: SCM Press, 1987), 6–13.

[26] McFague, *Models of God*, 19.

[27] McFague, *Models of God*, 72.

The Book of Revelation) leaves humans with no sense of responsibility for our shared home—the earth.

> [T]he king as benevolent patriarch encourages attitudes of passivity and escape from responsibility. In the triumphalist, royal model the victory has already been won on the cross and in the resurrection of Jesus Christ, and nothing is required of us.[28]

Rosemary Radford Ruether is, like McFague, an ecofeminist theologian. She is also a panentheist. According to her, the ecological crisis we find ourselves in can be traced back to when patriarchy began. The transcendent God of classical Judeo-Christian theism is a patriarchal male God who legitimizes the human dominance of nature.[29]

> Ecofeminism seeks to dismantle the whole paradigm of male over female, mind over body, heaven over earth, transcendent over immanent, the male God outside of and ruling over the created world—and to imagine an alternative to it.[30]

Ecofeminism points toward a holistic and inclusive theology and view of nature. It seeks a way that leads away from traditional, hierarchical, sexist, and environmentally destructive value-dualisms. Reuther's ecofeminist theology speaks of ecological and spiritual unity. Humans are part of nature, and if we destroy nature, we destroy ourselves. She is sympathetic to Whiteheadian process theology, which is a version of panentheism, and writes that everything is ultimately *relations*. There is no difference between energy and matter on a sub-atomic level. The only difference is due to different relational processes. Both ecology and sciences such as astrophysics show us that everything is connected in intricate relations, and ecosystems are dependent on diversity.[31] Reuther speaks of unity and diversity as two sides of the same coin. "Thus what we have traditionally called 'God' [...] and what we have called 'matter', the 'ground' of physi-

[28] McFague, *Models of God*, 69.

[29] Rosemary Radford Ruether, *Gaia & God* (San Francisco: HarperSanFrancisco, 1992), 2–4.

[30] Rosemary Radford Ruether, "Ecofeminist Philosophy, Theology, and Ethics: A Comparative View," in *Ecospirit: Religion, Philosophy, and the Earth*, ed. Laurel Kearns and Catherine Keller (New York: Fordham University Press, 2007), 85.

[31] Ruether, *Gaia & God*, 48, 248–249.

cal objects, come together. The disintegration of the many into infinitely small 'bits', and the One, or unifying whole that connects all things together, coincide."[32]

Catherine Keller, a process-panentheist, also writes about the human responsibility as caretakers for the earth. "When we mistake dominion for dominance, we fail in our responsibility as caretakers for the earth—*ipso facto* we abdicate 'dominion.' "[33] Although ecotheology is not necessarily based on Whiteheadian process theology, it has a close bond with this tradition.[34] Western ecotheology is inspired by the relationality and holistic philosophy of Whiteheadian process theology. Such process theology—which Keller adheres to—highlights the inherent value not only of human life but of all life. God and the world affect each other in a mutual relationship. Nature, with all its inhabitants—living or non-living—is valuable and has an inherent aim set by God. This kind of theology contributes to ethical and ecological thinking, with respect for the *diverse as inherently valuable*.[35]

Keller, McFague, and Ruether share a conviction that traditional Cartesian dualism has led us astray so that we no longer feel part of nature. If we are to take real responsibility for the well-being of the earth and other human beings, in-worldly value-dualisms cannot continue. The dualism of human-nature is destructive not only to nature but to us since we *are* nature. "We are nature with a concept of nature" as the poet Susan Griffin writes.[36] Keller, McFague, and Ruether all claim that *we are nature*, and we need a theology that acknowledges that. Panentheism does that, while classical theism fails in that regard.

> The "*en*" [in panentheism] asserts the difference of divine and cosmic, but at the same time makes it impossible to draw the line. For is not the line always already smudged? The smudge, the flux, "is" the *en*, the overlap, of

[32] Ruether, *Gaia & God*, 249.

[33] Catherine Keller, *The Face of the Deep: A Theology of Becoming* (Florence, United States: Taylor & Francis Group, 2003), 138, http://ebookcentral.proquest.com/lib/uu/detail.action?docID=181809

[34] Jay McDaniel, "Ecotheology and World Religions," in *Ecospirit: Religion, Philosophy, and the Earth*, ed. Laurel Kearns and Catherine Keller (New York: Fordham University Press, 2007), 21–44.

[35] McDaniel, "Ecotheology and World Religions," 26, 39–40.

[36] Susan Griffin, *Made from This Earth: An Anthology of Writings*, 1st ed. (New York: Harper and Row, 1982), 343. (McFague quotes Susan Griffin in *The Body of God*, 124).

divinity with world, of world with divinity. [...] the "in" channels the resonance between "deeps."[37]

According to what I here have called the panentheistic argument from environmental concern, a cosmos that is (or taken to be) inseparable from the divine will invoke more responsibility, more care, and a more profound feeling of holiness toward our environment than a cosmos that is not (or taken not to be) a part of God. In our time, we need to develop more responsibility, more care, and a more profound feeling of holiness toward our environment, because we *are a part of nature,* just as *nature is a part of God.* The inclusive and relational theology implied in panentheism contributes to a more relational and environmentally prosperous worldview.

Then, if panentheism, due to its less dualistic nature, entails a more environmental and ethically sensitive treatment of the world than classical theism, does it follow that pantheism—ontologically monistic—is even better? Environmental concern and human caretaking of both humans, animals, and plants are of utmost importance for a pantheist.[38] A pantheistic worldview highlights ontological God-world monism, a non-anthropocentric ethical concern for all things, and the shared responsibility for the well-being of all things.[39] So, ought a person who believes that environmental well-being is of the highest priority to be a pantheist rather than a classical theist or a panentheist? Before we can answer this question, we must critically analyze the claim that a panentheistic theology would entail a mentality that could contribute to a more environmentally flourishing world.

Panentheism, Environmental Concerns, and God-World Dualism

The panentheism that McFague argues for highlights that panentheism makes us (humans) more disposed to see the unity in creation rather than seeing it as a disposable object owned by us. The panentheistic argument from environmental concern states that *it would be better* if we

[37] Keller, *The Face of the Deep*, 219.

[38] Universal Pantheist Society 2022: Key Ideas/Pantheism and Earthkeeping, http://www.pantheist.net/earthkeeping.html (2022-05-13)

[39] Universal Pantheist Society 2022: Home/Are You A Pantheist? http://www.pantheist.net/ (2022-05-13)

conceptualized the divine in panentheistic terms. It would be better both for ourselves—with a more gender-equal society, and for the whole of nature and all its animals—because *we* would not treat and conceptualize nature and animals as being of lesser worth than us.

The question is whether a religious believer deeply engaged in environmental well-being ought to be a panentheist rather than a classical theist or a pantheist. First, there is no reason why a classical theist cannot be equally involved and interested in the environment as a panentheist. With that said, and all other things being equal, the religious believer with a deep engagement with the environment *has* pragmatic reasons to embrace panentheism over classical theism. I claim that panentheism has a pragmatic advantage relating to environmental concerns in its claim that all things are *part of* God and not merely created as separate from God. They affect each other. This makes the ontological God-world dualism of panentheism *less* dualistic than classical theism, which in turn, pragmatically, ought to invoke a more profound feeling of reverence for nature and other human beings. However, even classical theism can inspire us to think of human beings as equal or nature as inherently valuable. Nevertheless, all other things being equal, it is reasonable to hold panentheism as an inherently more "environmentally friendly" conception of God than classical theism.

Previously, I developed an understanding of ontological God-world dualism and the notion of in-worldly dualism. I stated that in-worldly dualisms or contraries are not always harmful. The problem is how in-worldly dualisms or contraries are conceptualized and *valued* and how this affects our shared reality and the possibility of promoting human and non-human flourishing.

Many would agree that human beings cannot treat other beings and nature, as such, however we please.[40] God, presumably, has an intention with the creation. God has revealed clues about that intention and about good and bad moral living, for example, in the Gospels or in nature itself. Classical theists believe that God is perfectly good and a perfect being. Many use theologies to justify immoral human behavior. This is unfortu-

[40] See e.g., Chad Meister, "Personalistic Theism, Divine Embodiment, and a Problem of Evil," *European Journal for Philosophy of Religion* 11, no. 2 (June 20, 2019): 119, https://doi.org/10.24204/ejpr.v11i2.2974; Charles Taliaferro and Chad Meister, *Contemporary Philosophical Theology*, 1 [edition] (London; New York: Routledge, Taylor & Francis Group, 2016), 117; Peter Byrne, "Omnipotence, Feminism and God," *International Journal for Philosophy of Religion* 37, no. 3 (1995): 145–165.

nate, but one could argue that the fault does not lie with classical theism.[41] If we believe that God is just and perfect, then we also believe that male domination, unjust actions, and environmental destruction are against the will of God.

Only because one is a classical theist or believes in Cartesian body-soul dualism, it is, of course, not necessarily the case that one does not care for the environment or other animals. Ecotheology is not exclusively for panentheists. Peter Byrne criticizes many feminist theologians for being *too* pragmatic and not taking metaphysical questions seriously enough. He writes that even if a classical understanding of God as omnipotent has resulted in a human glorification of dominant power, "it has yet to be shown that omnipotence in deity equals the exercise of domination over other personal agents."[42] Charles Taliaferro and Chad Meister similarly maintain that to believe that God is perfect requires that one carefully consider what the concept of perfection entails. Any notion that implies, for example, domination, must be a false notion of God's perfection. "[…] understanding God as perfect does not *ipso facto* make the concept of God patriarchal […]."[43]

So, if the problem is that we mistakenly attribute value to some in-worldly states of affairs while neglecting others, the problem is not the dualisms and contraries themselves but the way we understand and value them. If that is the case, can we not simply be better dualists? Ideally, yes. However, the way we speak of and conceptualize reality has a real effect on that very reality. We create reality by conceptualizing it in certain ways, and some ways are better than others. For example, it is not good to conceptualize human beings as if *male* human beings are the norm. Thus, some ways to conceptualize reality help promote human and environmental flourishing. Others hinder that effort.

Indeed, the problem is not the dualisms or contraries in themselves, and it is, of course, the case that we have in-worldly dualisms and contraries also within panentheism and pantheism. However, Grace Jantzen, Mary Daly, Pamela Sue Anderson, Beverly Clack, and others have shown us that the traditional classical theistic theology, which is ontologically

[41] Taliaferro and Meister, *Contemporary Philosophical Theology*, 116; Elime Judge-Becker and Charles Taliaferro, "Feminism and Theological Anthropology," in *The Ashgate Research Companion to Theological Anthropology*, ed. Joshua Ryan Farris and Charles Taliaferro (England: Ashgate Publishing Company, 2015), 86.

[42] Byrne, "Omnipotence, Feminism and God," 152.

[43] Taliaferro and Meister, *Contemporary Philosophical Theology*, 117.

dualistic, has contributed to harmful in-worldly value hierarchies.[44] The ontological God-world dualism entails a worldview where we have both spirit and matter—both God and the world. Since God is supreme in all aspects, those in-worldly entities associated with the spiritual (e.g., man, reason, and individuality) have been regarded as more divine—more valuable. And since the ontological God-world dualism of classical theism entails that the world is that which is *not* God, everything that has become associated with the material (such as woman, body, feelings, and relationality) has been regarded as non-divine—and hence less valuable. *This* is the problem with the classically theistic ontological God-world dualism. It is not problematic in itself but because it affects how we conceptualize other entities and beings in reality. It is problematic because of how it affects our very conception of reality.

If panentheism is true, the benefits of conceptualizing God in panentheistic terms are great. Assuming the belief that God created the world and "saw that it was good," the benefits of this far outweigh the cons if it turns out that classical theism is correct. Even if classical dualistic theism is right, no harm would come from us treating each other and the environment as inherently divine. The reverse cannot be said. If it turns out that panentheism is correct, the harms of not conceptualizing other beings and nature as part of God are significant. We need only look at the failed attempts to stop climate change and environmental exploitation such as in the Paris Agreement 2015 and the United Nations climate change meeting in Glasgow 2021.

McFague's main argument is that a model of God as embodied is better for us because then we would value physical entities such as bodies, the female, the male, plants, mountains, or polar bears, higher than we traditionally do today. Again, this is a pragmatic argument. McFague and most other feminist philosophers of religion do not argue that a classical theistic conception of God *ipso facto* makes the concept of God patriarchal, or that classical theism *ipso facto* contradicts care for the environment. Even so, it does appear always to be the case that ontological God-world dualism and body-soul dualism are used to legitimize patriarchal and hierarchal ideals.

[44] Jantzen, *Becoming Divine*; Mary Daly, *Beyond God the Father: Toward a Philosophy of Women's Liberation*, Beacon Paperback 488 (Boston: Beacon Press, 1974); Daly, *The Church and the Second Sex*; Anderson, "Feminist Challenges to Conceptions of God"; Clack, "Feminist Approaches to Religion.".

The classical theistic idea, found in Judaism, Christianity, and Islam, that human beings were created as "wardens" or "caretakers" of the earth ought to be equally fruitful in the pursuit of ecological well-being as panentheism or pantheism. A classical theistic conception of God can combine a belief in a strict ontological distinction between God and the world or between the body and soul and *still* emphasize equality, human and animal dignity, and ecological well-being. I see no reason why classical theism would contradict those values per se. Classical theism *and* panentheism adhere to ontological God-world dualism. If there is a problem with ontological God-world dualism in the risk that it affects how we conceptualize in-worldly states of being, this risk also applies to panentheism. However, due to the historical value hierarchy between body and soul, a hierarchy fueled by ontological God-world dualism, it is plausible to assume that a conception that highlights the inherent and equal worth of both *body and soul, matter and spirit* would result in more care and love for the bodily than a conception that highlights the superior value of soul/spirit would.

In panentheism, the distinction between God and the world is, to use the words of Keller, smudged. *Just as we cannot draw a line between us and nature, neither can we draw a line between God and the world.* When we realize this, we ought to regard nature not as a mere disposable dead object but as inherently valuable—possibly even divine. A panentheistic model emphasizes the divine immanence in nature and nature as part of the divine. A panentheistic model emphasizes less dualism—concerning both God-world and soul-body—than classical theism does. With a panentheistic and less dualistic worldview, we can see our place in nature not as detached masters but as co-inhabitants. Therefore, a panentheistic conception of God can contribute better to our environmental concerns than classical theism, because nature is more inherently valuable in panentheism than in classical theism. Hence, religious believers who are deeply engaged in the well-being of the environment ought to embrace panentheism rather than classical theism. Thus, we have a pragmatic argument favoring panentheism over classical theism. Again, this is not to say that a classical theist cannot be equally engaged in these questions.

Since I have argued that there are reasons to favor panentheism over classical theism, we can go back to whether pantheism, limited to environmental well-being, is even better than panentheism.

Pantheism, Environmental Concerns, and God-World Monism

> Nothing in the Universe exists alone; all matter and energy are united, interdependent, and interrelated. […] Man's arrogance in considering himself apart from and superior to other organisms is resulting not only in environmental abuse, but in deterioration of his own spirit.[45]

Given the pantheist belief that everything is divine and hence inherently valuable, there are good reasons to assume that a pantheistic worldview could contribute to a holistic and ethically respectful treatment of nature. Moreover, a *panpsychist* pantheist ought to inspire even more love, care, and respect for all of nature than a non-panpsychist pantheist. However, even though there are some pragmatic reasons to favor pantheism over panentheism and classical theism regarding environmental concerns, the reasons are significantly weaker than one might first think. The benefits of ontological God-world monism—essential to pantheism—cannot guarantee a lack of valued, harmful, in-worldly dualisms.

Pantheistic environmental ethics is essentially non-anthropocentric.[46] The inherent value of the universe comes from the fact that everything is divine. This is something that no naturalist can claim. Previously, we saw that Spinoza argues that the inherent value of the divine God-world does not depend on us—it comes from the divine unity. Since everything is, or is part of, the divine God-world unity, everything, according to the pantheist, is part of the moral community.[47] For this reason, our ethical considerations in general, and with particular respect to environmental ethics, cannot take only human beings as the moral ground for right and wrong. Pantheists instead want us to widen our understanding of what counts as the moral community—which is everything. The inherent value of all things comes not from its use or how well it suits our needs. It comes from the fact that it is divine in and of itself.

[45] Universal Pantheist Society 2022: Key Ideas/Pantheism and Earthkeeping. http://www.pantheist.net/earthkeeping.html (2022-05-13)

[46] John W. Grula, "Pantheism Reconstructed: Ecotheology as a Successor to the Judeo-Christian, Englightenment, and Postmodernist Paradigms," *Zygon* 43, no. 1 (2008): 174, https://doi.org/10.1111/j.1467-9744.2008.00904.x; Levine, "Pantheism, Ethics and Ecology," 131–133; Harold W. Wood, "Modern Pantheism as an Approach to Environmental Ethics," *Enviromental Ethics* 7, no. 2 (1985): 161.

[47] Levine, "Pantheism, Ethics and Ecology," 132.

Michael Levine writes that "[…] there is a prominent, if not prevalent, view, that its [pantheism's] implications (if it were true) would be a good thing for ecology, and for aspects of ethics having to do with the non-human (and the human)."[48] I claim that the non-anthropocentric ethics embedded in pantheism has these good implications for ecology, *even if it is not true*. However, as always, no philosophy or worldview can guarantee ethical behavior. Human beings are fallible by nature, and no matter what worldview we believe in or try to act by, we cannot guarantee ethically good behavior on our part. This is partly why empirical data does not give us the answer we seek. The best we can do is to find philosophical resources to promote a value system and ethics that, all other things being equal, would contribute to ethics that cause a good and flourishing world for both humans and non-humans.

With that said, pantheistic ethics grounded in the inherent value of everything entail what I call the *pantheist problem with value differentiation*. The pantheist problem with value differentiation is the challenge of coherently answering how an ontologically monistic God-world, where everything is divine, can be divided into different values. How can a pantheist coherently claim that the preservation of the jungles and animal diversity are *better* than deforestation, monocultures, and animal extinction? The pantheist needs to answer these questions because, otherwise, we cannot make sense of the pantheistic argument for environmental concern. If pantheism can answer the challenge of value differentiation, there are good reasons to take the pantheistic argument for environmental concern seriously.

If, as pantheism claims, everything is divine, if everything is part of the moral community, then everything is inherently valuable. If we hold this to be true, or the weaker stance, act as if this is true, there are good reasons to believe that a pantheistic ethical framework actually would contribute to a more environmentally flourishing and prosperous world than worldviews that do not hold everything as inherently divine and valuable, or if we act in accordance with worldviews that deny this. In the words of Harold Wood: "The identification of sacredness in the earth demands reverent behavior. In turn, such behavior necessitates a personal commitment toward living in greater harmony with the biosphere."[49]

[48] Levine, "Pantheism, Ethics and Ecology," 124.

[49] Wood, "Modern Pantheism as an Approach to Environmental Ethics," 160.

Value Differentiation

There are different kinds of pantheism. The Vedantic tradition of Advaita Vedanta, with Adi Shankara as a prominent figure, holds reality to be properly monistic. Only the Brahman exists. Everything that we think are separate entities and beings are *māyā*, illusions, because in fact, only the Brahman exists.[50] Pantheism in the tradition of Advaita Vedanta rejects the reality of anything but the Brahman even if the world, to us, appears to be constituted by many things.[51]

This sort of pantheism adheres to existence monism, the claim that there exists only one thing.[52] With such a strong monism, it is hard to see how we can make value differences at all. If all the things that we value as good and bad really are illusory phenomena without real existence it is hard to see how we coherently can argue in favor of environmental well-being rather than environmental exploitation.

However, there are other kinds of pantheism that do not ascribe to existence monism. The pantheism of Hindu philosopher Ramanuja, and the pantheism of Spinoza, claim the reality of in-worldly plurality.[53] For both Ramanuja and Spinoza, God takes different forms or modes.[54] Pantheism that does not ascribe to existence monism can instead ascribe to priority monism, the claim that "there is only one fundamental thing that is divided into parts and that the whole has priority over the parts."[55] One possible and to my mind, plausible reading of Spioza's pantheism is that it ascribes "a single substance [God], a plurality of derived modes [the plurality of beings and entities in the world], and the relationship between

[50] Bilimoria and Stansell, "Suturing the Body Corporate (Divine and Human) in the Brahmanic Traditions," 239, 252.

[51] Wood, "Modern Pantheism as an Approach to Environmental Ethics," 160.

[52] Bilimoria and Stansell, "Suturing the Body Corporate (Divine and Human) in the Brahmanic Traditions," 251; Puligandla, "God and Ultimate Reality: An Analytical Interpretation of Śaṇkara's Philosophy," 615; Sharma, *The Philosophy of Religion and Advaita Vedānta*, 1–2; Sucharita Adluri, "The World as the Body of God: Ramanuja on What Is Ultimately Real," in *Models of God and Alternative Ultimate Realities*, ed. Jeanine Diller and Asa Kasher (Dordrecht: Springer, 2013), 626.

[53] Sharma, *The Philosophy of Religion and Advaita Vedānta*, 2; Valera and Vidal, "Pantheism, Panentheism, and Ecosophy," 10.

[54] Adluri, "The World as the Body of God: Ramanuja on What Is Ultimately Real," 626.

[55] Valera and Vidal, "Pantheism, Panentheism, and Ecosophy," 10.

the substance and the modes is not a relationship of parts and whole because it would imply that modes are ontologically prevalent."[56]

Some pantheists argue that the divinity of the pantheistic God can be more significantly present in certain situations and less so in others. Grace Jantzen argues for example that the pantheistic God was more significantly present in Jesus Christ than in the *New York Times*.[57] If this way of thinking is coherent with pantheism, it certainly cannot be the existence monistic version, such as the pantheism of Shankara. Like Jantzen, Wood also attributes diversity and value distinction in the world. Human beings are unique, and we should use our unique ability to exercise empathy and compassion with the rest of creation.

> Pantheism confirms the uniqueness of humanity, and its ethics derives from a simple expression of human abilities for empathy, compassion, and a mystical oneness with the rest of the natural world. Pantheist ethics has as its goal a closeness with nature that enables humankind to better participate with, rather than dominate nature, a relationship with nature equivalent to traditional religion's relationship with God. It is a closeness based not upon imitation, but upon reverential communion.[58]

In this sort of pluralistic pantheism, God is the one necessary substance or entity that makes up the entire universe, but we can understand God as everything without sacrificing the reality of the different modes. Different aspects of the one divine reality can be distinguished. If some aspects or modes of God can be understood to be more revelatory of the divine will or telos than others, then we can understand some in-worldly phenomena to be more significantly divine and valuable than others. We can do this while still adhering to non-anthropocentric ethics. For example, if environmental flourishing, environments with prosperous animal diversity, peace, and compassion are more significantly revelatory of the inherent teleology of the pantheistic God-world, then we can coherently argue that these phenomena indeed are better than environmental exploitation, animal cruelty, oppression, war, and so on. A pantheist takes the divine inherently valuable unity as the grounding metaphysical principle for her ethics and value judgments—not an egoistic anthropocentric one.

[56] Valera and Vidal, "Pantheism, Panentheism, and Ecosophy," 12.

[57] Grace M. Jantzen, *God's World, God's Body* (London: Darton, Longman and Todd, 1984), 98–99.

[58] Wood, "Modern Pantheism as an Approach to Environmental Ethics," 161.

Panpsychist Pantheism

If we accept this conclusion, we can see that there are reasons to prefer not only pantheism in general but *panpsychist* pantheism. Panpsychism stems from the Greek words *pan* (all) and *psyche* (the mental). There are many different forms of panpsychism and different ways to understand the claim that everything is mental.[59] For my purposes here, it suffices to settle for a wide interpretation of panpsychism, claiming that *everything, on a fundamental level, either has mentality/psyche, subjectivity, consciousness, or at least experience.*

Since a pantheist can attribute different values to different aspects of the world without being an anthropocentric egoist, we have reason to believe that a panpsychist pantheist who attributes significant divine value to everything mental/psychic, subjective, conscious, or at least experiencing contributes to an even better environmental ethics than a non-panpsychist pantheist. This is because a panpsychist pantheist attributes these significantly divine values to *everything*—including mountains, trees, rivers, frogs, and snails. A non-panpsychist pantheist who does not attribute mentality/psyche, subjectivity, consciousness, or at least experience to everything need not claim that mountains, trees, rivers, frogs, and snails are significantly valuable and divinely revelatory. A panpsychist thus attributes *more* significantly valuable and divinely revelatory entities to the world than a non-panpsychist pantheist.

With that said, even the panpsychist pantheist must answer why environmental well-being and preservation are better than destruction. Why is a flourishing garden with prosperous and healthy animals better than a dried-up garden with starving animals if *everything*—including the dead leaves, dried-up earth, and starving animals are inherently mental/psychical, subjects, conscious, or at least experiencing and thus significant divine expressions of the divine pantheistic telos?

[59] See a description of the nuances of the different forms of panpsychism in Joanna Leidenhag, "Deploying Panpsychism for the Demarcation of Panentheism," in *Panentheism and Panpsychism*, ed. Godehard Brüntrup, Benedikt Paul Göcke, and Ludwig Jaskolla (Brill, mentis verlag, 2020), 65–90. See also William, E. Seager, Philip Goff, and Sean Allen-Hermanson, "Panpsychism," in *The Stanford Encyclopedia of Philosophy*, ed. Edward N. Zalta, Winter 2017 (Metaphysics Research Lab, Stanford University, 2017), https://plato.stanford.edu/archives/win2017/entries/panpsychism/; David Ray Griffin, *Reenchantment Without Supernaturalism: A Process Philosophy of Religion*, Cornell Studies in the Philosophy of Religion (Ithaca, N.Y: Cornell University Press, 2001).

Regardless of what type of pantheism we adhere to, if we cannot coherently argue that certain aspects of the divine are *more valuable and significantly divine* than others, we again have no reason to think that a prosperous planet is preferable to a destroyed planet. If we conceptualize everything as *equally* valuable, we will not gain the pragmatic benefits of rejecting ontological and in-worldly dualism.

Unfortunately, even if we accept that we coherently can argue for value differences within pantheism, the value difference can be used to argue in favor of harmful in-worldly value hierarchies that cause harm to the environment. Neither classical theism nor panentheism *necessarily* entail that we conceptualize reality to produce environmentally harmful in-worldly dualism. Similarly, a worldview based on ontological God-world monism cannot guarantee a flourishing world of harmony.

As far as I can see, a pantheism of the existence monist kind that rejects the reality of everything, but one single indistinguishable category cannot coherently make sense of value distinctions. Instead, a more pluralistic version of pantheism that does not reject the reality of different in-worldly modes of being (but still adheres to ontological God-world monism) appears to be the more advantageous position if we as pantheists want to argue that pantheism entails good environmental ethics. In other words, even if no pantheism can guarantee the lack of environmentally harmful in-worldly value hierarchies, we still have reason to consider a panpsychist pantheism that allows the reality of in-worldly plurality.

There are good reasons to think that a pantheistic conception of God, whether panpsychist or not, could contribute to non-anthropocentric environmental ethics in a way that classical theism cannot. However, due to the pantheist problem of value differentiation and the remaining risk of environmentally harmful in-worldly value hierarchies, pantheistic ethics is not obviously better than a pan*en*theistic one. The benefits of ontological God-world monism are less significant than one might have thought.

Conclusions

I have explored and critically analyzed pragmatic arguments for panentheism and pantheism regarding environmental well-being in this chapter. There are pragmatic reasons to prefer panentheism over classical theism because of the *less* dualistic nature of panentheism. Furthermore, we have reason to think that pantheism—which rejects ontological God-world dualism—could contribute to more profound environmental ethics than classical theism and panentheism since it regards all aspects of the world as sacred and divine. If we accept the claim that life, consciousness, and experience are inherently valuable, I also suggested that a *panpsychist* pantheist position entails an even stronger focus on the *whole* world than a non-panpsychist pantheist does. Even if panpsychist pantheism is false, a world and a theology based on the conviction that it *is* true would contribute to a non-anthropocentric and non-egoistic ethics that would benefit environmental flourishing. With that said, pantheism's benefits are small compared to panentheism's benefits since they both emphasize the divinity of the physical, embodied, natural world.

Bibliography

Adluri, Sucharita. 2013. The World as the Body of God: Ramanuja on What Is Ultimately Real. In *Models of God and Alternative Ultimate Realities*, ed. Jeanine Diller and Asa Kasher, 625–632. Dordrecht: Springer.

Anderson, Pamela Sue. 2007. Feminist Challenges to Conceptions of God: Exploring Divine Ideals. *Philosophia* 35 (3–4): 361–370. https://doi.org/10.1007/s11406-007-9083-7.

Aquinas, Thomas, Summa Theologiae. Accessed Jun 07, 2022., from http://summa-theologiae.org/question/04505.htm

Bilimoria, Purushottama, and Ellen Stansell. 2010. Suturing the Body Corporate (Divine and Human) in the Brahmanic Traditions. *Sophia* 49 (2): 237–259. https://doi.org/10.1007/s11841-010-0183-7.

Bishop, John, and Ken Perszyk. 2016. Concepts of God and Problems of Evil. In *Alternative Concepts of God: Essays on the Metaphysics of the Divine*, ed. Andrei A. Buckareff and Yujin Nagasawa, 106–127. Oxford: Oxford University Press. https://doi.org/10.1093/acprof:oso/9780198722250.001.0001.

Brierley, Michael W. 2004. Naming a Quiet Revolution: The Panentheistic Turn in Modern Theology. In *In Whom We Live and Move and Have Our Being: Panentheistic Reflections on God's Presence in a Scientific World*, ed. Arthur Peacocke and Philip Clayton, 1–15. Grand Rapids, Mich: Eerdmans.

Buckareff, Andrei A. 2022. *Pantheism*. 1st ed. Cambridge: Cambridge University Press. https://doi.org/10.1017/9781108558266.

Byrne, Peter. 1995. Omnipotence, Feminism and God. *International Journal for Philosophy of Religion* 37 (3): 145–165.

Clack, Beverley. 2021. Feminism, Religion and Practical Reason. In *Cambridge Elements: Elements in the Philosophy of Religion*, ed. Yujin Nagasawa, 1st ed. Cambridge: Cambridge University Press. https://doi.org/10.1017/9781108859653.

———. 2017. Feminist Approaches to Religion. In *The Routledge Handbook of Contemporary Philosophy of Religion*, ed. Graham Oppy. London: Routledge Taylor and Francis Group.

Clayton, Philip. 1997. God and Contemporary Science. In *Edinburgh Studies in Constructive Theology*. Edinburgh: Edinburgh Univ. Press.

Clayton, Philip, and Arthur Peacocke, eds. 2004. *Whom We Live and Move and Have Our Being: Panentheistic Reflections on God's Presence in a Scientific World*. Grand Rapids, Mich: William B. Eerdmans Publishing Company.

Culp, John. 2017. Panentheism. In *The Stanford Encyclopedia of Philosophy*, ed. Edward N. Zalta. Metaphysics Research Lab, Stanford University. https://plato.stanford.edu/archives/sum2017/entries/panentheism/.

Daly, Mary. 1974. *Beyond God the Father: Toward a Philosophy of Women's Liberation*. Beacon Paperback 488. Boston: Beacon Press.

———. 1999. *The Church and the Second Sex*. Boston, Mass: Beacon.

Eyring, Veronika, Nathan P. Gillett, Krishna M. Achutarao, Rondrotiana Barimalala, Marcelo Barreiro Parrillo, Nicolas Bellouin, Christophe Cassou, et al. 2021. Human influence on the climate system. In *Climate Change 2021: The Physical Science Basis. Contribution of Working Group I to the Sixth Assessment Report of the Intergovernmental Panel on Climate Change*, ed. Valérie Masson-Delmotte, Panmao Zhai, Anna Pirani, Sarah L. Connors, C. Péan, Sophie Berger, Nada Caud, et al. Cambridge University Press.

Frankenberry, Nancy. 2018. Feminist philosophy of religion. In *The Stanford encyclopedia of philosophy*, ed. Edward N. Zalta. Metaphysics Research Lab, Stanford University. https://plato.stanford.edu/archives/sum2018/entries/feminist-religion/.

Gregersen, Niels Henrik. 2004. Three Varieties of Panentheism. In *In whom we live and move and have our being: Panentheistic reflections on god's presence in a scientific world*, ed. Philip Clayton and Arthur Peacocke, 19–35. Grand Rapids, Mich.: William B. Eerdmans Publishing Company. 0–8028–0978-2.

Griffin, David Ray. 2001. Reenchantment Without Supernaturalism: A Process Philosophy of Religion. In *Cornell Studies in the Philosophy of Religion*. Ithaca, N.Y: Cornell University Press.

Griffin, Susan. 1982. *Made from This Earth: An Anthology of Writings*. 1st ed. New York: Harper and Row.

Grula, John W. 2008. Pantheism Reconstructed: Ecotheology as a Successor to the Judeo-Christian, Englightenment, and Postmodernist Paradigms. *Zygon* 43 (1): 159–180. https://doi.org/10.1111/j.1467-9744.2008.00904.x.

Jantzen, Grace. 1999. *Becoming Divine: Towards a Feminist Philosophy of Religion*. Bloomington, Ind: Indiana University Press.

Jantzen, Grace M. 1997. Feminism and Pantheism. *The Monist* 80 (2): 266–285.

———. 1984. *God's World, God's Body*. London: Darton, Longman and Todd.

Judge-Becker, Elime, and Charles Taliaferro. 2015. Feminism and Theological Anthropology. In *The Ashgate Research Companion to Theological Anthropology*, ed. Joshua Ryan Farris and Charles Taliaferro. England: Ashgate Publishing Company.

Keller, Catherine. 2003. *The Face of the Deep: A Theology of Becoming*. Florence, United States: Taylor & Francis Group. http://ebookcentral.proquest.com/lib/uu/detail.action?docID=181809.

Leftow, Brian. 2016. Naturalistic Pantheism. In *Alternative Concepts of God: Essays on the Metaphysics of the Divine*, ed. Andrei Buckareff and Yujin Nagasawa, 1st ed., 64–87. Oxford: Oxford University Press.

Leidenhag, Joanna. 2020. Deploying panpsychism for the demarcation of panentheism. In *Panentheism and Panpsychism*, ed. Godehard Brüntrup, Benedikt Paul Göcke, and Ludwig Jaskolla, 65–90. Brill, mentis verlag.

Levine, Michael P. 1994a. *Pantheism: A Non-Theistic Concept of Deity*. London, United States: Taylor & Francis Group. http://ebookcentral.proquest.com/lib/uu/detail.action?docID=169344.

———. 1994b. Pantheism, Ethics and Ecology. *Environmental Values* 3 (2): 121–138.

Mander, William. 2020. Pantheism. In *The Stanford encyclopedia of philosophy*, ed. Edward N. Zalta. Metaphysics Research Lab, Stanford University. https://plato.stanford.edu/archives/spr2020/entries/pantheism/.

McDaniel, Jay. 2007. Ecotheology and World Religions. In *Ecospirit: Religion, Philosophy, and the Earth*, ed. Laurel Kearns and Catherine Keller, 21–44. New York: Fordham University Press.

McFague, Sallie. 1987. *Models of God: Theology for an Ecological, Nuclear Age*. London: SCM Press.

Meister, Chad. 2019. Personalistic theism, divine embodiment, and a problem of evil. *European Journal for Philosophy of Religion* 11 (2): 119–139. https://doi.org/10.24204/ejpr.v11i2.2974.

Puligandla, R. 2013. God and Ultimate Reality: An Analytical Interpretation of Śaṇkara's Philosophy. In *Models of God and Alternative Realities*, ed. Jeanine Diller and Asa Kasher, 615–623. Dordrecht: Springer.

Rogers, Katherin. 2000. *Perfect Being Theology. Reason and Religion*. Edinburgh: Edinburgh University Press.

Rosemary, Radford Ruether. 2007. Ecofeminist Philosophy, Theology, and Ethics: A Comparative View. In *Ecospirit: Religion, Philosophy, and the Earth*, ed. Laurel Kearns and Catherine Keller, 77–93. New York: Fordham University Press.

Rubenstein, Mary-Jane. 2021. *Pantheologies: Gods, Worlds, Monsters*. New York: Columbia University Press.

Ruether, Rosemary Radford. 1992. *Gaia & God*. San Francisco: Harper San Francisco.

Seager, William E., Philip Goff, and Sean Allen-Hermanson. 2017. Panpsychism. In *The Stanford Encyclopedia of Philosophy*, ed. Edward N. Zalta. Metaphysics Research Lab, Stanford University. https://plato.stanford.edu/archives/win2017/entries/panpsychism/.

Sharma, Arvind. 1995. *The Philosophy of Religion and Advaita Vedānta: A Comparative Study in Religion and Reason.* Hermeneutics University Park, Pa: Pennsylvania State Univ. Press.

de Spinoza, Benedict. 2018. *The Ethics (Ethica Ordine Geometrico Demonstrata)*. Translated by R.H.M. Elwes. Mineola, New York: Dover Publications.

Sprigge, Timothy. 1997. Pantheism. *The Monist* 80 (2): 191–217.

Stenmark, Mikael. 2019. Panentheism and Its Neighbors. *International Journal for Philosophy of Religion* 85 (1): 23–41. https://doi.org/10.1007/s11153-018-9687-9.

Taliaferro, Charles, and Chad Meister. 2016. *Contemporary Philosophical Theology*. 1st ed. London: New York: Routledge, Taylor & Francis Group.

Universal Pantheist Society. 2022a. Key Ideas/Pantheism and Earthkeeping. Accessed May 13, 2012, from http://www.pantheist.net/earthkeeping.html

———. 2022b. *Home/are you a pantheist?* Accessed May 13, 2022, from http://www.pantheist.net/

Valera, Luca, and Gabriel Vidal. 2022. Pantheism, Panentheism, and Ecosophy: Getting Back to Spinoza? *Zygon* 0 (0): 1–19. https://doi.org/10.1111/zygo.12800.

Williams, Thomas. 2013. Introduction to Classical Theism. In *Models of God and Alternative Ultimate Realities*, ed. Jeanine Diller and Asa Kasher. Dordrecht: Springer.

Wood, Harold W. 1985. Modern Pantheism as an Approach to Environmental Ethics. *Enviromental Ethics* 7 (2): 151–163.

Yalcin, Martin O. 2011. American Naturalism on Pantheism. *American Journal of Theology & Philosophy* 32 (2): 156–179.

CHAPTER 4

Landscapes of the Unconscious and the Longings of Nature

Elaine Padilla

Trees, as symbols of our longings for mutually beneficial forms of interconnectivity, arise out of the depths of the unconscious. A tree like that of Myrna Báez's *El Mangle* (below), which is depicted in water, can express nature's longing for what Carl Jung likens to a mystical harmony potentially attained through creative processes similar to that of human consciousness entering a mother's womb and arising to conscious awareness.[1] My emphasis

I want to thank The Louisville Institute for a year-long research grant that made this study possible, and my colleague and friend A.W. Barber, Professor of Asian Studies, University of Calgary in Canada, for the many conversations on Buddhism that have informed some of the insights contained in this essay.

[1] See, for instance, Carl G. Jung, *Psychology of the Unconscious: A Study of the Transformations and Symbolisms of the Libido: A Contribution to the History of the Evolution of Thought*, trans. and intr. Beatrice M. Hinkle (London: Kegan Paul, Trench, Trubner & Co. Ltd., 1946), esp. 136–139.

E. Padilla (✉)
Philosophy and Religion, LatinX/Latin American Studies, University of La Verne, La Verne, CA, USA
e-mail: epadilla@laverne.edu

T. J. Hastings, K.-W. Sæther (eds.), *Views of Nature and Dualism*,
https://doi.org/10.1007/978-3-031-42902-6_4

in this essay is on the depths that Robert S. Corrington (primarily in *Nature's Sublime*) compares to a dark matrix that I perceive *El Mangle*'s painting to represent.[2] To me, the opacity of this archetypal tree image, of which entwined root systems are engulfed by oceanic waters, is symbolic of the eternal longings of nature arising through the human unconscious. The tree's grounding opacity intertwines selves with the dark bodies that have suffered the wounds of the colonial impulse, and its potencies or energies can provoke bloodied seeds of hope to activate one's innermost creative activities. A piece of art beckons beauty within selves and for selves to radiate the beauty of interconnectedness; it becomes an *artwork-in-the making* of relationality, holding together *co-artisans of themselves* in union with others, and within communities where the whole can flourish.

El Mangle by Myrna Báez (1997, Museo de Arte de Ponce)

The maternal can resemble creative processes of nature that churn underneath surfaces seemingly at rest, where contrasting energies are conducive for human awakenings. This aesthetic state of disruption awakens

[2] Robert S. Corrington, *Nature's Sublime: An Essay in Aesthetic Naturalism* (Lanham, MD: Lexington Books, 2013).

the hoping self to the self that has suffered the wounds of oppression, as found in the principles of "poetics of relation" elaborated by Afro-Caribbean philosopher Edouard Glissant.[3] This essay will highlight what can be inferred as an agitated unconscious arising from the Caribbean depths in order to reconsider how an ancient bodily wisdom infused by the enslaved of the colonial era, with the effects still felt today, is resurrected for the purposes of hope to overcome past wounds and to enact fuller expressions of relationality.

If one perceives in an artwork such as *El Mangle* nature's beckoning to human consciousness to enter into the depths of nature in ways where relational bonds can become expansive and non-tribal in the sense of remaining balkanized as with Glissant, this piece of art can serve as an example of how, for Corrington, artwork can be a cipher of the radiating and ancient wisdom of the *not-yet-being* in nature. As any tribal bonds, including those of African ancestry, are liberated from their colonial bondage, the repressed relational elements in the unconscious suffer the loss of horizons once viewed as the world. Interconnectivity navigates through a *grounding opacity* with the potential for de-essentializing selves, as described by the aesthetic thought of another Caribbean philosopher, Francisco José Ramos. Depths that extend via the root systems of *El Mangle* become navigational linkages for selves to become aware of their migration, wide interconnections, and impermanence. Dark bodies, dark souls, African ancestry, *any*body, and even coloniality can undergo processes of not-yet-being, encounter a grounding opacity, and navigate through passageways of contrasting interconnectivities, so as not to repeat violent histories.

Landscapes like Báez's *El Mangle*, pulsating the eternal longings of nature through its somber colors, can resemble wombs of hope*ful* creative activity for selves to become artists who, once awakened to the unconscious of nature, can take on the task of becoming a work of art. Being aware of a moment's presence ever so transient yet contrasting starkly enough with colonial realities can leave a memorable ancestral footprint. A hope suffused with its stubborn desire for non-balkanized and expansive interconnectivities, for which one's deep abysses long, can stir selves toward an aesthetic praxis. This distantly near and fragile trace, entangled

[3] Edouard Glissant, *Caribbean Discourse: Selected Essays*, trans. J. Michael Dash (Charlottesville, VA: University Press of Virginia, 1989). See also *Poetics of Relation*. Trans. Betsy Wing (Ann Arbor: The University of Michigan Press, 1997).

in the deepening and widening roots, suffers the wound of uncertainties that pose an inquiry on socialities—this, to me, is an initial step toward what Jung calls a "mystical harmony" that can lead to breaths of flourishing in all that there is. In this essay, I'll provisionally delineate this flourishing interconnectivity by hinting at how art can serve to create better life conditions based on the aesthetic principles of the above-mentioned Caribbean thinkers and Corrington.

Dark Tree Longings

Trees as symbols of a collective unconsciousness can be images that signify both feelings of alienation ailing societies and individuals, and various expressions of longing for and paths toward flourishing harmonies and mutually beneficial co-existence. According to Jung, while tendencies and desires smothered by social conditioning can be buried in the unconscious, there also lay subjacent energies, value intensities, or life instincts that are symbolized by primordial images such as a tree. These, which he calls archetypes, have yet to reach "the threshold of consciousness" and are in a continuous state of "*creative activity*."[4]

Furthermore, origins as past "lives" yet to be discovered in the root systems are embedded in the symbolic material of the tree. Yet, more than resembling a search for personal archives or lineage, something like an archetypal tree can reveal a tendency that sets itself apart from the personal unconscious and arrives at its most "impersonal" components even as its categories are inherited. Especially when contemplating on trees with visibly entangled root systems like *El Mangle*, the archival material latent in the tree can become the content of consciousness arising from multiple origins. For instance, archetypes that give rise to tree images, for Jung, "constitute a common psychic substrate of a suprapersonal nature which is present in every one of us,"[5] because these tree symbols have been nurtured by a global stock from the East and West, South and North. Furthermore, as contents of the collective unconscious, archetypes are similar to Plato's *eidos*,[6] such that "treeness" is archaic and primordial, as

[4] See Morris Philipson, *Outline of a Jungian Aesthetics* (Northwestern University Press, 1963), 47–48; quoted from Carl G. Jung, *Two Essays on Analytical Psychology* (New York: Meridian Books, 1956), 278.

[5] Carl G. Jung, *The Basic Writings of C.G. Jung*, trans. R.F.C. Hull (Princeton, NJ: Princeton University Press, 1990), 300.

[6] Jung, *The Basic Writings of C.G. Jung*, 301.

well as not fully graspable. As the archetypal tree unfolds from its "embryonic germ-plasm" state of entangled relations, an original way of conceptualizing oneself can result. The tree—an archetype image of oneself beyond one's own horizons, as one whose origins are entangled with those of others, and whose roots are also grounded in the eternal—can awaken the imagination toward creative acts aimed at the flourishing of the self in relation to itself and others alike.[7]

Nonetheless, questions can arise with regards to a colonizing uniformity or sameness when wrestling with issues of social marginality. What kind of aesthetic pathway would there be in such instances when a priori translations of an "impersonal" imaginary have already been employed as a tool for debilitating the creative components of some segments of society, and so repressing their impulses and potential for flourishing, in particular for those of African descent?

A partial response to how an aesthetic pathway can activate creativity in ways that further liberate meaning from colonizing imageries comes from Corrington, who would ground the embryonic content of the tree in the dark substratum of the womb of nature or the dark matrix. Whereas for Jung, the process of delving into the unconscious resembles re-entering the maternal womb, in Corrington's model, an archetypal image like the trees of *El Mangle* would arise from an intricate interaction between the human unconscious and the unconscious of nature, in which ancient wisdom is infinitely infolding. A longing for flourishing forms of life cannot be exhausted in this interaction since "sacred folds" in nature are "intensified semiotic fields that fold in on themselves over and again increasing in depth and power of meaning with each infolding."[8] So a creative advance would be an outgrowth comparable to a fetal development of the aesthetic processes of *nature naturing* capable of undergoing multiple rebirths. Drawing from Schopenhauer, Corrington describes *nature naturing* as a "unified momentum 'underneath' the phenomenal world that we experience through the senses," yet it has no form and no telos, and "plays no favorites in nature."[9] Similar to how Jung abstracts the archetypes from the images that these come to represent, Corrington's understanding of the unconscious of nature challenges sign-meanings segmented by present

[7] Philipson, *Outline of a Jungian Aesthetics*, 53; see Jung, *Two Essays on Analytical Psychology*, 108.

[8] Corrington, *Nature's Sublime*, 4.

[9] Corrington, *Nature's Sublime*, 38.

global imaginaries, and so the way that longings can cross temporalities and spaces. In other words, just as Jung separates out global archetypes from the individual images that represent them,[10] the global sign-meanings within Corrington's unconscious of nature challenge the local signifiers isolated from the whole.

Essentially, though, Corrington progresses several steps deeper into the collective unconscious than Jung when *darkening* the enlivening processes of creative activity of the unconscious as it pertains to the infinite interrelationality of nature. What for Jung is "suprapersonal," in the sense of an ancestral archaic stock, is progressed by Corrington into the deepening of the unconscious through the pathways of the dead as the seedbed of an infinite relational soil. As if buried in the depths of nature, by adopting a Hindu-like stance, selves are reborn through the unconscious. Similar to sign-meanings arising from quasi-reincarnating rebirths, one creates as if resurrecting the longings of those who have passed on. Unconscious processes of time and eternity, space and place, or of past lives and places, are woven together with the yet to be defined in terms of communal bonds—for in nature lie subjacent interrelated selves. In this regard, an aesthetic process that entails contemplating on a painting, with a tree's entwined root systems that ground a collective unconscious through the sand and into the watery depths of nature, is opaque due to its vast past interconnectivities. The long-past relations that have endured as desirable traits (and not as a result of an essence)—individual as well as generic traits, attitudes, and experiences—become part of the substrate of nature's ancient wisdom.

This ancient wisdom can arise as the stubborn longings of nature pulsating through the lungs of trees, continuously nourished by those who once dreamed of flourishing worlds. As I see it, the human unconscious, upon contemplating multiple entangled roots like those of *El Mangle*, can unearth hope for the *not-yet-being* as an excess of "flourishing" (for the whole) already embedded in nature's energies. Such energies are useful for an upsurge of regenerative impulses toward mutually beneficial relations. As for Corrington, the process of the unconscious of nature, in pouring "itself into the human collective unconscious and its archetypes,"[11]

[10] For Jung, "The archetype is essentially an unconscious content that is altered by becoming conscious and by being perceived, and it takes its colour from the individual consciousness in which it happens to appear." Jung, *The Basic Writing of C.G.* Jung, 301.

[11] Corrington, *Nature's Sublime*, 67.

regenerates a surplus charge that lifts up the self "above the power of its regnant sign series and hold open free for creative and novel semiosis."[12] The symbolic regains new meaning from nature's vast stock of signs and significations, which to me, incorporates the sedimented longings beyond selves that have found no rest. Meaning, embedded in the depths of nature are the energies—desires, the will, the impulse—to overcome constricting and repressive relationalities perceived as permanently conditioning the imagination through, for example, dominant images. This would be another way of describing the Jungian process whereby contents aiding consciousness can widen horizons for the development of self-uniqueness and for assimilation into "a plan of life" that to me resembles the way in which archetypes can play a role in the "*creative activity.*"[13]

With regards to an artwork, the natural matrix or maternal unconscious can act as passage into consciousness by exposing the self to the powers and potencies of a relational ideal that, for Corrington, would be embedded in the work of art. This ideal holds the potential for rotating the human unconscious "toward the light of the *not yet*, toward an open and expanding future that beckons it onward toward a vision of wholeness that can't be replicated by anything that is a mere object within the vast infinity of *nature natured*."[14] *Nature natured* in art, as Corrington defines it, acts as a self-object such as "a sacred text, a history of (claimed) revelation, a powerful narrative…or anything that opens out the potencies of nature as appearing within key orders of the world."[15] The emphasis is on the dynamic activity that results in the vision of representations of that which has yet to be that can turn selves toward that potential for relational flourishings.

Subsequently, and more specific to a work of art like Báez's painting of *El Mangle*, when comparing the concept of "womb" to the unconscious of nature, as in Corrington's model, the tree as an archetypal image can act as a reservoir of meaningful relations because human consciousness enters a potential rebirthing process—one in which sacred folds fecund with potentialities and energies arise from nature with an increase of depth and power (intensity). The bonds of nature are not only vast and expansive,

[12] Corrington, *Nature's Sublime*, 86.

[13] See, for example, Jung, *The Basic Writings of C.G. Jung*, 107–108, 117, and 143–145; and his ideas on *Two Essays*, 278 quoted in Philipson, *Outline of a Jungian Aesthetics*, 48.

[14] Corrington, *Nature's Sublime*, 65.

[15] Corrington, *Nature's Sublime*, 61–62.

complex and multiple, but also unendingly potential. Each time consciousness arcs back away from its levels of consciousness into its unconscious, the human unconscious can become more fully aware of an infinite relationality embedded in the depths of nature (and not the human consciousness alone). Via the tree, human consciousness can awaken anew to its "connection with universal powers and structure that are part of shared communal energy and trans-individual form" for the benefit of the various overlapping communities.[16]

The opacity of the dark womb is due to its inexhaustible wealth of potencies and energies, longings for the *not-yet-but-possible* bonds, renewed hopes, and stubborn desires for conscious and bodily expressions of an unending vast array of interconnectivities latent in nature. Thus, beyond the human self, as Corrington notes, because nature "harbors an astonishing array of communities 'within,'" the commitment remains to "a strong a ecological framework that affirms the intrinsic worth of non-human species."[17] This moving away from the conscious back into the unconscious, that then arises back to consciousness, breaks free from tribal bonds in unforeseeable ways because of nature's immeasurability that awakens selves to the *sign*ificance of nature as such. With the deepening into the womb of nature, which is useful in overcoming the human impulse toward employing dominant masks and exclusive personifications, can also come expressions of interconnectivity with which to counter the prevalent androcentricism. The human who becomes conscious of her widespread entanglement can be an active communal member easing and blocking practices that deplete resources and for some that desecrate the land. Such relational bonds at a level of human consciousness would demand the continued assessment of relationality being turned exclusive and organized into hierarchical categories that thwart flourishing interconnectivities and that in the end perpetuate social marginalization.

Nature, therefore, pulsates continuously, eternally longing for the activation of more mutually beneficial forms of relationality in the human psyche with planetary implications. Archetypal images such as trees can provoke a spark toward such mystical harmony. This burst of energy that fuels ongoing rebirths of relational bonds is a positive aspect in Jung's model, especially when bolstered with an emphasis on nature's unconscious and its relational quality beyond the human (as what is and what

[16] Corrington, *Nature's Sublime*, 75.
[17] Ibid., 87.

will be) in relation to the *whole*ness of nature implied in Corrington's ecstatic and aesthetic naturalism.

Nonetheless, a challenge remains when contemplating on a collective unconscious that has suffered the wounds of colonization such as those still oozing in the Caribbean. For many, being drawn into the wombic depths of nature can agitate the unconscious and resurface wounds caused by past and ongoing oppressions. A sense of flourishing ways of *being* in the world revive the self, yet it must first confront origins pillaged from one's world that obscure the sight of a horizon with which to wed world and infinity. It would mean encountering the abysmal unruliness that evokes viscerally the feeling of an existence forever tainted, at least seemingly so, by the darkestmost events in history—events that, even when unrepeatable, are pregnant with pain and visions of a truncated telos. This too would be a needed step in the therapeutic processes of perceiving a tree like that of *El Mangle,* an archetypal image of the eternal longing in nature for interconnectivity.

Ethno-Poetic Agitations

A scrutinizing look at human histories can tell us that the actualization of Jung's mystical harmony brushed as *El Mangle* in the collective consciousness might be seen when transposing the painting beyond the canvas's frame; for not all interconnectivities have resulted in the flourishing of selves, communities, or ecologies. Much harm has been done as people groups have crossed oceans and have settled, as vampirizing economies have spread their tentacles into natural resources and annihilated entire ecosystems, and as dominant nations have created co-dependent bonds that threaten the wellbeing of the global population. Thus one might ask, how would an archetypal tree soaked with the Caribbean imaginary like that of Báez's *El Mangle* serve as reminder of colonizing pasts, yet be a liberative source of a stubborn hope in the process of actualizing nature's longings for interconnectedness? A path to flourishing forms of relationality entails wrestling with former forms of injustice, the blows of colonization, and the need to shatter ongoing shackles of oppression as the collective memory encounters the unutterable flux of past waters. As contemplated through the aesthetic eyes of Afro-Caribbean philosopher Glissant, in his poetics of relation, the Caribbean waters and their deepening root systems reveal an agitated unconscious that can result in a *counter-poesis.*

In Jung, archetypal images can spur creative activity, yet how can one explain such instances when an inexplicable counter-movement suffuses the body's sympathetic system, an effect in the viscera that can seem opposite to the desire for actualizing beauty? For example, when contemplating the landscape of *El Mangle,* one can feel as if the painting offers a horizon dimmed by the contradictions of life's longings suffering the wounds of dominating bonds. For Glissant, this tree would depict an ancient cry prevalent among people groups of the African diaspora that continuously overflows the bounds of any awakening, especially when colonization caused its underground currents to churn up with an "agitation," of which the Caribbean landscapes serve as reminders. This "agitated unconscious"[18] surfaces in the exposed sands (surfaces) enveloping the entangled root systems once the surging seas have clashed against them (depths)—these surges also continuously mutate the contours of the islands, at times violently in an instant.

If any longing for the beautiful were to become manifest, it cannot be by simply putting to rest these waters by means of lulling memories, silencing the voices of history. The tree regenerates a sense of hope amidst the many colonizing and violent undercurrents it suffers. This understanding of the tree would be in agreement with Corrington, for whom archetypes reflecting nature's unconscious resemble a "vibrating reality that is a harmony of contrasts,"[19] the rich field of struggle with a regenerative quality of death and life from which beauty emerges. For Glissant in particular, in a work of art the landscape is a *character* depicting ceaseless disruptions to numbness or lethargy, from the inside out, and again, from the outside in (the deaths of the ego for the rebirth of a new being).[20] But in partial agreement with Jung's notion of an *arche*, an agitated unconscious is also swollen by sandy and oceanic movements that can resuscitate an ancient Caribbean longing for an ethic of intercollective becomings (what Jung would call a "mystical harmony") that has yet to be actualized for

[18] This term "agitated unconscious" is employed by J. Michael Dash when describing Glissant's poetics of the unconscious in the Introduction to Glissant's book *Caribbean Discourse* (xxi). To this term, I later add, "of the Caribbean" because to me Glissant, while speaking mostly as a Martinique author, seeks to shed light on the Caribbean "agitated existence" in general (See, for example, *Caribbean Discourse*, 9).

[19] Corrington, *Nature's Sublime*, 161.

[20] Glissant, *Caribbean Discourse*, 146.

numberless individuals of African descent, those whose dark bodies bear the marks of a slave trade.

In the Caribbean, as Glissant keenly points out, undercurrents are symbolic of processes of erasure at the hands of colonialism uniquely and individually carried out by the English, French, Dutch, and Spanish.[21] What remain in place are economic dependencies, exploitations of the tourist industry, linguistic assimilations, loss of historical memory, and cultural denials or homogenization. In order for life to be tolerable in the Caribbean, its consciousness exists as a domesticated, calm surface. An unconscious repression at a personal level perpetuates itself when bedazzled by the disguise, like visions of light sparkling through the production of progress and development. The present-day Caribbean skirts around what is being perceived, and simply verifies or authenticates what is presumed to be normal in the existing co-dependent system. The result is the inner resistance and inner struggle between the onslaught of sameness and the search for intercollectivity. In light of Glissant's descriptions of the unconscious, this inner struggle can be called the agitated unconscious of the Caribbean.

For Glissant, to understand the Caribbean unconscious one must listen to the silences of its history characterized by numberless ruptures such as "the brutal dislocation" of the slave trade.[22] Deep within, lay sedimented the longings for returning to "the point of entanglement" from which the Afro-populations have been "forcefully turned away."[23] These he identifies, not with a nostalgia for a singular origin, as much as with a form of *neurosis* that foregrounds a primordial relation. The depths of Caribbean histories are characterized by both, an abyss of neurosis or sense of "relocation (in the new land) as a repressive phase," and sense of a continuum, because the subterranean convergence of this neurotic abyss is "primarily the site of multiple converging paths."[24] The Caribbean unconscious can resemble its landscapes tracing the underside of all its history because: its roots "extend in all directions in our world through its network of branches" forming cross-cultural relationships[25]; of "the agitation of its

[21] Glissant, *Caribbean Discourse*, 5.
[22] Glissant, *Caribbean Discourse*, 61.
[23] Glissant, *Caribbean Discourse*, 26.
[24] Glissant, *Caribbean Discourse*, 65–67
[25] Glissant, *Caribbean Discourse*, 67.

beaches, forgetful of all who climbed its coconut trees" and for being the "ultimate frontier, visible evidence of our past wonderings and our present distress"[26]; and because its sea "explodes the scattered lands into an arc," enrooting and diffracting, harmonizing in errantry, opening all elements to the unforeseeable.[27]

In particular, the Caribbean unconscious becomes agitated at each stitch of interwoven past with a potential future whereby the self undergoes *counter-poesis*. The dark soul of nature gravitationally spirals the Caribbean downward and into the depths of that which is opaque because it has been denied and erased, while simultaneously establishing its *non-being* as potentiality, as the yet to exist. Harmonies of contrasts ensue when spiraling down. As if immersing into the deep longing waters through the multiple roots of *El Mangle*, the depths of the self-darken the line of descent with each twist and turn of linear histories, as multiple zones of contact recombine. *El Mangle* displays a dim horizon due to the counter-movements of nature's oceanic depths. For Glissant, by undergoing a violent descent that appears to offer no path to recovery, the stone of time does not "stretch into our past" and calmly carry us into the future, but rather implodes "in us in clumps, transported in fields of oblivion where we must, with difficulty and pain, put it all back together if we wish to make contact with ourselves and express ourselves."[28]

Being drawn into the archetypal womb of Báez's *El Mangle* results in *counter-poesis* because delving into an infinite hopefulness can be swollen by an ancestral blood violently being robbed of a present that runs through its entangling root systems. An ancestral *arche* agitates the unconscious, for the breathlessness of the Maroon longs within the root systems of *El Mangle*—dying, fleeing, and when alive, building their dwellings without rooftops. Since it is a bodily hopefulness, the mangrove (conglomeration of Caribbean *manglares*) would then be no dreamland to settle in, but what Glissant would describe as a "knotted mass of somber greens" (swamp-like) where the Maroons once ago found death as much as an open-air refuge.[29] Its purplish depths are thus a submarine unity of decomposed remains of enslaved Africans—the shackled, weighed down,

[26] Glissant, *Caribbean Discourse*, 11.
[27] Glissant, *Poetics of Relation*, 33.
[28] Glissant, *Caribbean Discourse*, 106.
[29] Glissant, *Caribbean Discourse*, 10.

pursued, forced to hide, entangle histories at the deepest levels of the unconscious with their "seeds of an invisible presence."[30]

With a similar Hindu flavor as found in Corrington's thought, but with a unique tragic tone, harmonies of contrasts erupt from within *counter-poetically* for regeneration of selves and more greatly for global ecologies. Histories reincarnate through the agitated unconscious, desperately pressing against the membranes of the bowels of the ego, seeking to birth novel frameworks and dynamics, hoping afresh, claiming anew, painstakingly propelling the self to move past mere survival, so as to build realities with the architectonics of the cosmos, and, when stepping into the unknown wilderness, to grow soiled wings that free ecologies from a victimizing past.

While the collective unconscious might seem to resemble a land of neurotic ghosts, the emphasis here is on Caribbean landscapes that reflect processes of an unconscious *poesis* and *counter-poesis* that stems from a space that is an anti-space that for Glissant is "limited to the point of gnawing away at one's being, but diverse enough to multiply it into infinity."[31] By means of lament, this longing to enact liberation can result in an irreducible interweaving of ancestral voices, including those of non-humans. Entangled histories spread horizontally and deepen in the subterranean convergences of cultures and identities, mobilities amongst peoples and the non-human, communicating multiple origins, hopes, and potential becomings.

Hence, entrance into the unconscious of nature radiating through *El Mangle* can turn subversive its agitation on account of the resistance to violence that the depths of nature have also buried. The fragile landscapes of the Caribbean weave together strange and denied beliefs, forbidden desires, and contrived imaginations with the syncopated rhythms of nature that, though threatened, survive as deferred and stubborn dreams of the Caribbean sedimented in the dense and opaque unconscious (*Antillanitè*). The "submarine roots" of a "cross-cultural relationship" provide passageway into subversive densities of irreducibility of the self and of the other,[32] what Glissant calls "ethno-poetics" or non-homogenizing and non-homogeneous ways of being future humans. Hence the

[30] Glissant, *Caribbean Discourse*, 67.
[31] Glissant, *Caribbean Discourse*, 159.
[32] Glissant, *Caribbean Discourse*, 67.

tragic seeds of contaminated survival that the slave trade carried are likewise made manifest through a hidden awareness of the stubborn unreason of the Caribbean unconscious (reincarnating the slave) with its capacity for metamorphoses.[33]

Traces of Excess

A potential for *counter-poesis* lies in the opacity of the eternal longings of nature churning within the sedimented wounds of violent histories that, at present, continue to suffuse a stubborn desire through the seemingly truncated hopes for flourishing interconnectivities. The point at which the unconscious submerges into the dim waters can result in flourishing metamorphoses when the Caribbean consciousness frees itself from its emotional and mental shackles. When contemplating on a painting like *El Mangle*, for instance, an initial step already discussed in relation to Glissant would be to enroot oneself in the dark sands bloodied with the ancestral vitality and polluted by neo-colonial trade systems that remain in place in the Caribbean. For Afro-Caribbean philosophers, these waters, perceived as having a magnetic pole that runs through the former slave trade route, cannot be skipped. Nonetheless, the unconscious is drawn deeper. Upon immersion in the darkened matrix that grounds it, the unconscious passes through a site-event of unknowing. This is conducive to prolonged awareness of self-irreducibility that can result in metamorphosis and, through ethno-poesis, a further process of de-essentialization.

In Jungian analysis, such opaque site-event draws from the archetype of the "shadow." Jung describes it as a "tight passage, a narrow door, whose painful constriction no one is spared who goes down to the deep well."[34] The opacity of the oceanic waters of *El Mangle* can be compared to the "dark psyche" acting as a deep mirror of the self. To darkly face oneself upon entering this opaque space-moment resembles what for Caribbean philosophers, such as Francisco José Ramos, involves a process of emptying that sheds off the layers of an "essential," assimilated, domesticated, or "pre-packaged" way of being human.

This Jungian shadow can be for Corrington symbolic of the *sublime* in nature, which, although referred to with adjectives such as "clearing," it is not devoid of opacity. Earthquakes caused by encounters with the sublime

[33] Glissant, *Caribbean Discourse*, 3.
[34] Jung, *The Basic Writings of C.G. Jung*, 317.

compare to an experience of being shipwrecked that results in the unconscious spiraling “down a black hole” where the world “looms large and threatening.”[35] In an encounter with the sublime or opaque elements of the regenerative *poesis*, a world can seem infinitely expansive. In this encounter with the sublime, boundaries are washed away, and a world's limiting horizon loses its status as the world. This can be an aesthetic moment's event where painting, for instance, extends far beyond the frame of its canvas. When one realizes that a horizon is not the same as “world,” one can experience a sense of vertigo, fear at the loss of world-meaning. But nature does not abandon itself, and so its impulses are toward awakening to an array of other horizons that selves must come to grasp. Without the many horizons coming to their aid, selves can fall into an abyss of meaninglessness.[36] More positively stated, selves in getting ahold of the potential for becoming anew, by perceiving an expanded world-horizon, can gain a sense of newness despite the loss.

Furthermore, the opacity of selves entering the shadows of an expanding world, seemingly without bounds, is instrumental in a liberative self-awareness as it de-clutters the space for visions of the *not-yet-but-possible* upsurging beauty from within nature. In a more Buddhist tone than found in Corrington, like that of Ramos,[37] the darkestmost purple potentialities and energies brushed over the twisted veins of *El Mangle*, momentarily awaken the ego to its *non-being anywhere* (non-subordination to any form of being, including non-being) by means of an overflow of vacuity or excess of emptiness of being (*desbordamiento of sunyata*).[38] All things appear as they are, and as that which they are not. And so the “I am,” in

[35] Corrington, *Nature's Sublime*, 170.

[36] Corrington, *Nature's Sublime*, 169.

[37] Following the aesthetics of thought of Francisco José Ramos, I weave in the three basic principles of Buddhism into my arguments: dissatisfaction (*dukkha*), impermanence (*anicca*), and insubstantiability or lack of substance (*anata*). See *Estetica del pensamiento II: La danza del laberinto: Meditacion sobre el arte y la accion humana* (San Juan, PR: Editorial TalCual, 2003), 347.

[38] He would call this aesthetic form of awakening, navigational or the naontological, which resembles forms of maritime travel that have no destination. It is defined as that by which one neither “is” nor “is not,” and does not intend “to be.” It holds no specific location in that it cannot be subordinated to anything that “is” and “is not” or any condition under which it can appear, because it is related to “el todo” which in English can be said to be “the all of everything.” For him the navigational state resembles being swollen by vacuity (vacuously full) or by what he understands to be the Buddhist concept of sunyata. Ramos, *Estetica del pensamiento II*, 27.

being "non-being" (as potentiality or energy) with that opaque ground that is likewise related to everything, as if in union with *El Mangle*'s root system, experiences its capacity for non-being (one's own death and rebirth)—not only its own, but the intercollective *non-beingness* of the universe. Rather than being in shackles, might what is painted in *El Mangle*'s root systems be perceived as another way of being ligamented? Might we bring to *image* anew an already availed mystical harmony and interconnectivity within nature, oneself, and others including the non-human and the cosmic?

Being in mystical harmony or in union with the multiple root systems, as portrayed in the archetypal *Mangle*, can be described as a navigation or passage that free-flowingly links selves. Selves navigate through the waterways and the root systems created by the mangroves interconnecting one element with another via a dynamic flow. The mystical harmony of Jung, in becoming migratory, can free selves from the multitudinous harmful entwinements. For instance, Ramos would argue that by freeing *El Mangle* from the obligation to name itself, the aesthetic thought (that he speaks of) can navigate through an opaque passageway through which it can partake of infinity without censorship. The ego becomes vacuously full, meaning, it indeterminately instantiates its unconscious as free from preconceived notions of being and non-being (like that of a dominant ancestry). In freeing itself from what is imposed by society on selves, it can transverse from the inside out (or migrate in and out of the self) and interrelatedly with the whole of the universe in concrete ways moment by moment.

In the aesthetic thought of Ramos, the ego, awakened to its own impermanence, deepens by likewise giving itself through the root systems horizontally, rather than by floating above the world as if ultimately transcendent. In union with the mangroves, the ego passes over or migrates. In delivering itself to this mode of momentary existence, being *in* via can travel as *yet-being* simultaneously with being (that which is yet to be in what appears to be)—the shipwreck of everything that is an essential being. The traversing inherent to impermanence resulting in a no-thing or that which no one can pre-determine can gain meaning from that which gives itself. It receives at each precise moment of givenness the totality of existence as the totality of pluriverses within.[39] And

[39] I borrow the term "pluriverse" from Ernesto Cardenal to emphasize the element of multiplicity existent in the cosmos. See Ernesto Cardenal, *Pluriverse: New and Selected Poems*, ed. Jonathan Cohen (New York: New Directions, 2009).

without becoming subject to any form of identification or personification, it can catch a glimpse of being giftedness and generosity, thankfulness and vulnerability, and as widespread interconnectivity anew time and again.

Lastly, if transposing the potentially regenerative processes of *El Mangle* into what Corrington views as the radically new potencies of nature bursting through human consciousness, non-tribal content deepened in nature's womb can arise in human consciousness. As with flashes of the "*not-yet* conscious within the potencies of nature,"[40] interconnecting powers can radiate through the watery depths of nature depicted in *El Mangle*. In perceiving the world-horizon from a new vantage point, the artist comes to represent on the canvas the yet to become relationship that is churning in nature. In Jungian language, this would be the point at which the painting shines forth beauty in the making from its contact with the eternal realm of the archetypes—a beauty which for Corrington, as already stated, would be the sublime in nature.

For the one contemplating on the artwork, a sharpened awareness of the beautiful can result via an opening at a conscious level to the non-tribal or non-balkanizing content embedded in the innumerable natural complexes. The human, as one complex among them, deepens into nature's clearing to encounter that source of harmony of contrasts which is larger than human—an energy or potency that pulsates like a "micro-burst of pure expanding energy" and splits open stale and cemented meanings of being human to clear the path for novel ways of being *one* self (or "who one is") within community.[41] For Corrington, through such pulsations of energy one can be filled with a sense of internal unity even when amidst "swirls" of conscious existence. By reaching down into the unconscious, one befriends infinite depths, out of which to radiate beauty and *become* a work of art.

This "friendship" with opacity, though an uncomfortable one since it can be experienced as a horizon being swept from under one's feet, is needed for cultivating beauty in healthy relations with others in communities.[42] The *not-yet-being* clears the path for communal forms of awareness—awareness of oneself and others and of mutual

[40] Corrington, *Nature's Sublime*, 96.
[41] Corrington, *Nature's Sublime*, 122.
[42] Corrington, *Nature's Sublime*, 136.

awareness—and so to a complex sense of community that welcomes contrasts and impermanence. The finite and adaptable content—the vast array of combinations of colors, brush techniques, schools of art—of the "always-opening potency of the *not-yet-being*"[43] can be forged democratically as an antidote to monochromatic utopias that come to light when one is awakened to injustices and oppressions, and one's tendency toward homogenization and domination. The sacred power of energy can be distributed throughout the individuals in a community, as with multiple tones and shapes on a canvas, and can serve to create a plan for society that adapts through time to the various social and/or natural environments.[44] That which is at the heart of the consciousness of nature can be made manifest in the democratic processes of the human communities.

Becoming Art

To conclude, an archetypal image like a tree depicted in a painting can bolster our aesthetic processes. In depicting the deepest longings in nature and in us for interconnectedness, it can awaken us to our deepest desires for becoming our most flourishing forms of relationality. These longings lie subjacent in our unconscious enrooted in the depths of the womb of nature, with a ground that is infinitely relational (like a continuum). In this regard, works of art like Myrna Báez's *El Mangle* can aid in countering the effects of coloniality in the human psyche and societies, in particular for people of African ancestry. More widely put, as an aesthetic process, drawing meaning from the symbolic can move selves toward meaningfulness in relationships across multiple borders, and activate the inner creativity needed for selves to radiate from within their beauty—becoming works of art that expands selves beyond perceived world-horizons, and away from mere social conditioning, passivity, and personal inertia (and possible alienation).

As described in this essay, such aesthetic processes mean deepening oneself into the dark womb of nature where relationality is infinite, and a grounding opacity interconnects past and present selves with *not-yet-being* a work of art. These processes can aid a potential break in cycles of

[43] Corrington, *Nature's Sublime*, 97.
[44] Corrington, *Nature's Sublime*, 109.

violence. The *interwoven relational*, as depicted in the root systems of *El Mangle*, counter-poetically serves as the opaque ground for experiencing an agitation in the unconscious due to a harmony of contrasts caused by encountering the "other-than" the selves who remains alive in the depths of the twisted roots of the mangroves. Glissant describes it as the "agitated existence" characteristic of the Caribbean islanders whose ancestors had obstinately endured the slave trade systems,[45] yet explicitly, in a way similar to Corrington's understanding of aesthetics that favors democratic interpretative processes. Linking one with the other, hoping with one another, being interwoven, on the one hand, counters systems of balkanization in the Caribbean that perpetuate oppression, and on the other, prevents wounds and feelings of alienation from turning the oppressed into an oppressor.

In this manner, aesthetic processes involving the unconscious can be grasped anew. For while the fragile reality of the Caribbean will not cease to disturb a dormant consciousness, the innerworkings of fragmentation by their different "social, political, and economic regimes," in the words of Glissant, can be the landscape out of which an artist can create the possibility for "bonds of unity in the future."[46] The artist who seeks to activate the collective life in reality can shift it to consciousness, presenting the possibility of a collective effervescence to which the artist might aspire. Tapping into the fragility of the Caribbean landscape of diverse cultures, languages, ideas, races, and political and religious systems can result in listening to how these root systems communicate among continents and through time interconnecting multitudinous ancestries.

Consequently, this is how *El Mangle* can *make life*,[47] as Corrington would describe it. A form of loss to a stagnant and essential world would occur, without a total loss of the literal world of water and waves, roots and sand, air and sky as such. As opacity clears the path for the novel, consciousness becomes aware of the elasticity of horizons being expanded by worlds that spread in all directions and intertwine numberless elements. The opening can result in an "emancipatory reenactment...whereby the not yet opens out a hidden potency from the past and brings it forward

[45] See Glissant, *Caribbean Discourse*, 9.
[46] Glissant, *Caribbean Discourse*, 235.
[47] Corrington, *Nature's Sublime*, 178.

into the present moment of creativity."[48] When an ancient wisdom grounded in the Caribbean waters can become pluriversal, the rays of our longings for liberation can be embodied anew time and again—selves become icons of *art-in-the-making*.

In this aesthetic process of emancipation, for Ramos, nature's grounding opacity navigationally interconnects in time at the speed of an instant. What is left with each appearance is only a trace of the glory of *El Mangle* that dissolves all appearances. Since the phantasmagorical presence is a mere reminder of a far greater desire that cannot exterminate longing—infinitude—justice as a concept will thus come in many faces, movements, and will need to continue to create new pathways, grow new root systems, and become interconnected anew.

With each life experience accompanied by feelings of dissatisfaction, desperation to capture that which appears can result in manifesting negative and positive forms of identification with desire; it can also result in selfish substantiations such as greed, oppression, and violence. In agreement with Ramos, with each moment of satisfaction accompanied by dissatisfactions, the point would be to celebrate with some form of melancholic enjoyment the beauty that is fanned by an unlimited pleasure. With an awareness of that which is always beyond any ego, there might also come the positive suffering of the ecstatic limit. We can ache with a Bacchic delirium each time that our unconscious longings for flourishing interconnectivities come to rest on some totality of the pluriverse being made partly manifest in art as in the *hope*-filled depths of *El Mangle*.

Bibliography

Cardenal, Ernesto. 2009. *Pluriverse: New and Selected Poems, ed. Jonathan Cohen*. New York: New Directions.

Corrington, Robert S. 2013. *Nature's Sublime: An Essay in Aesthetic Naturalism*. Lanham, MD: Lexington Books.

Glissant, Edouard. 1989. *Caribbean Discourse: Selected Essays.*, trans. J. Michael Dash. Charlottesville, VA: University Press of Virginia.

———. 1997. *Poetics of Relation*. Trans. Betsy Wing. Ann Arbor: The University of Michigan Press.

[48] Corrington, *Nature's Sublime*, 184.

Jung, Carl G. 1946. *Psychology of the Unconscious: A Study of the Transformations and Symbolisms of the Libido: A Contribution to the History of the Evolution of Thought.*, trans. and intr. Beatrice M. Hinkle. London: Kegan Paul, Trench, Trubner & Co. Ltd.

———. 1956. *Two Essays on Analytical Psychology.* New York: Meridian Books.

———. 1990. *The Basic Writings of C.G. Jung*, trans. R.F.C. Hull. Princeton, NJ: Princeton University Press.

Philipson, Morris. 1963. *Outline of a Jungian Aesthetics.* Evanston: Northwestern University Press.

Ramos, Francisco José. 2003. *Estetica del pensamiento II: La danza del laberinto: Meditacion sobre el arte y la accion humana.* San Juan, PR: Editorial TalCual.

CHAPTER 5

Ferd Toward a Joyful Change: Nature, Mountaineering Philosophers, and the Dawn of "Higher" *Friluftsliv* Education

Dag Erik Wold

Introduction

The Norwegian philosopher Arne Næss's opening chapter in one of his best-known books, *Ecology, Community and Lifestyle* (1974), is entitled "The gravity of the situation." Næss expresses a clear warning about the environmental situation.

In the Norwegian landscape, wind turbines were not present when Næss wrote his book; however, dams dried-up, and piped rivers were present and caused disgust and sorrow, mainly among those who hiked and climbed the Norwegian mountains. "Has Norway anything to tell the world?" Næss asks, "I don't know anything else other than the classical Norwegian *friluftsliv*, free-air-life."[1] Næss became internationally known

[1] Arne Næss 1994a, "The Norwegian Roots of Deep Ecology," in *Nature the True Home of Culture*, ed. Børge Dahle (Oslo: Norges idrettshøgskole, 1994), 15.

D. E. Wold (✉)
Department of Physical Education, Volda University College, Volda, Norway
e-mail: dag.erik.wold@hivolda.no

T. J. Hastings, K.-W. Sæther (eds.), *Views of Nature and Dualism*,
https://doi.org/10.1007/978-3-031-42902-6_5

as one of the founders of the deep ecological movement, which had close ties to the *friluftsliv* philosophy, as it was chiseled out especially from the late 1960s and throughout the 1970s. In climbing and philosophy circles around Næss, an interest in ecophilosophy was established from the mid-1960s. It functioned as an academic exercise with an activist aim, especially as Gandhi-inspired nonviolence campaigns were initiated in relation to hydroelectric development in the Norwegian wilderness.[2] Central to this movement was the nature-friendly *friluftsliv*, which at the time was having a philosophical, academic, political, and educational resurgence.

Næss's deep ecology was a protest against Cartesian dualism, and the goal was that a new and less anthropocentric view of nature would contribute to deep social change; moreover, this concept is undeniably needed today; however, we should consider whether deep ecology in general or ecophilosophical *friluftsliv* in particular is still necessary and relevant.

This article aims to explore the understanding of the relationship between humans and nature in the early Norwegian "higher"[3] *friluftsliv* pedagogy and identify the main challenges related to this understanding. This topic is of interest today because the nature crisis requires suggestions of viable paths for change. In addition, my study can provide renewed knowledge of improving educational frameworks for a more sustainable *Bildung*.[4] My task does not include studying humans' relationships with nature as a general topic in *friluftsliv* pedagogy. Instead, I focus on how humans can develop a deep, joyful love of nature through simple living in nature.

The relationship between humans and nature is an important topic in *friluftsliv* education, which has emphasized civilizational criticism and nature-friendly education. Of particular interest here is the early views of *friluftsliv* pedagogy in proximity to the advent of ecological concern, mainly from the 1960s and 1970s, a period that can also be described as the early years of *friluftsliv* education for adults. Here, we find that a *ferd*

[2] The Mardøla (1970) and Alta (1979–81) protests are the best-known campaigns in Norway because of the extensive nonviolent actions that occurred.

[3] "Higher" refers to the fact that the *friluftsliv* pedagogy largely developed as didactics in the mountains and teaching for adults, both in academic and nonacademic institutions.

[4] The notion of *Bildung* embraces several understandings. In this article I take a broad approach, which can imply terms like "formation," "self-formation," "growth," "cultivation," and "self-development."

methodology was developed.[5] *Ferd* methodology consists of many different discourses, such as safety, health, and leadership. Although *ferd* spans a wide range of topics, my concern is to explore *ferd* methodology within one particular context for *friluftsliv*, that is, climbing, or better described as mountaineering, which was the field of practice for most ecophilosophers. Næss proposed a new Norwegian verb instead of climbing (*klatring*), namely *fjelling*, to describe being in and close to the mountains and not just focusing on vertical movement.[6]

My methodological approach is a historical–systematic reconstruction.[7] Within the humanities, systematic reconstruction aims to analyze material for elucidating a particular problem—in my case, the human–nature relationship in *friluftsliv* pedagogy. Such a reconstruction also has a critical component by problematizing findings in the material. Further, my systematic reconstruction is historical, based on Norwegian ecophilosophers' writings. My main sources are Næss's writings from the 1970s, particularly his work *Ecology, Community and Lifestyle*. In addition, I will use the *friluftsliv* philosopher and professional mountain guide Nils Faarlund and other ecophilosophers' texts from the journal *Mestre fjellet* from 1968 to 1999.[8] My historical–systematic reconstruction focuses on the theme of *ferd*, which is discussed in the context of humans' relationship with nature.

First, I will identify and elaborate on the concept of *friluftsliv* developed by the mountaineering philosophers, where I find "nonconventional goal direction" a crucial concept.[9] This chapter provides a necessary basis for my second step, where I will investigate the cultural construction of *ferd* by analyzing two historical *ferds*. Furthermore, I will discuss and

[5] The Old Norse word *ferd* (pronounced faerd) is derived from *fara*, which means to travel. In conjunction with *friluftsliv*, it usually means a long journey in nature. *Ferd* is proposed as something between the more common terms *tur* (hike) and "expedition."

[6] In a greeting to Faarlund, Næss wrote that he would like a verb that focuses on the mountain and not the people and their nagging. Næss stated that the translation *mountaineering* is usable. See Arne Næss, "Litt om fjelling," in *Om å gripe fjellet—og bli grepet av Fjellet. En vennebog om friluftslivet og Nils Faarlund*, ed. Torbjørn Ydegaard (Sorø: Nordisk Forum for Vejledning i Natur- og Friluftsliv, 1987), 37–39.

[7] For such an approach see Knut-Willy Sæther, *Naturens skjønnhet. En studie av forholdet mellom estetikk, teologi og naturvitenskap* (Oslo: Cappelen Damm Akademisk, 2017), 14ff.

[8] Faarlund was the main editor of the epic journal *Mestre fjellet* [Mastering the Mountain], which was released at irregular intervals from 1968 to 1999.

[9] "Nonconventional goal direction" is a notion I have constructed, which I believe covers Næss's view in Arne Næss and David Rothenberg, *Ecology, Community and Lifestyle. Outline of an Ecosophy* (Cambridge: Cambridge University Press, 1976), 305.

problematize *ferd* and deep encounters with nature as components in *friluftsliv* education, with the notions of free nature, durability, and reflection. Finally, I address how *friluftsliv* and *ferd* methodology can be related to a broader picture for change and sustainable interactions with nature. Altogether, this will provide new perspectives on understanding humans' relationships with nature and how this fosters a deep connection with and love for nature in education.

Mountaineering Philosophers and the Concept of *Friluftsliv*

Næss is essential for identifying the idea that *friluftsliv* as a pedagogical practice has something to offer in developing a close love for nature. Næss's *friluftsliv* was above all different forms of climbing—an activity he practiced almost all his life. His theoretical contribution to the *friluftsliv* discourse is particularly related to the simple, playful, and reflective *friluftsliv*, as well as concrete climbing ethics. Even though Næss participated in some of the most famous climbing expeditions among Norwegians and developed ethics and philosophy related to this form of *friluftsliv*, the simple *friluftsliv* is often associated with small social settings. Næss spoke and wrote with passion about a *friluftsliv* with no more equipment than you can fit in your backpack, with a strong emphasis on the encounter with nature. If we add up all short and long visits, he spent several years at his simple cabin called Tvergastein on the mountain plateau called Hallingskarvet.[10] Since his own theoretical contribution to *friluftsliv* pedagogy is somewhat vague, it is often elaborated by his mountaineering friends,[11] in particular Faarlund.[12] In *Ecology, Community and Lifestyle*, Næss allowed Faarlund to speak in the chapter entitled "*Friluftsliv*—exuberance in nature." Nevertheless, Næss contributed substantially to the

[10] Tvergastein (1506 m) is situated more than 5 km from the nearest road in a mountainous area in southern Norway, beneath the southern cliffs of a mountain plateau called Hallingskarvet. Næss had the cabin built in 1937. At times, he brought piles of books and would stay there for weeks at a time. Tvergastein is important for understanding Næss's life and his teachings.

[11] Peter Wessel Zapffe (1899–1990), Sigmund Kvaløy Setreng (1934–2014), and Nils Faarlund (b. 1937) were all connected to Næss in different ways through mountaineering. The first two he knew through the Department of Philosophy at the University of Oslo.

[12] Later, Bjørn Tordsson (b. 1953) was among those who outlined *friluftsliv* pedagogy in Scandinavia.

philosophical foundation of *friluftsliv*. Næss can be described as a pluralist and possibilist, favoring diversity, and he rarely offered a uniform proposition or final definition. He avoided explicit definitions of many of his terms. Instead, he encouraged gradually coming to our own understanding of keywords.[13] In his Ecosophy T ("T" stands for *Tvergastein*), he saw the Norwegian *friluftsliv* tradition and simple cabin life as a joyful and inspiring example of the many paths to societal change for a better world.

The *friluftsliv* movement from the 1960s was associated in part with the international "green wave." In particular, nature conservation became an important topic within the emerging international mountaineering community, with people in Norwegian academic and philosophical circles leading the way. Deep ecology had its heyday in the 1970s and 1980s. As it gradually declined in Norway, it became better known internationally. In present-day Scandinavia, we find appeals to deep ecology primarily in a few educational institutions and nature-protection organizations.[14] "Ecophilosophy" may be a more apt word than "ecosophy" in Norway today. Sigmund Kvaløy Setreng[15] explains why ecophilosophers wanted to hold on to "ecoPHILOsophy" and not use Næss's "ecosophy," because "philo," as *love for*, was important for them.[16] According to Kvaløy Setreng, "We wanted to keep a clear signal of involvement with *love*. We would seek understanding and work politically, continuously inspired by love of nature and love for the human partaker in nature. We associate 'ecosophy' more with a purely intellectual pursuit."[17]

When Næss wrote about *friluftsliv* in the 1970s, he focused on the *idea of freedom*, the *deep joy of solitude*, and the need to explore wild nature—values that can be found in the early romantic days of *friluftsliv* (from the middle of the nineteenth century) manifested by Norwegian celebrities such as Henrik Ibsen (1828–1906) and later by Fridtjof Nansen

[13] Peter Reed and David Rothenberg. *Wisdom in the Open Air: The Norwegian Roots of Deep Ecology* (Minneapolis: University of Minnesota Press, 1993), 66.

[14] Espen Gamlund, "Hva er galt med dypøkologien? Noen kommentarer til Arne Næss' Økosofi T," *Norsk Filosofisk Tidsskrift* 47 (4) 2012: 229–242.

[15] Kvaløy took the farm name Setreng when he managed his uncle's farm, which he had promised his uncle. In Norway, using the farm name as a surname is not unusual, and it can be seen as a way of linking one's identity closer to the place one feels at home.

[16] Sigmund Kvaløy Setreng, "To økofilosofer i Norge; deres begynnelse og en del til." *Norsk filosofisk tidsskrift.* 37:1–2 (2002), 117–126.

[17] Reed and Rothenberg, *Wisdom in the Open Air: The Norwegian Roots of Deep Ecology*, 119–120.

(1861–1930). *Friluftsliv* was now understood to embrace some values other than competition and achievement orientation.[18] This redefinition is evident in the following quote: "Conventional goal direction: to get there, to be skillful, to be better than others, to get things done, to describe in words, to have and to use new and fancy equipment—is discouraged."[19] In several areas, and not least in the "sport" of climbing, competition in the 1970s was viewed with growing skepticism. Faarlund claimed: "Because competition between people permeates our culture, we need life forms that contain more nuanced issues 'than better than'/inferior to."[20]

The mountaineering philosophers were unhappy with much of the attitude toward nature and mountains in expedition climbing and the aggressive activity they had seen between WWI and WWII in climbing in the Alps.[21] They were worried about an increasingly competitive mentality in climbing. At a seminar entitled "Climbing: Why? How?" at Høgfjellskolen ("the Norwegian School of Mountaineering") in 1969, Faarlund saw Scandinavia as the last hope for autotelic climbing, but he also referred to the positive elements of Yosemite climbers' philosophy and climbing ethics. Faarlund described their climbing philosophy as an "ecological" path between that of the modern USA (utility, efficiency, speed) and the "North American Indians" (way of life, few means, close contact with Nature), as inspired by John Muir.[22] Næss wanted to promote climbing as a contemplative attitude to the mountain and referred to the monks in the Himalayas and their aesthetic and religious experience of the mountain. The *wild* mountain was important for these climbing philosophers; therefore, bolts and other climbing equipment should not be left behind. In general, in the mountain landscape, the majestic nature should not be made cozy (cf. no trail markings at Hallingskarvet). Hence, in line with the critique above from the mountaineering philosophers, I coin the term "nonconventional goal" for describing a fundamental quality of *friluftsliv*.

[18] André Horgen argues that, beginning with the dawn of Norwegian mountaineering in the 1870s, a cultural resistance took place in The Norwegian Tourist Association against British sports ideal of ascending the summit. See Horgen, "Friluftsliv, idrett eller Sport? Mening og begrepsbruk i Den Norske Turistforenings årbøker, 1868–1979," *Historisk tidsskrift* 101 (3) 2022: 213–228.

[19] Næss and Rothenberg, *Ecology, Community and Lifestyle. Outline of an Ecosophy*, 179.

[20] Nils Faarlund, "Nei til konkurranseklatring," *Mestre fjellet* 2:2 (1971). My translation from Norwegian.

[21] Norsk Alpincenter, "Klatring—hvorfor, hvordan?" *Mestre fjellet* 3–4 (1969): 25–26.

[22] Norsk Alpincenter, "Klatring—hvorfor, hvordan?" *Mestre fjellet* 3–4 (1969): 26.

An Alternative to Competition: The *Ferd* as Cooperation

Identifying the nonconventional goals leads us to *ferd* as a practice that differs from competing. In the following sections, I will unpack the notion of *ferd* by emphasizing our encounter with nature and its pedagogical foundations.

One of the first remarkable *Bildung* travels, now called *ferd* by Faarlund, went to the "holy" mountain of Stetind in northern Norway in 1966.[23] Having plenty of time and a lot of wind and rain (what Næss called "Spinoza weather"), Faarlund disturbed Næss in his reading with challenging questions about ecology, science, Gandhi, and Spinoza's ethics. Ideas were brought back to the Department of Philosophy at the University of Oslo. In 1969, these ideas were developed into the "Nature and Humans seminar," resulting in the founding of the Co-working Groups for the Protection of Nature and the Environment.[24] At that time, several watercourses were under threat in Norway. Mardøla waterfall, one of the highest waterfalls in Europe, was chosen as the first camp for direct action and anti-violence demonstrations in the summer of 1970.[25] This mountain camp was a coalition of (snm) members, academics, young students, local farmers, carpenters, and mountaineers, mainly led by Kvaløy Setreng, who was educated by Næss in Gandhi's nonviolent philosophy over the

[23] In 1965, Faarlund obtained the latest from the climbing community in California: "Chouinard-forged" chrome-molybdenum steel pitons that could be repeatedly used. He was therefore able to tempt Næss to join a new *ferd* to Stetind along with Kvaløy Setreng and members of Tindegruppa (the student climbing group that Faarlund co-founded at NTH in Trondheim in 1959). Here, as early as the summer of 1966, Faarlund challenged Næss, a former "participant" in the positivist Vienna Circle, to confront the natural sciences. Faarlund found the journey to Stetind that summer a kind of hotbed of ecophilosophy that would later develop in slightly different directions (source: conversations with Faarlund).

[24] It was also called *the Ecophilosophy Group*, written as (snm). Lowercase letters and parentheses were used as a reminder to the members that the organization should never become an end in itself. It was a circle of cooperating subgroups spread over Norway dealing with environmental problems, including an ecophilosophy group and practically oriented groups, such as the "nonviolent direct-action group." (Setreng, "To økofilosofer i Norge; deres begynnelse og en del til,")

[25] Approximately 500 people joined the Mardøla action, with participants from France, America, and Holland. Setreng wrote: "We lost the waterfall but we started a movement" (Sigmund Kvaløy Setreng, "The Universe Within," *Resurgence*. 106 (1984): 8.

years as his student at the University of Oslo from 1958 to 1966.[26] The tent camp was in many ways a place where cooperation, learning, and deep joy were central to the participants, completely in line with Næss's ideas of a life—"simple in means, rich in ends,"—and with nonconventional goals. Those values were also evident later in nature protection camps, such as in the Sàpmi heartland, Alta, in 1980–81.

In the following section, my analysis of the notion of *ferd* will be followed by two concrete *ferds*: the *Tseringma* and the *Azourki*.

Mother of all **Ferds**: *The Anti-Expedition to Tseringma*

In 1971, Næss, Faarlund, and Kvaløy Setreng went on a "Grand Tour"' (*Bildung-reise*) to the Himalayas Rolwaling Valley on a two-month-long "pilgrim's journey." They were going to climb, but the goal was not conquest. On the contrary, they would demonstrate this by returning before reaching the summit, due to respect for the locals' religious feelings for the sacred mountain Jomo Tseringma—"The holy mother of the lasting life."[27] In this way, they would convince the authorities that the sacred mountains of the Sherpas should be protected from climbing. In consultation with the lama in the Rolwaling Valley, they agreed that mountaineering should not be done on Tseringma above 6000 meters.[28]

In a report from this *ferd*, entitled "Rock climbing in a new way," Kvaløy Setreng wrote about mountaineering as an opportunity to achieve contact with nature without a conquering mentality, in line with Næss's critique of conventional goals.[29] In the village of Beding, which had not been visited previously by expeditions, they spent a lot of time adapting to daily life, agriculture, and the Tibetan Buddhistic religious life and rituals. They described the Sherpa community as "a life in harmony with nature," an "equilibrium society," and a "society that had managed to survive through centuries despite a harsh climate and scarce resources—and to do so with an unprecedented cultural surplus."[30]

[26] Examples of specific advice that were attempted to be followed were: to seek the center of the conflict and treat opponents with the utmost respect, and to be nonviolent in language, judgement, and action.

[27] The Holiest of all sacred mountains for the Sherpas—"the ultimate symbol of man's cohesion with and dependence on nature" (Kvaløy 1973b 10).

[28] Nils Faarlund, "Et 50 års minne fra Himalaya," *Hallingdølen*, October 12, 2021, 25.

[29] Kvaløy 1973b, 10.

[30] Kvaløy 1973b, 10.

The *ferd* inspired several articles and lectures, especially from Kvaløy Setreng, who later made journeys to the valley called "the Plow Furrow,"[31] where he got much of the inspiration for the distinction between complicity and complexity. He recognized the complexity of the valley's spiritual, social, and cultural life and compared it to the complexity of other pre-industrial natural societies. He pointed out that the natural ecological complexities of the environment are subverted by the Industrial Growth Society (IGS). One of the main arguments for Kvaløy Setreng for the Life Necessities Society (LNS) was that industrial communities were a tremendous extra burden on the natural environment due to the extreme need to transport goods and services. The Sherpa and old peasant communities in Norway were ideal for his ecophilosophy of strictly decentralized societies.

The inspiration from this *ferd* also developed into an almost uncompromising requirement to use only nature-friendly materials, such as wooden skis and woolen clothes. The equipment had to be simple and versatile. The journal *Mestre fjellet* ran many articles on the topic, and eventually, descriptions of how to make the equipment yourself were published. Moreover, as expressed by Faarlund, "[m]aking your own hiking equipment is a form of liberation struggle."[32] In *friluftsliv* education in Norway in the 1970s and 1980s, a combination of *friluftsliv* and craft-making was evident. Typically, students made their own wooden skis, backpacks, and other hiking equipment using natural materials.

According to Næss, the way sacred mountains are cultivated varies in different religions and cultures. While in Buddhist culture, it is considered unworthy to climb a sacred mountain, at least if one does not have the right attitude, he sees in Chinese culture that ascending mountains is a kind of pilgrimage all the way to the top. The hiking pattern may differ, but the notions of the dignity of the mountain are common, which brings us to the appearance of safety attitudes toward mountain hikes in the ecophilosophical *friluftsliv* environment after the Tseringma *ferd*. Much like the views of the ecophilosophers, climbers in general ought to approach the mountain with open-minded, friendly thoughts, and a benevolent attitude. Before the ascent, the climbers should carry out traditional rituals and let the monks bless the expedition. According to Næss, the right attitude toward the mountain is to "linger on the cliffs," which reflects his typical predilection for curiosity and dwelling in "wonder."

[31] Sigmund Kvaløy Setreng, "The Universe Within," 1–18.

[32] Nils Faarlund, "Klær og redskap for friluftsliv om vinteren," *Mestre fjellet*, 29 (1980): 20.

The idea of returning before reaching the summit, or to avoid the summit point, is later found in the educational ideals of Høgfjellskolen and later in academic *friluftsliv* education. Respect for the sacred was transformed into consideration for the dangers in the mountains and learning to be humble in nature. Gandhi's challenge, "the way is the goal," became a prevailing doctrine in the *friluftsliv* movement.[33]

Approximately 50 years later, Faarlund says Rolwaling's *ferd* gave him inspiration for a pedagogy with nature as the great educator, described as "grandparent pedagogy"—a practice where values, traditions, and knowledge are transferred between generations in natural cultures.[34] Among naturalists and in primary industries, older generations have taught children through participating in life and work. They have safely been able to make their own experiences. Faarlund refers to Konrad Lorenz's statement that *nature is immediately understandable*—as opposed to using excessive indoor teaching in classrooms with technical aids.[35] Outdoors, one learns about the weather by being in the weather. Faarlund is skeptical of Edmund Hillary's English School for the Sherpas and calls it a destruction of their culture. He claims an unfortunate development from participation in real life to "a spectator school" following the European pattern.[36]

The Tseringma *ferd* was followed by new expeditions to the Himalayas. In the autumn of 1979, businessman Arne Næss Jr. (professor Næss's nephew), Faarlund, and several of the members of Tindegruppa and from the Høgfjellskole environment set off on the Numbur expedition.[37] When the expedition occurred, the ecophilosophical-inspired pedagogy was already well-developed and incorporated in institutions such as

[33] Nils Faarlund, "Læresetninger," June 2021, https://www.norgeshogfjellskole.no/laeresetninger/

[34] Petter Erik Leirhaug, Per Ingvar Haukeland, and Nils Faarlund, "Friluftslivsvegledning som verdidannende læring i møte med fri natur," in *Friluftslivspedagogikk*, eds. Linda Hallandsvik and Jannicke Høyem (Oslo: Cappelen Damm Akademisk, 2019), 15–32.

[35] Nils Faarlund, "Friluftslivets naturmøte som veg til et liv i lage," *Mestre fjellet*, 44 (1999): 22.

[36] Nils Faarlund, "Hillary og Sherpaene: Vennetjeneste eller kulturdrap?" *Mestre fjellet*, 25 (1977): 23.

[37] Faarlund called it "the Norwegian Himalayan expedition" to Shorong Himal (Nils Faarlund, "Erfaringer fra Den norske Himalayaekspedisjonen," *Mestre fjellet* 30 (1981). "Erfaringer fra Den norske Himalayaekspedisjonen," *Mestre fjellet* 30: 5–12). The Nepalese government named the most sacred mountain for the Shorong sherpas Numbur, whereas Sherpas named it Shorong Yulla (Nils Faarlund, "Toppen er et sted på vegen," in *Bergtatt*, ed. Norman Kjærvik (Oslo; Grøndahl Dreyer, 1992).

Høgfjellskolen. In addition, *friluftsliv* education took place at the Norwegian School of Sport Sciences and at several of Norway's teacher training schools. The Numbur expedition clearly showed how the *ferd* philosophy was still contrasted with the more common imperialistic expedition philosophy. The expedition was in line with the *ferd* philosophy by creating heterogeneous groups—broadly composed of climbers with a strong competitive instinct and performance focus, together with climbers with nonconventional goals. These major ideological differences led to disagreements and conflicts over goals and means during the expedition. Faarlund wanted the Numbur expedition to continue along the lines of "the way is the goal" in keeping with Tseringma.[38]

Despite the quarreling, for Faarlund's part, the Numbur expedition met some of his goals, even though he would not call it a *ferd*.[39] Faarlund mentions the encounter with Sherpa culture, great nature experiences, and the time to develop a sense of togetherness with free nature. He is thankful that Næss Jr. allowed the expedition to use nature-friendly and durable materials in the equipment. However, more problematic was the fact that this expedition was preparation for a later trip to Mount Everest. The Everest expedition had mostly ordinary "expedition frames" and the summit as the ultimate goal, contradicting the values of *friluftsliv* and *ferd*.

In 1985, real ideological differences among Norwegian expeditioners emerged, this time for a larger audience—in Norwegian newspapers.[40] This is when Næss Jr. gathered a group of mountaineers, including several of the participants from the Numbur expedition (but without Faarlund), to fulfill the dream of climbing Mount Everest. Faarlund and Kvaløy Setreng strongly opposed the expedition, as they considered it part of the imperialistic international climbing industry that had destroyed the local Sherpa culture of the Khumbu Valley after 1953.[41] However, the *ferd* to

[38] See Ralph Høibakk, Høibakk, "Numbur 1979, Egotrip, pilgrimsferd og samlivstrening," in *Norsk Fjellsport 1983*, 225–236. Norsk Tindeklub.

[39] Nils Faarlund, "Den norske Himalaya ekspedisjonen 1979. En optimistisk meddelelse fra første leir over tregrensen," *Mestre fjellet* 31–32 (1983): 43.

[40] Næss Sr. stated that the expedition had "sold itself to the mass media and other sponsors, and Faarlund called it a banal status trip that put the Sherpas at risk." (Martin Nilsen, "Banal status-trim," *Dagbladet*, April 20, 1985, 8, and Anne-Sissel Skånvik, "Solgt seg til massemedia," Dagbladet, April 20, 1985), 9.

[41] The Everest expedition became a hot topic in *Mestre fjellet* 1986 (no. 34–35). Næss agreed with Kvaløy Setreng and Faarlund but was careful not to criticize his nephew publicly in the newspapers. He also followed the expedition up to about 5000 (Thorbjørn Ydegaard, "Om Færden som metode," *Mestre fjellet* 36 (1986): 9–10.

Azourki 12 years earlier, far from the Himalayas, followed up the nonconventional goals and virtues from the Tseringma *ferd* and gave *ferd* a format and content that had an impact on the new *friluftsliv* education.

Ferd *to Azourki:* Ferd *Methodology and* Friluftsliv *Pedagogy Takes Shape*

The 1973 journey to Azourki in the Atlas Mountains of Northwest Africa consisted of Nils Faarlund, Helga Faarlund,[42] Atle Tellnes,[43] and Per Lund. They went by train from Norway to Marrakech in Morocco, and onward with vehicles and mullets.[44] Azourki is also the name of the Sacred Mountain of the Berbers, a mountain people living in the forest border at a height of approximately 2000 m.[45] Like the Sherpas, they had—until then—preserved their culture free from the influence of modern technology and industrialization.[46] The journey was seen as a part of the work at Høgfjellskolen and was a continuation of the journey to Tseringma. The purpose of the *ferd* was mountain skiing 3000–4000 meters above sea level, studying the Berbers's way of life, and continuing the work with *friluftsliv* and ecoculture studies.[47]

Faarlund seemed to be disappointed that meeting the Berbers engendered different experiences than expected. Despite a traditional peasant culture without the heavy influence of industry, they encountered an

[42] Faarlund' s wife, Helga, had a vital role administrating Høgfjellskolen' s assignments, and the students there described her role as very inspiring and important for their student life in Hemsedal (Torbjørn Ydegaard (ed.), *Nils Faarlund Samlede Verker VII, Veglederstudenter* (København: Books on Demand), 2016.

[43] Tellnes was among the first (*friluftsliv*) students at NIH, wrote one of the first master theses there in 1978, and started *friluftsliv* education in Bø, Telemark in 1982 (see Nils Faarlund, "Takk for "vakre handlinger" for vegledning i friluftsliv!" in *Friluftsliv, kystkultur og vakre løsninger*, edited by Gunnar Tellnes (NAKULHEL-forlaget, 2018), 266–279.

[44] In retrospect, the mode of travel and equipment to Tseringma was intended, but not nature friendly. However, this became better on the Azourki *ferd* (Faarlund, "Takk for "vakre handlinger" for vegledning i friluftsliv!").

[45] Nils Faarlund, "Ferden som arbeidsmetode i friluftsliv," *Mestre fjellet* 18 (1974): 6.

[46] Atle Tellnes has systematically studied the Berbers's use of resources, which he finds largely sustainable, but also threatened. See Atle Tellnes "Friluftslivs-seminar Agder folkehøyskole 1974," *Mestre fjellet* 18 (1975): 9–16.

[47] Norsk Alpincenter, "Nytt fra Høgfjellskolen," *Mestre fjellet* 3–4 (1972): 38.

While the journey to Tseringma was spontaneously created by Professor Næss, the journey to Azourki was deliberately designed and carried out on the basis of method assessments (Faarlund, "Ferden som arbeidsmetode i friluftsliv," 6.

attitude toward nature close to the anthropocentric attitude he knew from Europe, and not what he had experienced in the Sherpa culture in Beding.[48] However, the Azourki *ferd* gave new vigor and ideas to his and Tellnes's pedagogical thinking. Subsequently, the *ferd* methodology spread to different *folkehøgskoler*,[49] primary schools, teacher training schools, and *friluftsliv* organizations.[50]

My analysis above shows that the main goal of the *ferd* is to familiarize oneself with free nature and close-to-nature cultures. Such a goal—or even a virtue—has remained in the *friluftsliv* education, which emphasizes that the consequence of this deeper knowledge will lead to committed friendships with nature, as well as the local cultures.[51] Further, the understanding gained promotes a joyful change from shallow acquaintances to a deeper love and care for nature. By emphasizing the relationship between humans and nature, the experiences from the *ferd* described above and previous *ferds* may—with relevance for *friluftsliv* education—be summarized in the following main points:

- Participation and co-responsibility: A *ferd* should not have a fully prepared program beforehand. Instead, the participants are encouraged to take initiative, to discuss, and to organize themselves in interest groups. Friendly gatherings may take place through artistic activities. The ideal is noncompetition and cooperation.
- Duration: A *ferd* that lasts a week is considered too short. The group should have sufficient time to become aware of nature's rhythms, such as sunrise, sunset, and avalanches.[52] Time and space for crafts and artistic activity are important.

[48] Norsk Alpincenter, "Nytt fra fjellet," *Mestre fjellet* 15 (1973): 23.

[49] Nordic "Folk Academy"/"folk high schools" (*folkehøgskole*) are based on the ideas of education and *Bildung* by the Danish poet and priest N. Grundtvig (1783–1872).

[50] Tellnes, "Friluftslivs-seminar Agder folkehøyskole 1974," *Mestre fjellet* 18 (1975): 9–16.

[51] Faarlund, "Ferden som arbeidsmetode i friluftsliv."

[52] Enough time was also considered essential for the safety of the group, especially if it is a new environment one is not used to. Gradual adaption due to circadian rhythm, oxygen pressure, heat, and cold is necessary to prevent injury and illness (Faarlund 1974, 7). A basic (safety) principle on *the study journey* at Høgfjellskolen, and in other *friluftsliv* educational studies, started with the near and familiar and gradually moved toward the distant and unknown.

- Slowness: The traveling to the journey's start should be slow enough so that one has time to prepare well, both for approaching nature and culture. It is essential to grasp the soul and rhythm of the place.[53]
- Being together as a group: Traveling as an educational path is not limited to distant places[54] or monthly stays. The educational point is to bring together a small group and choose a location that actualizes the goals one has set. This should include *friluftsliv* and (participatory) observation of alternative life forms that differ from today's technoculture. Being together should be characterized by a simple life in tents, cabins, igloos, or snow caves.

When *ferd* becomes a part of adult education in *friluftsliv*, it takes on a fairly structured form based on the recommended method. That pattern starts with preparatory work (a "*ferd* council"), then the *ferd* itself, and closes with post-*ferd* reflections and tasks ("afterthoughts"). In line with the principle of not rushing, one should have sufficient time in all of these phases. On a *student ferd*, as a part of the journey's daily rhythm, the *ferdråd* ("*ferd* council") should gather the participants around the campfire in the evening for each member of the group to give a summary of the day. The students are strongly advised to keep a diary for personal reflections. As with the journey to Azourki, the mode of travel to the journey's point of departure is preferably train, bus, or boat. Natural materials and durable equipment are preferred.[55]

My analysis thus far, starting with identifying the concept of *friluftsliv* by emphasizing nonconventional goals, is built upon the main characteristics and virtues of *ferd*, exemplified by Tseringma and Azourki. Based on the material above, I will now discuss and problematize what kind of understanding of the relationship between humans and nature is evident in this early *friluftsliv* pedagogy.

[53] Means of transport that allow time to readjust mentally (Faarlund, "Ferden som arbeidsmetode i friluftsliv.") As early as 1974, we can read skepticism about airplanes. Also, due to pollution, one should choose a train or a boat. In the final part to the base camp, one could use a bicycle, a rowboat, or a pack animal.

[54] Faarlund admits that such long-range travel can take a toll on resources and can negatively affect foreign cultures and that it also makes it difficult to speak out against others who have expedition wishes (Faarlund 1974). Just as Næss defended his long journeys, Faarlund thought that the end justifies the means.

[55] See Norges Høgfjellskole, *Mestre fjellet* 43 (1996).

Ferd and Deep Encounters with Nature: Discussion and Problematization

The idea of a direct connection between humans' encounters with nature and environmentally friendly attitudes and behavioral change has been criticized for being overly optimistic and oversimplifying the complex picture of environmental issues.[56] My analysis shows that ecophilosophers have responded to this by emphasizing that our encounters with nature must be sufficiently deep. They seem to be well aware that their project is more than just creating an individual cognitive understanding: It is about developing a strong emotional connection with nature over time.

In the following, I will problematize how to understand these deep encounters and what can be considered "sufficiently deep." I will do this through the notions of free nature, duration, and reflection. These notions cover what the mountaineering philosophers had in mind in their understanding of *ferd* and *friluftsliv* as education or *Bildung*. Finally, I will discuss the *ferd* further, and how its holistic philosophy and methodology can be related to a broader picture for a joyful change and a sustainable interaction with nature.

Free Nature

According to Næss and Faarlund, a prerequisite for deep commitment seems to be deep experience and joy. "Cross-country skiing without a machine-made track is great. Skiing in ski resorts is not."[57] I interpret Næss as saying that the experiences of nature in these two examples are qualitatively different. There is a question of the *depth of* the encounter with nature. In *Mestre fjellet*, the term "untouched nature" is used back in the very first issue.[58] Some of the inspiration to do this probably came from Peter Wessel Zapffe[59] who emphasized radical conservation of nature

[56] See, for example, Klas Sandell and Johan Öhman, "An educational tool for outdoor education and environmental concern," *Journal of Adventure Education and Outdoor Learning* 13 (1) (2013): 36–55.

[57] Hans Kolstad, *Besinnelse, Natufilosofiske essays* (Oslo: Humanist forlag, 2007), 131.

[58] See Nils Faarlund, "Vi må lære å bruke naturen—uten å forbruke den," *Mestre fjellet* 1 (1968): 28–29.

[59] See note 12.

and felt a strong bond with the untouched wilderness.[60] It was probably the high mountains that were in the mountaineering philosopher's mind. However, below the alpine timberline and all the way down to sea level, *friluftsliv* in "untouched nature" is a problematic term, because nature in all its variety is more or less affected by humans. Kvaløy Setreng saw the cultural landscape in Rolwaling Valley, with its houses and fences, as direct expressions of life that give lasting protection because it is embedded in nature while at the same time being built on centuries-long harvesting of knowledge.[61] In this sense, human-touched nature is something desirable to preserve. Increasingly, "untouched" was replaced by *free* nature—understood as where the natural rhythms are undisturbed and unbroken. Hence, a slalom facility with artificial snow and floodlights is a good example of a place where humans break the natural rhythms.

Moreover, the notion of free nature may also be problematic in light of the Anthropocene, because pollution in general has in one way or another affected most places on the globe. However, it may be better for us to consider nature to be as *free as possible*. Areas with less disturbance from mechanical or artificial sounds and where we can see, for example, swimming fishes or hear the whispering sound of the wind or the stream trickle can provide the potential for deep encounters with nature. The first non-violent actions were about protecting rivers, that is, water that is allowed to flow freely. If unregulated, a river is the ultimate expression of free nature: "nature that is allowed to be itself." The water level drops and rises with the weather, time of day, and season, and the sound of running water has its distinctive natural rhythm. According to Kvaløy Setreng, we not only have to observe running water from a distance, but can also choose to enter the river, throw ourselves in, and physically feel the temperature and the currents.[62]

But *how free* does nature have to be? Zapffe was uncompromising on this and argued as early as 1934 that mountain areas with signs are something completely different from those without signs, and he expressed it this way: "It is the first signpost that is devastating, not the twentieth or

[60] Dag O. Hessen, "Die norwegishe apparatlantenschafft—Zapffe som naturverner," in *Barske Glæder, og andre temaer fra et liv under åpen Himmel*, by Peder W. Zapffe (Oslo: Cappelen Damm, 2012), 145–153.

[61] Sigmund Kvaløy, "Rytme, natur og menneske," *Mestre fjellet* 16 (1973b): 10.

[62] Sigmund Kvaløy Setreng, "Økokrise-glimt fra det norske økofilosofiske forsøket," in *Den uoverstigelige grense. Tanke og handling i miljøkampen*, eds. Svein Gjerdåker, Lars Gule and Bernt Hagtvet (Oslo: Cappelen, 1991), 102–116.

thirtieth."[63] For Zapffe, the deepest value of wild untouched nature is that nature (and the Earth) lives well without people. I find an elitist element in Zapffe's argument that the mountain should be for the few who have earned it through their own efforts. Næss is more moderate and states that a child's encounter with nature does not have to involve large areas of wilderness. A small flower meadow, a little forest, or a stream may be enough and prevail for deep, joyful play.[64] Inspired by Spinoza, Næss says: "Man is born as a good person and born to joy. We need to find our child's joy, but with the insight and action of an adult."[65] According to Næss, Spinoza speaks of different concepts of joy, and of interest here is joy as *hilaritas*. This is the joy that affects the entire soul and body. *Hilaritas* affects the entire person, and thus all that we perceive—our entire reality. According to Næss, Spinoza believes that *hilaritas* is the only affect of which there can never be too much. Næss translates *hilaritas* as "a bright mind" but believes it is more about a disposition in humans than an affect, corresponding to Spinoza's *inner calm* and compatible with a strong desire for activity and the joy of unfolding.[66]

For Næss, it is significant to experience the whole of nature through concrete encounters with individual creatures.[67] In an encounter with a wolverine in the mountains, we do not meet an abstract species of wolverine, but the species as an individual, which is an expression of the whole. Descartes's reduction of animals to machines and soulless creatures clearly contrasts with such encounters in nature.[68] Inspired by Martin Buber, Faarlund describes encounters with free nature as *I–Thou* encounters in *friluftsliv*.[69] "The sense of nature"—which Næss also applies to landscapes, especially those that are not dominated by human activity—is known as Næss's Ethical Holism.[70] The ecophilosophers would say that we do not go *out*, like something "alien" outside, but *into* nature, like something

[63] Hessen, "Die norwegishe apparatlantenschafft—Zapffe som naturverner," 150.

[64] Arne Næss, "Friluftslivet i våre tanker," *Mestre fjellet* 41–42 (1994b): 19–22.

[65] Øystein Dale, "Med fjell som venner - en samtale med Arne Næss," *DNTs Årbok* 2002, 61. My translation.

[66] Per Ingvar Haukeland and Arne Næss, *Dyp Glede: Med Arne Næss inn i dypøkologien* (Oslo: Flux forlag, 2008).

[67] Haukeland and Næss, *Dyp Glede: Med Arne Næss inn i dypøkologien.*

[68] Arne J. Vetlesen. *The Denial of Nature. Environmental philosophy in the era of global capitalism* (Abingdon: Routledge, 2015).

[69] Faarlund "Friluftslivets naturmøte som veg til et liv i lage," 21–26.

[70] Kolstad, *Besinnelse, Naturfilosofiske essays.*

domestic—something in us.[71] Hence, our latent ability of identification is closely tied to deep encounters with nature: We tend to see ourselves in everything alive.[72]

Further, free nature is freedom from distractions that can impoverish nature encounters. In particular, noise can distract from this joy and rhythm. A major challenge in outdoor life today is finding silent places with no mechanical noise where only nature's own sounds are predominant. Næss would say that wind is a harmonious sound, whereas snowmobiles are false tones in the mountains. Another challenge related to sound during the journeys described earlier is distractions from the other group members. It seems that the interaction with the other expedition members may have both led to "noise" and disturbed them in listening to the voice of the mountain. Næss has spoken warmly about being in nature in solitude. When hiking and climbing, he and Zapffe agreed not to talk but only point and smile to avoid spoiling the experience with words.[73] Therefore, it is somewhat surprising that group size has not been addressed more explicitly in the earliest statements about the qualities of *ferd*. However, this became one of the central values later.[74]

Duration

To achieve a deeper encounter with nature, the *ferd* methodology indicates that it is necessary to have hikes or camps that last several days and preferably several weeks. Inspired by the Tseringma *ferd* and his life at Tvergastein, Næss addresses the question of adequate time for adjustment: "It takes time for the new milieu to work in depth. It takes time for the sensitivity of nature to fill the mind."[75] Næss warns that if "too much technique and apparatus is placed between us and nature, nature cannot possibly be reached."[76] As I understand it, a fundamental reason for the

[71] Haukeland and Næss, *Dyp Glede: Med Arne Næss inn i dypøkologien.*

[72] Reed and Rothenberg, *Wisdom in the Open Air: The Norwegian Roots of Deep Ecology*, 73.

[73] Dale, "Med fjell som venner—en samtale med Arne Næss," 59–61.

[74] According to an instruction book by Faarlund on Nordic-mountain skiing (1973), the tour group should not be too numerous (7–8 participants). During a seminar for educators of *friluftsliv* (*vegledere*) in 1979, led by Faarlund, several agreements arose concerning the best practices: group size should be small (3–7 participants). (Norges Høgfjellskole, "Krav til vegledere i friluftsliv," *Mestre fjellet*, 13–14 (1980): 14.

[75] Næss and Rothenberg, *Ecology, Community and Lifestyle. Outline of an Ecosophy*, 179.

[76] Næss and Rothenberg, *Ecology, Community and Lifestyle. Outline of an Ecosophy*, 179.

ecophilosopher's desire for duration is the importance of captivating the uniqueness of the place and having time to meet its "residents," even including the smallest plants. Just as one needs to befriend nature, one also needs time to become friends with close-to-nature cultures. As described earlier, when *ferd* becomes part of the Høgfjellskole pedagogy, the *ferd* preferably goes to places where one can also meet local people who represent cultures living close to nature for generations. In *Mestre fjellet*, we read that there may also be such valuable places in the Nordic countries (e.g., in encounters with the Sami people, Inuits, or coastal peasants). The idea was that these places and inhabitants could teach students about cultures that stood in contrast to the prevailing culture, affected by—what is pejoratively described as—Cartesianism and technoculture. In this way, Næss stresses the significance of place and argues for an intertwined relation between humans and nature, in contrast to a subject–object dualism that distinguishes the self from its geographical surroundings.[77]

At the same time, a deep encounter with place and its residents based on duration and dwelling may be criticized for romanticizing the natural life of indigenous people, especially the "noble wild" of the Himalayas. Although Sherpa culture appears to be exemplary, Kvaløy Setreng wrote: "But Beding is not the paradise of romantics. The people must work hard from childhood, and they are plagued by certain diseases in the same way as Norwegian medieval society."[78] Faarlund, 50 years later, is clear that "we cannot heedlessly copy their way of life but as exemplary patterns of life in which tone and rhythm are given by the human and natural world."[79] Indeed, when it comes to the journey to Azourki, we see that they do not necessarily look blindly for some Arcadian idyll: "[T]he local's brutal behavior towards free nature, animals, and trees shocked him."[80]

In light of the Tseringma *ferd* and the notion of duration, we find a discourse about safety that is closely related to the view of nature in

[77] Arne Næss, "An Example of Place: Tvergastein," in *The Selected Works of Arne Næss, eds.* Alan R. Drengson and Harold Glasser (Dordrecht: Springer, 2005), 339–359. The importance of place in ecophilosophy is elaborated by such as Knut-Willy Sæther, "Naturen som sted: En eksplorativ studie av. økofilosofi, økoteologi og kristen økospiritualitet," in *Rom og sted: Religionsfaglige og interdisiplinære bidrag*, eds. Knut-Willy Sæther og Anders Aschim, 189–206 (Oslo: Cappelen Damm Akademisk, 2020).

[78] Sigmund Kvaløy, "Rolwaling—et livssamfunn i likevekt," *Mestre fjellet* 15 (1973a): 12.

[79] Faarlund, "Et 50 års minne fra Himalaya," 25.

[80] Faarlund, "Takk for "vakre handlinger" for vegledning i friluftsliv!"

friluftsliv pedagogy. It is an uncompromising connection between safety and the mountaineering ecophilosophers' friendship-with-nature-thinking. By moving too fast, without active sensing and being mindful, travelers do not have time to create sufficient friendships or pick up on nature's warnings. Although "speed is safety" is a commonly used expression in steep terrain, Himalayan expeditions through the ages contain tragic examples of fatal accidents due to a lack of will and the ability to listen to "the voice of the mountain."

Safety and feeling safe are also about feeling socially safe. One strength of the *ferd* methodology is that both duration and distance from "unfree" nature can give experiences meaning through the possibility of perceiving oneself in relation to the outside world. The conditions for making this happen are better in a slow and long-term *ferd* than through short-term programs. Developing confidence with new hiking companions takes time and focus, and trusting social interactions is one of the characteristics of what we can describe as a good *ferd*. The participants "take off the mask from everyday life," become more honest with each other, and get closer to life's big questions, including a love that transcends the human world.[81]

Reflection

In the context of *ferd*, experience is more than doing. In my analysis, I have examined reflections on behaviors toward nature and fellow human beings that are systematized through daily *ferdråd*[82] reflections. In particular, Faarlund states that if the journey's experiences are to mean anything for the participants' lives and involvement in society, impressions and thoughts need to be discussed in the plenary during and after the journey. Thus, *friluftsliv* may be seen as an intermediary between nature and society with *ferd* as the method.[83] Learning in *friluftsliv* involves

[81] See Helga S. Løvoll and Knut-Willy Sæther, "Awe experiences, the sublime, and spiritual well-being in Arctic wilderness," *Frontiers in Psychology*, 13 (2022):973922, doi: 10.3389/fpsyg.2022.973922.

[82] The word *ferdråd* consists of *ferd* and *råd*, which can mean both a council and advice, and has become an established concept in *friluftsliv* pedagogy as a compulsory and democratic gathering where everyone in the same *ferd* group meets. Tradition dictates that the participants gather in a circle, on the same level, and everyone can express their background, experiences, and expectations in relation to the *ferd*. A common understanding of goals and potential risks is desirable.

[83] Ydegaard, "Om Færden som metode," 11.

mediation that alternates between individual experiences in nature, group experiences, and academic reflection. These interactions are essential for constructing an academic theory of *friluftsliv*.[84] Here, I find resources from John Dewey's theory of reflection, experience, and meaning helpful for shedding light on this learning context.[85] Dewey showed how strong sensory experiences, which are concentrated within a meaningful context, are educational, in the sense of *Bildung*. Reflective experience is a central concept in Dewey's pedagogy, where he insists on the strong bonds between embodied experience and thinking.[86] For Dewey, learning is about interaction and how the group can learn involvement, respect, participation, and social skills. A diversity of perspectives helps us to find good solutions.[87] A day's cycle of *ferdråd* in the morning, discussions at lunch, and gatherings around the campfire in the evening, with articulation and group discussions, is in line with Dewey's holistic view, where being, doing, and knowing are unified in his model of aesthetic experience.[88] The daily experiences are thematized, and with sufficient time they can lead to deeper philosophical conversations with the ambition of generating conversations on meaning, values—including nature's intrinsic value—and the future. This process can be seen as part of an ecopolitical *Bildung* in the pedagogy of *ferd*. Dewey connects education and democracy in a way that is "more than a form of government, it is primarily a mode of associated living, of conjoint communicated experience."[89]

[84] Helga Synnevåg Løvoll, "Naturen som Rom. Filosofisk samtale gjennom erfaring i natur," in *Rom og sted: Relgionsfaglige og interdisiplinære bidrag*, eds. Knut-Willy Sæther and Anders Aschim, 189–206 (Oslo: Cappelen Damm Akademisk, 2020), 195.

[85] John Dewey, *Democracy and Education* (New York: The Free Press/Macmillan Publishing Co, 1916).

[86] John Quay and Jason Seaman, "Outdoor studies and a sound philosophy of experience," in *Routledge International Handbook of Outdoor Studies*, eds. Barbara Humberstone, Heather Prince, and Karla A. Henderson (New York: Routledge), 40–48.

[87] Per Ingvar Haukeland and Hanne Lund-Kristensen 2019. "Den livskraftige barnehagen. En økopedagogisk-filosofisk tilnærming," *Studier i Pædagogisk Filosofi* 8 (1): 69–89.

[88] See Quay and Seaman for Dewey's *aesthetic experience* to describe the experience of what is going on in the immediate present. Quay and Seaman, "Outdoor studies and a sound philosophy of experience," 43.

[89] John Dewey, *Democracy and Education* (New York: The Free Press/Macmillan Publishing Co, 1966 (1916)).

Ferden Further: Toward a *Friluftsliv* as a Holistic Interaction with Nature

In my final section, I will examine how humans' encounters with nature—according to *friluftsliv* and particularly the *ferd* methodology elaborated above—can be related to a broader picture for change and sustainable interactions with nature.

In line with Dewey, there is a strong argument for the value of "adventure" internationally. In *ferd* pedagogy as a way of working in the 1970s and beyond, some elements differ from the Anglo–American approach. *Ferd* pedagogy has a strong ecophilosophical justification (e.g., nonconventional goals, learning from local culture and indigenous peoples, and the value of dwelling). Adventure, as a way of working in outdoor practices, has a long history but has been based on an education with a more anthropocentric perspective than the *ferd* tradition. One example of the adventure approach is the Outward Bound movement. This movement focuses on physical training, survival training, risk, and (inter)personal growth.[90] However, it need not be entirely one way or the other. In fact, due to the influence of both Muir and Næss, we see more holistic approaches recently in adventure practices in the UK/US.[91]

The philosopher Espen Gamlund asks why the deep ecology movement has lost momentum. Most of the world's population lives in big cities, and Gamlund is skeptical about whether urbanists today will agree that biodiversity increases their happiness and quality of life.[92] I agree, to some extent, with Gamlund's suggestion that a problem with deep ecology's idea of identification is that it requires a relationship with free nature. He believes it does not apply to everyone (including in Norway), something Næss and the other ecophilosophers have taken for granted. In addition, the distance to nature is too far for many, and it is constantly getting farther due to ongoing urbanization. For Gamlund, ecophilosophy can be perceived as contextual in time (pre-oil) and space (pre-urban). In contrast, he emphasizes that the diversity of life forms is meaningful to those who spend much time in nature.

[90] Anette R. Hofmann et al. *Norwegian friluftsliv. A Way of Living and Learning in Nature* (Münster: Waxmann, 2018).

[91] One example is the Sierra Club, founded in California 1892, and the "Leave no trace" movement.

[92] Gamlund, "Hva er galt med dypøkologien? Noen kommentarer til Arne Naess' Økosofi T."

A question remains about the feasibility of nature-friendly and long-distance *ferds* and *friluftsliv*. These activities require much travel for large parts of the population, especially in other countries that are not as wealthy and as sparsely populated as Norway. Næss uses the journey to Tseringma to illustrate the "joyful mountain life," with plenty of time to dwell by the cliffs. At the same time, he encouraged more people to commute between the mountains and the city, with their main stay being in the mountains. This view was likely perceived as naive and elitist even in its time. But as Næss expresses to his audience in *Mestre fjellet*, there should be great possibilities for those aged 15–25 and 50–100 years.[93] In many ways, Norwegians' penchant for life in their cabins corresponds to this desired mountain encounter. Næss was fascinated by the traditional Norwegian cabin culture, not least his own simple and rich cabin life at Tvergastein. However, the cabin culture in modern (oil-rich) Norway has developed in a nonsustainable and nature-unfriendly direction. Today, many new "mountain lodges" and "cabin towns" are far from such a mountain encounter and simple living.

At the same time, Næss was aware that *friluftsliv* could never become a common and general way of living in modern society, but considered *friluftsliv* to be one of many different ways or ecosophies.[94] For Faarlund, *friluftsliv* and *ferd* are important as a part of the cultural heritage in Norway, pointing backward to the historical importance of people in urban areas regularly having (deep) relations with nature. Hence, "free nature" never becomes old-fashioned.[95] In a city like Oslo, *Marka* (the Forest) embraces the city. Therefore, it is a short distance for the inhabitants to a relatively "free nature," where you can attain almost complete solitude, silence, and "a sense of freedom." Faarlund expresses this with *friluftsliv* as a way home to these values and to a world in which human dignity and nature dignity are brought to life.[96]

Although there are many great thoughts about joyful play and deep encounters, *ferd* may have become, in practice, an overly serious and theoretical project. When analyzing literature on *friluftsliv* and *ferd*, which often takes the form of practical skills, knowledge, and nature wisdom,

[93] Arne Næss, "Om å være på fjellet," *Mestre fjellet* 1–2 (1972): 4–5.

[94] Gunnar Breivik, "'Richness in Ends, Simpleness in Means!' on Arne Naess's Version of Deep Ecological Friluftsliv and Its Implications for Outdoor Activities," *Sport, Ethics and Philosophy*, 15, no. 3 (2021).

[95] Nils Faarlund, *Friluftsliv—en dannelsesreise* (Oslo: Ljå Forlag, 2015).

[96] Faarlund, *Friluftsliv—en dannelsesreise*.

one can lose sight of heartfelt joy. The relevant question is whether or not the strong tradition of freedom in *friluftsliv* has been lost along the way. Even if the intention of the ecophilosophers is clear that this nature-centered freedom is about a transformative joy, it might have had an opposite effect in leading to fear or apathy. Nevertheless, ecophilosophers must first be recognized for being very early adopters of a partly Eastern wisdom,[97] in that change must go through the heart, genuine joy, and activity in nature.

Conclusion

In this *ferd* of exploring the relationship between humans and nature in the context of *friluftsliv* developed by the mountaineering philosophers, I have found a strong emphasis on ***deep encounters*** with nature in the *ferd* methodology. In my analysis of two particular *ferds*, I found that the love of and friendship toward nature required what they called ***free nature***. However, achieving such a goal of deep encounters with nature is challenging because of human intervention, pollution, and a lack of landscapes with a wilderness character. Further, my analysis of the long-distance *ferds* shows that deep encounters are not only related to nature but also to the travelers' interactions with each other and the local culture. In addition to the dimension of ***place***, the journey encompasses an ***individual dimension*** in that the *ferd* participants need time to achieve natural sensitivity. A lack of awareness of nature, hurriedness, and disrespect for the sacred in nature are great dangers in the mountains. For this reason, it is important to allow sufficient time for all phases of a *ferd*. This point has been embedded in the pedagogical way of working with *ferds*. Allowing time for inquiry, reflection, and interaction with the others in the *ferd* group is key to this experiential form of learning.

Underlying the complex notion of *ferd* in *friluftsliv*, where free nature, duration, and reflection are crucial, nonconventional goal direction is a *ferd* virtue nurtured through cooperation, in contrast to competition and commercialism. Respect and humility, not only for mountain peaks, but for everything that lives in the mountains (i.e., the whole ecosystem), points us toward a holistic interaction with nature, both in attitude and in practice.

[97] With Eastern wisdom I refer here to the descriptions earlier in this article of the mountaineering philosopher's inspiration from Gandhi and the Tibetan Buddhism of the Sherpas.

Bibliography

Breivik, Gunnar. 2021. 'Richness in Ends, Simpleness in Means!' on Arne Naess's Version of Deep Ecological Friluftsliv and Its Implications for Outdoor Activities. *Sport, Ethics and Philosophy* 15 (3): 417–434. https://doi.org/10.1080/17511321.2020.1789719.

Dale, Øystein. 2002. Med fjell som venner - en samtale med Arne Næss. *DNTs Årbok* 2002: 59–61.

Dewey, John. 1916. *Democracy and education*. Vol. 1966. New York: The Free Press/Macmillan Publishing Co.

Faarlund, Nils. 1968. Vi må lære å bruke naturen - uten å forbruke den. *Mestre fjellet* 1: 28–29.

———. 1971. Nei til konkurranseklatring. *Mestre fjellet* 2: 2.

———. Norsk skiinstruksjon III, *Fjellskiløping*. Foreningen til Ski-Idrettens Fremme/Barnas Skifond, Landsrådet for skoleidrett, 1973.

———. 1974. Ferden som arbeidsmetode i friluftsliv. *Mestre fjellet* 18: 6–7.

———. 1977. Hillary og Sherpaene: Vennetjeneste eller kulturdrap? *Mestre fjellet* 25: 23.

———. 1980. Klær og redskap for friluftsliv om vinteren. *Mestre fjellet* 29: 20.

———. 1981. Erfaringer fra Den norske Himalayaekspedisjonen. *Mestre fjellet* 30: 5–12.

———. Den norske Himalaya ekspedisjonen 1979. En optimistisk meddelelse fra første leir over tregrensen. *Mestre fjellet* 31-32 (1983): 43–46.

———. 1992. Toppen er et sted på vegen. In *Bergtatt*, ed. Norman Kjærvik, 42–56. Oslo: Grøndahl Dreyer.

———. 1999. Friluftslivets naturmøte som veg til et liv i lage. *Mestre fjellet* 44: 21–26.

———. 2015. *Friluftsliv—En dannelsesreise*. Oslo: Ljå Forlag.

———. 2018. Takk for "vakre handlinger" for vegledning i friluftsliv! In *Friluftsliv, kystkultur og vakre løsninger*, ed. Gunnar Tellnes, 266–279. NAKULHEL-forlaget.

———. 2021a. "Læresetninger." https://www.norgeshogfjellskole.no/laeresetninger/. downloaded July 10, 2023.

———. "Et 50års minne fra Himalaya." Hallingdølen, October 12, 2021b.

Gamlund, Espen. 2012. Hva er galt med dypøkologien? Noen kommentarer til Arne Naess' Økosofi T. *Norsk Filosofisk Tidsskrift* 47 (4): 229–242.

Haukeland, Per Ingvar, and Arne Næss. 2008. *Dyp Glede: Med Arne Næss inn i dypøkologien*. Oslo: Flux forlag.

Haukeland, Per Ingvar, and Hanne Lund-Kristensen. 2019. Den livskraftige barnehagen. En økopedagogisk-filosofisk tilnærming. *Studier i Pædagogisk Filosofi* 8 (1): 69–89.

Hessen, Dag O. "Die norwegishe apparatlantenschafft–Zapffe som naturverner." In *Barske Glæder, og andre temaer fra et liv under åpen himmel*, by Peder W. Zapffe, 145–153. Oslo: Cappelen Damm, 2012.

Hofmann, Anette R., Karsten Gade Rolland, Kolbjørn Rafoss, Herbert Zoglowek, and Norwegian friluftsliv. 2018. *A way of living and learning in nature.* Münster: Waxmann.

Horgen, André. 2022. Friluftsliv, idrett eller sport? Mening og begrepsbruk i Den Norske Turistforenings årbøker, 1868-1979. *Historisk tidsskrift* 101 (3): 213–228.

Høibakk, Ralph. 1983. Numbur 1979, Egotrip, pilgrimsferd og samlivstrening. In *Norsk Fjellsport*, 225–236. Norsk Tindeklub.

Kolstad, Hans. 2007. *Besinnelse, Natufilosofiske essays.* Oslo: Humanist forlag.

Kvaløy, Sigmund. 1973a. Rolwaling–et livssamfunn i likevekt. *Mestre fjellet* 15: 11–12.

———. 1973b. Rytme, natur og menneske. In *Mestre fjellet*, vol. 16, 10.

Leirhaug, Petter Erik, Per Ingvar Haukeland, and Nils Faarlund. 2019. Friluftslivsvegledning som verdidannende læring i møte med fri natur. In *Friluftslivspedagogikk*, ed. Linda Hallandsvik and Jannicke Høyem, 15–32. Oslo: Cappelen Damm Akademisk.

Løvoll, Helga Synnevåg. 2020. Naturen som Rom. Filosofisk samtale gjennom erfaring i natur. In *Rom og sted: Relgionsfaglige og interdisiplinære bidrag*, ed. Knut-Willy Sæther and Anders Aschim, 189–206. Oslo: Cappelen Damm Akademisk.

Løvoll, Helga Synnevåg, and Knut-Willy Sæther. 2022. Awe experiences, the sublime, and spiritual well-being in arctic wilderness. *Frontiers in Psychology* 13: 973922. https://doi.org/10.3389/fpsyg.2022.973922.

Nilsen, Martin. 1985. Banal statustrim. *Dagbladet* 20: 41.

Høgfjellskole, Norges. 1980. Krav til vegledere i friluftsliv. *Mestre fjellet* 29: 13–14.

———. 1996. *Mestre fjellet* 43.

Alpincenter, Norsk. 1969. Klatring–hvorfor, hvordan? *Mestre fjellet* 3–4: 25–27.

———. 1972. Nytt fra Høgfjellskolen. *Mestre fjellet* 3–4: 38.

———. 1973. Nytt fra fjellet. *Mestre fjellet* 15: 23.

Næss, Arne. 1972. Om å være på fjellet. *Mestre fjellet* 1–2: 4–5.

———. 1976. Økologi, samfunn og livstil. In *Utkast til en økosofi.* Oslo: Universitetsforlaget.

———. 1987. Litt om fjelling. In *Om å gripe fjellet–og å bli grepet av fjellet. En vennebog om friluftslivet og Nils Faarlund*, ed. Torbjørn Ydegaard, 37–39. Sorø: Nordisk Forum for Vejledning i Natur- og Friluftsliv.

———. 1994a. The Norwegian Roots of Deep Ecology. In *Nature the True Home of Culture*, ed. Børge Dahle, 15–18. Oslo: Norges idrettshøgskole.

———. 1994b. Friluftslivet i våre tanker. *Mestre fjellet* 41–42: 19–22.

Næss, Arne, and David Rothenberg. 1992. *Ecology, Community, and Lifestyle. Outline of an Ecosophy.* Cambridge, Cambridge University Press. (Translated & edited by David Rothenberg).

Næss, Arne. 2005. An Example of Place: Tvergastein. In *The Selected Works of Arne Næss*, ed. Alan R. Drengson and Harold Glasser, 339–359. Dordrecht: Springer.

Personal conversations with Nils Faarlund.

Quay, John, and Jason Seaman. 2018. Outdoor Studies and a Sound Philosophy of Experience. In *Routledge International Handbook of Outdoor Studies*, ed. Barbara Humberstone, Heather Prince, and Karla A. Henderson, 40–48. New York: Routledge.

Reed, Peter, and David Rothenberg. 1993. *Wisdom in the Open Air: The Norwegian Roots of Deep Ecology.* Minneapolis: University of Minnesota Press.

Sandell, Leif, and Johan Öhman. 2013. An Educational tool for outdoor education and environmental concern. *Journal of Adventure Education and Outdoor Learning* 13 (1): 36–55. Taylor & Francis Online. https://doi.org/10.1080/14729679.2012.675146.

Setreng, Kvaløy Sigmund. 1984. The universe within. *Resurgence.* 106: 1–18. Permission for publication on OpenAirPhilosophy generously granted by Resurgence, http://resurgence.org.

———. 1991. Økokrise-glimt fra det norske økofilosofiske forsøket. In *Den uoverstigelige grense. Tanke og handling i miljøkampen*, ed. Svein Gjerdåker, Lars Gule, and Bernt Hagtvet, 102–116. Oslo: Cappelen.

———. 2002. To økofilosofer i Norge; deres begynnelse og en del til. *Norsk filosofisk tidsskrift* 37: 117–126.

Skånvik, Anne-Sissel. 1985. Solgt seg til massemedia. *Dagbladet* 20.

Sæther, Knut-Willy. 2017. *Naturens skjønnhet. En studie av forholdet mellom estetikk, teologi og naturvitenskap.* Oslo: Cappelen Damm Akademisk.

———. 2020. Naturen som sted: En eksplorativ studie av økofilosofi, økoteologi og kristen økospiritualitet. In *Rom og sted: Relgionsfaglige og interdisiplinære bidrag*, ed. Knut-Willy Sæther and Anders Aschim, 189–206. Oslo: Cappelen Damm Akademisk.

Tellnes, Atle. 1975. Friluftslivs-seminar Agder folkehøyskole 1974. *Mestre fjellet* 18: 9–16.

Vetlesen, Arne Johan. 2015. The denial of nature. In *Environmental philosophy in the era of global capitalism.* Abingdon: Routledge.

Ydegaard, Torbjørn. 1986. Om Færden som metode. *Mestre fjellet* 36: 9–11.

———, ed. 2016. *Nils Faarlund Samlede Verker VII, Veglederstudenter.* København: Books on Demand.

Zapffe, Peter W., and Barske Glæder. 2012. *Og andre temaer fra et liv under åpen himmel.* Oslo: Cappelen Damm.

CHAPTER 6

An Overview of Natural, Human, Philosophical, and Theological Dualisms

Mark Graves

Introduction

Dualism divides something into two contrasted aspects. Typically in studying a dualism, one would examine the two aspects of something and their contrast. However, the focus of this chapter is on the division itself and the places and ways that division occurs. Investigating how divisions take place across different types of dualism illuminates commonalities across various dualisms, suggests common structures to them, and yields insights into ways they may relate. The domains examined and divided in this chapter are nature, human existence, metaphysical realms, and the theological relationship between God and creation. A systems approach to exploring these dualities identifies a few common types of division, such as between biological and mental, and the relative emphasis placed on the different divisions classify and helps organize the various dualisms.

One reason the divisions have apparent commonalities across domains of discourse is the interrelation among these four domains—nature,

M. Graves (✉)
School of Psychology, Fuller Theological Seminary, Pasadena, CA, USA
e-mail: markgraves@fuller.edu

T. J. Hastings, K.-W. Sæther (eds.), *Views of Nature and Dualism*,
https://doi.org/10.1007/978-3-031-42902-6_6

human existence, metaphysical realms, and the theological relationship between God and creation—especially from a human perspective. One aspect of nature is human existence, which includes mental activity, such as examining nature, including human nature, and studying metaphysics and theology. Identifying and characterizing the something to be divided is a metaphysical investigation, but the only access humans have to that something is through human experience (including thought), which in turn depend upon historical and cultural social constructs and mental processing occurring in an embodied neurobiological platform, i.e., an aspect of nature.

Examining the division of something generally appears to depend upon having a place to stand in which to observe the division from a somewhat objective position. However, in this case, the something being examined is all-encompassing with highly interdependent aspects that preclude an external or "etic" perspective. One could react to the expanse by considering it all as socially constructed, and thus a construct of human thought and culture, but that fails to engage the theological and scientific dimensions which purport to exceed human construction and imagination. A third option takes both objective and subjective perspectives seriously and begins with human experience and interpretation of reality as having practical effects separate from their human study. I take the third approach and begin with human scientific and religious perspectives on nature, examine some historically and philosophically significant divisions there, and consider their metaphysical and theological implications.[1]

The scientific and religious perspectives on nature define an ontological foundation for science and a narrative characterization of Creation.[2]

[1]The third perspective draws upon the pragmatic philosophies of C.S. Peirce (semiotic metaphysics) and Josiah Royce (objective idealism). Kelly A. Parker, *The Continuity of Peirce's Thought* (Nashville: Vanderbilt University Press, 1998); Kelly A. Parker, "Josiah Royce: Idealism, Transcendentalism, Pragmatism," in *The Oxford Handbook of American Philosophy*, 2008, https://doi.org/10.1093/oxfordhb/9780199219315.003.0006; Donald L Gelpi, *The Gracing of Human Experience: Rethinking the Relationship between Nature and Grace* (Collegeville, Minn.: Liturgical Press, 2001); Mark Graves, *Mind, Brain, and the Elusive Soul: Human Systems of Cognitive Science and Religion* (Aldershot, Hants, England; Burlington, VT: Ashgate, 2008).

[2]This chapter focuses primarily on ontology, i.e., the study of being or existence, as a component of the broader study of metaphysics, i.e., the study of reality, which also includes study of causation, modality, and the nature of time and space.

Ontologically, I use emergent monism as a tool to organize and study various dualisms.[3] The novel method begins with a monist perspective, examines where strong emergence appears to take place (i.e., the emergence of causal powers), then considers the type of dualisms to occur if these transitions were postulated dualistically as a division into two realms. Thus, the hard division of dualism is situated within softer monistic alternatives (e.g., dual-aspect or emergent monism), using the monistic realm as a place to stand to examine the postulated dualism. This approach organizes a variety of historical dualist positions; introduces a novel and systemic way to study various monisms, dualisms, and their relationships; and initiates identifying disparities in the impact those dualities have on Western (WEIRD; Western, Educated, Industrialized, Rich, and Democratic) academic theology and science discourse (e.g., Cartesian dualism appears ingrained in philosophical presuppositions while vitalism is typically interpreted metaphorically).

The present investigation begins with a human perspective on creation, making use of the religious metaphor of creation as the Book of Nature. A scientific and religious analysis of nature suggests chapter transitions for the Book of Nature, which identify places significant to human existence and human experience of our existence. Those cosmological, evolutionary, and cultural developments of human existence within nature highlight possible ontological distinctions, where significant dualisms might be postulated. Those ontological (and causally significant) divisions, corresponding to the natural chapter transitions from a human perspective, are limited by human experience but may nevertheless identify aspects of the reality in which humanity developed and evolved. Systems theory characterizes human existence and human experience (e.g., of nature) as distinct levels of systems, where each level and transition correspond to the chapters and

[3] Elsewhere I argue for an emergent monism as ontological foundation or phenomenological perspective. Graves, *Mind, Brain, and the Elusive Soul*; Mark Graves, "The Emergence of Transcendental Norms in Human Systems," *Zygon* 44, no. 3 (2009): 501–532; Mark Graves, *Insight to Heal: Co-Creating Beauty amidst Human Suffering* (Eugene, OR: Cascade Books, 2013); Mark Graves, "Places of Information Generation: Bridging Pannenberg's Logos and Deacon's Emerging Semiosis," *Theology and Science* 14, no. 3 (2016): 305–324, https://doi.org/10.1080/14746700.2016.1191880; Mark Graves, "Emergent Models for Moral AI Spirituality," *International Journal of Interactive Multimedia and Artificial Intelligence* 7, no. 1. Special Issue on AI, Spirituality, and Analogue Thinking (2021): 7–15, https://doi.org/10.9781/ijimai.2021.08.002

transitions of the Book of Nature and the corresponding structure of human nature. Attending to the chapter transitions as the primary focus of investigation yields a framework for examining the dualisms they may indicate across the natural, human, philosophical, and theological aspects of reality. This investigation considers various philosophical positions where the transitions are over or underemphasized to characterize historically significant dualisms and monisms, respectively. The overemphasis of a chapter also serves as a metaphoric root for metaphysical positions, and the overemphasis of some transitions serves metaphorically to describe possible God-world divides.

Book of Nature

How can one understand an evolving cosmos in a theological context? Drawing upon the study of emergence by scholars and scientists like Nancey Murphy, Philip Clayton, and Terry Deacon, I suggest a model for emergent Creation that honors Christian traditions while fully engaging scientific findings.[4] The religious metaphor of a Book of Nature analogous to Scripture has ancient roots. Augustine writes,

> May the sacred pages be a book for you, so that you hear this; may the globe of the earth be a book for you, so that you may see this. In these books [of scripture] only those who know the letters read; in the whole world, even a fool can read.[5]

Ancient, medieval, and modern theologians recognize the value of reading the Book of Nature (though always with caution to prioritize Scripture), and scientists and philosophers have identified several places

[4] Nancey C. Murphy, "Nonreductive Physicalism: Philosophical Issues," in *Whatever Happened to the Soul?: Scientific and Theological Portraits of Human Nature*, ed. Warren Brown, Nancey Murphy, and H. Newton Malony (Minneapolis, Minn.: Fortress Press, 1998); Philip Clayton, *Mind and Emergence: From Quantum to Consciousness* (New York: Oxford University Press, 2004); Terrence W. Deacon, *Incomplete Nature: How Mind Emerged from Matter* (New York: W.W. Norton, 2011); Terrence W. Deacon, "Emergence: The Hole at the Wheel's Hub," in *The Re-Emergence of Emergence*, ed. Philip Clayton and Paul Davies (Oxford: Oxford University Press, 2006), 111–150.

[5] St. Augustine, *Enarratio in Psalmum* XLV 6–7.

where nature's transitions appear significant to human existence, sufficient to identify chapters in the human edition of the Book.[6]

Beginning. The Book of Nature currently appears to begin with time, space, and energy and continues to unfold reality into existence with expanding space-time, development of matter, and cooling temperatures. Cosmologists organize a general structure for cosmological development over about 13.8 billion years with iterative stellar processes of element formation leading to planets such as Earth. Although the elements appear to be a surprisingly small fraction of the matter and energy in the Universe, they are particularly relevant for telling the Creation story to humans.

The elements developed after the Big Bang with gravity forming protostars from nebulae and with generations of stars forming and exploding to create more elements and the ingredients for planets. In particular, the variation in electric charge due to the relative absence of electrons in the atom's outer shell will become very significant to the human story. One could deepen the metaphor of Book and consider the creation of elements—and formation of planets where they can interact—as the beginning of a new chapter.

Physical. As planets form and cool, elements have time and space to interact and to begin forming molecules. The molecules combine into larger structures and begin to take on new emergent properties, such as surface tension or energy storage, which then also begin to interact and become regularized. The pages of the ancient book are faded and difficult to read, but somehow nucleic acids and RNA appear to form along with amino acids, translational processes, and permeable membranes. Eventually, among the myriad complex molecules, a new molecule emerges called DNA.

[6] As examined in the following, several researchers have distinguished what simplifies to four levels of inquiry for the scientific study of human systems; and to capture both the casual relationships that give rise to the physical laws and aspects of spirituality, one needs two additional boundary levels. Ludwig von Bertalanffy, *Perspectives on General System Theory: Scientific-Philosophical Studies* (New York: G. Braziller, 1975); Arthur Robert Peacocke, *Theology for a Scientific Age: Being and Becoming—Natural, Divine, and Human* (Minneapolis: Fortress Press, 1993); Clayton, *Mind and Emergence*; Graves, *Mind, Brain, and the Elusive Soul*; Graves, "The Emergence of Transcendental Norms in Human Systems"; Graves, *Insight to Heal: Co-Creating Beauty amidst Human Suffering*.

DNA currently exists within a complex system of molecular processes that very rapidly and robustly transcribe long, well protected strands of nucleic acid; translate them into folded proteins; and transport those proteins to other locations for structural and energetic activity. In particular, the regularization of nucleic structure and genetic transcription enables nucleotide bases to readily substitute for each other within the long nucleic acids. The variation enabled by multiple nucleotide bases and their chemical complementarity will also become very significant to the human story and what centuries of thinkers considered the magic of Life—opening up another language for creation sufficiently different to warrant a third chapter.

Biological. The collections of proteins surrounding and maintaining nucleic acids and other supportive structures began new ways of interacting and sharing molecules and processes. Cell structures become more robust, and coordination between cells increases to create multicellular organisms. Those organisms continue to increase coordination with cells becoming more specialized and developing new ways to exchange encoded molecular processes. Bacteria, fungi, and other organisms continue to evolve into plants and early animals with some cell types evolving even more complex and specialized ways of coordinating their tissues into complex adaptive structures. Eventually, neural structures develop which could coordinate complex interactions and respond to numerous environmental changes.

The neuron can receive complex electrochemical and other molecular inputs and process those signals into reliable outputs. Assemblies of neurons coordinate those signals across synaptic clefts into new activities and can adapt based upon input signals and feedback to maintain a memory for the organism and respond to new situations in novel ways. The neuron also becomes significant to the human story and creates a new way of communication within embodied brains sufficient to warrant a new chapter. The linearization of a complex network of biological activity into the pages of a book highlights that the Book of Nature, which humans read, is also very much read from a human-centric position. Considering the Book of Nature as a 1382 page volume (with an estimated 10 million years per page), we now find ourselves within the last 50 or so pages of the book.

Psychological. Animals extend biology into increased sentience and sophisticated behaviors—creating more fluid ecological niches, which subsequently influence continued evolution. Animals engage in competitive and cooperative behaviors—finding mates and increasing genetic diversity and phenotypic fitness. Ecological structures create an incentive for species to become more social and create more sophisticated means of cooperation and communication. As the communication increases in abstraction, the sounds or signs used in communication gain the capacity for definition by social convention—and form symbolic language.

Language requires sophisticated neural and social structures to form and propagate. Although aspects of language exist in non-human animals, the ability to create abstract symbolic representations and to combine them through complex grammars appears specific to humans.[7] Symbolic language enables a new chapter in the Book of Nature. Indeed, it enables the very metaphor of books and chapters. Perhaps we can update the ancient metaphor of book from a modern printed book to a contemporary collection of hypertext pages where instead of new linear chapters, some obscure link in a large network of sites takes us to a new, previously inconceivable realm of the Internet where new kinds of continued creation can take place.

Cultural. Language and human culture co-evolve—with diverse spoken and written languages and novel interpretations of human experience. Human culture extends animal sentience and behavior into a realm of rituals and ideas that can propagate as human history and enable abstract reason. Culture includes the formation and exchange of linguistic and other tools, such as books, and creating civil, economic, and governance structures to maintain and institutionalize social relationships. Complex social, historical, and cultural interactions lead to emergence of ethical values, such as justice, and projection of social relations onto presumed unseen and unexplainable forces. A transition into what some call the Axial Age about 2300–2800 years ago leads to new levels of abstraction with historical religions and new economic systems emerging.[8] One could consider the Book of Scripture as emerging within the Book of Nature and incorporating deeper revelations into human language. This raises the question whether a new chapter or realm enters human existence that extends or transcends human language and culture.

The prior transitions between realms depended upon a construct that emerged in the earlier realm and became foundational for the processes

[7] Terrence W. Deacon, *The Symbolic Species: The Co-Evolution of Language and the Brain* (New York: W.W. Norton, 1997).

[8] Robert Neelly Bellah, *Religion in Human Evolution: From the Paleolithic to the Axial Age* (Cambridge, Mass.: Belknap Press of Harvard University Press, 2011).

and causal activity of the subsequent realms. The biological anthropologist Terry Deacon's work on emergence suggests the construct has some memory and regulatory capacity that enables it to affect other constructs as well as a structured place of variability incorporating a relational absence, he calls *absentials*, which prevents reduction from the higher level constructs (and identifies the places that chapters break in the metaphorical book).[9] The emergent realms depend upon an essential absence that functions as a switch between complex states of the lower-level constructs. An evolving cosmos appears to depend upon an essential emptiness—which is open to relationship—at its places of greatest creativity. The empty places in an electron's outer shell enable chemical relational bonds through shared electrons; the empty places in the structure of DNA allow for substitution of nucleotide bases (and their relationship with complementary bases); the empty places between neurons at synaptic junctions enable neurons to vary their relationships with other neurons; and the unspecified references of symbolic language gain definition through linguistic discourse and social convention. A new emergent realm that transcends culture would likely depend upon an essential relational absence and openness within complex cultural structures enabling new kinds of causal activity and creativity.

Transcendent. One cannot identify the key bridging construct between foundational and emergent realms without familiarity with the constructs of the emerging realm. DNA is just another molecule to a physicist, but biologists have an interest in reproduction and evolution that requires DNA to deeply understand. Neurons are just another cell to a biologist, but psychologists have an interest in behavior and learning and that study benefits from deeper neuroscientific insights. What are the constructs of a transcendent realm that would transcend culture?

- Money in a modern economic system substitutes for value and frequently has strong causal influences on culture and human behavior.
- Mathematics is a language that has proven helpful for understanding an enduring cosmos.
- Religion with its rituals and prayer has a readiness for relationship that might help identify a key bridge construct.

[9] Deacon, *Incomplete Nature: How Mind Emerged from Matter.*

Other possible transcendent constructs might include the Greek transcendentals of Truth, Beauty, and Goodness. Perhaps the transcendent realm is what we refer to by spirituality.

At this point, the story ends. The Book of Nature has no more pages for us to read. Perhaps God knows what is next, but we do not. But if the Book of Nature is a story of very large, complex unfolding dynamics told from a human-centric position, where one chapter builds upon the last, then perhaps we are called to continue the creation story as what the Lutheran theologian Phil Hefner calls created co-creators.[10] When Jesus said people cannot serve both God and money (Matthew 6:24), perhaps we were given a choice as to what realm we create. The story of the evolving cosmos ends, but we can continue trying to understand how the next realm might continue, and what we might do to help co-create it.

Human Systems

How can one characterize the chapters and chapter transitions in a systematic way? Scientific study creates understanding of the natural realm (and metaphoric Book of Nature). As the book is read from a human-centric perspective, the chapters and chapter transitions we identify informs humanity not only about nature and how we perceive nature but also on how, as humans, we *can possibly* perceive nature. As a human being, our dependence upon elements, DNA, neurons, and symbolic language serves as a reliable foundation for a scientific, philosophical, and theological understanding of nature.[11] Shifting focus from cosmology to ontology, these chapter divisions constrain the types of ontology that would be meaningful to humans. An ontology that does not allow for elements might be theoretically possible, but it could not be used to explain nature from a human perspective. Additional constructs may be important to cosmology or ontology, but I claim that a focus on the chapter divisions provides essential insight into nature, including humanity, especially

[10] Philip J. Hefner, *The Human Factor: Evolution, Culture, and Religion* (Minneapolis: Fortress Press, 1993).

[11] If one wishes, one could examine a scientific field, such as cosmology, deemphasizing some theories or findings, historical influences, philosophical perspectives, or cultural contexts. But if one attempts to exclude the existence of the atomic elements, then something is fundamentally missing from the study of cosmology, from a human perspective.

because we have no choice but to understand nature from a human perspective.[12]

Human systems theory formalizes the chapters in the Book of Nature as levels of systems.[13] My claim in this paper is that different ontologies variously emphasize the continuity between levels (monism) or separation between levels (dualisms). Historically, systems theory operated in a monist framework (either more physicalist or idealist, depending upon the theorist). Elsewhere, I have proposed using theories of emergence to characterize the distinction between levels (briefly discussed later in this chapter).[14] In the present study, the levels of human systems indicate both ontological claims about the nature of human existence (as described scientifically and philosophically by studies at each level of human existence) and phenomenological claims about the possible ways humans can experience their world. The ontological and phenomenological perspectives interrelate. Nature unfolded in a particular way that results in the existence of humanity; and given current human nature, we have phenomenological constraints on how we can experience and study nature.

Using systems theory to characterize human existence and experience as interrelated clarifies this investigation's starting point of human scientific and religious perspectives on nature. The interrelated framework of

[12] Thomas Nagel and others argue convincingly that our awareness and understanding of reality is heavily conditioned by our embodiment and experience, thus we would be hard pressed to imagine nature from the perspective of another animal, much less reality in the abstract. Thomas Nagel, "What Is It Like to Be a Bat?," *Philosophical Review* 83, no. 4 (1974): 435–450.

[13] Systems theory (originally developed in the 1940s) focuses on the complexity and interdependence of relationships often using information and decision-making/control concepts. A *system* is a collection of interacting elements that form an integrated whole. The elements coordinate their interactions along functional divisions to yield functional capabilities for the system. The system as a whole has an organization and the activity of its interrelated elements brings about local order, i.e., they both reduce local entropy. As a whole, the system has continuity of identity, and its behavior necessarily and sufficiently depends upon the independent activities of its elements. Those elements may themselves be systems with their own elements and subsystems. Because a system as a whole accomplishes something that none of its elements accomplishes alone, a system does not reduce to its elements nor their independent interactions. Bertalanffy, *Perspectives on General System Theory: Scientific-Philosophical Studies*; Lars Skyttner, *General Systems Theory: Perspectives, Problems, Practice*, 2nd ed. (Singapore; River Edge, N.J.: World Scientific, 2006).

[14] Graves, *Mind, Brain, and the Elusive Soul*; Graves, "The Emergence of Transcendental Norms in Human Systems."

human systems integrates findings from natural and social sciences as well as humanities scholarship into a cohesive framework that supports scientific and theological investigation. The framework incorporates the needs of theologians to situate human nature and morality within a larger story of Creation; acknowledges the social construction of the human-centric model through science and the humanities, as needed by social scientists and other scholars; and satisfies the objective bias of natural scientists by locating spirituality within social systems that are grounded in neurobiological processes. In particular, the framework incorporates theological investigation of human spirituality as transcending human culture while emerging from it within a social-historical-linguistic context of religious traditions and builds a bridge between those investigations and the broader scientific investigations of cosmology and evolving nature.

In summary, the present investigation focuses on the transitions common to theological, metaphysical, ontological, human existential, and phenomenological realities. Those transitions are formalized using human systems theory, which enables study of various dualisms (and monisms) with only secondary consideration of the realities in which they occur. The remainder of this chapter considers how philosophical theories about humanity and nature (ontological and metaphysical) are clarified by examining how they account for the transitions among the system levels associated with chapter transitions.

Philosophical Perspectives on Transitions

How significant of a distinction are the transitions between levels? The phenomena emerging in each chapter are significant for humans and are characterized by systems for each corresponding level. However, various philosophical perspectives emphasize certain aspects and may consider some transitions as more or less significant than others. I claim some dualist perspectives result from overemphasizing the differences in two levels of phenomena as opposed to the emergent monist position that considers them in mutually intertwined developmental and evolutionary context. A dualist believes that the transition divides reality into distinct realms (but must postulate some way of bridging them if they co-exist). In many cases, the examination of nature from the perspective of different chapter transitions clarifies commonalities between phenomena previously considered belonging to different dualistic realms.

Conversely to dualism's overemphasis on a chapter transition, some characterizations of monism underemphasize the chapter transitions and consider one chapter (or level) as predominant and the others as secondary derivatives. From a monist perspective, there is one realm, and everything is constituted within that realm. Various monisms occur depending upon what is considered predominant within the realm, with some flavors of monism emphasizing each chapter. Two examples are physicalism and idealism, which essentially overemphasize the physical or transcendent chapter, respectively. In addition, the overemphasized constructs of a particular level (or chapter) can also serve as a root metaphor for various metaphysics, as described in a later section.

Monist Perspectives

Two significant monisms that prioritize the phenomena and processes of one systems level over others are physicalism and idealism.[15] In addition to the claim about what is fundamental to existence, in some formulations, the positions also have reductive characterizations that claim the phenomena of the other levels either do not exist or exist only as a derivative of either the physical or ideal.

Physicalism. When physicalists describe the world, they prioritize the processes created in the physical chapter in explaining the phenomena of the later chapters. Physicalist acknowledge the other phenomena exist (in some way) but as aggregations of physical objects that may require additional explanations or other properties to fully characterize. Although these aggregated properties may be needed to explain the other phenomena and may constrain additional aggregated objects, they and their interactions can still be characterized physically.

Most physicalists acknowledge the need for explanations for other phenomena, but they differ on whether the aggregations have particular ways of influencing or constraining other objects. Physical reductionists claim reduction of all phenomena and their interactions to the mechanistic interactions between atoms and molecules, often described

[15] Not all monisms prioritize the phenomena and processes of one level over others. Emergent monism and multilevel monism claim that there are both significant similarities and differences between the levels and attempt to neither overemphasize nor underemphasize the transitions. Dual-aspect monism emphasizes the systems of two levels (often physical and transcendent, or physical and psychological/mental) and deemphasizes the transitions between the remaining levels.

philosophically as efficient causation and in physics as Newtonian mechanics.[16] As a countering position, nonreductive physicalists characterize the world in terms of physical phenomena but explicitly reject the complete reduction of other phenomena to physical properties and relationships.[17] The key distinction appears to be whether the apparently causal influences between phenomena of other levels are completely described or not by the causal powers that emerged among purely physical phenomena.

A useful distinction on causation from theories of emergence is between weak and strong emergence. In the position of *weak emergence*, emergent structures may constrain lower-level structures and emergent categories are required to explain causal processes, but causal processes do not emerge, while the *strong emergence* position claims that ontologically distinct levels arise over time characterized by their own distinct laws or regularities and causal forces.[18] Physicalists reject the strong emergence position and claim that some type of physical causation ultimately suffices to explain all phenomena. In general, for nonreductive physicalists, these other phenomena may constrain other phenomena (and their physical constituents), while for physical reductionist, only additional explanations are needed.

Idealism. When idealists describe the world, they identify an essential aspect of phenomena as ideas (ἰδέα) or form (εἶδος) with historical implications for abstract thought (via universals) and spirituality (via Neoplatonism). These ideas are considered necessary to explain the world, which would otherwise be unintelligible or unknowable. In the characterization of chapters, "ideas" emerge as transcendent phenomena from human recognition of universality within certain constructs of symbolic language. Ted Peters identifies a "single univocal meaning" as an abstraction from symbolic discourse and makes an important, subtle, and clarifying distinction in this context between the meaning of symbol and an idea,

[16] Careful construction of the position involves the supervenience of other phenomena on purely physical events. Jaegwon Kim, *Supervenience and Mind: Selected Philosophical Essays* (New York, NY, USA: Cambridge University Press, 1993).

[17] Murphy, "Nonreductive Physicalism: Philosophical Issues."

[18] Mark A. Bedau, "Weak Emergence," in *Philosophical Perspectives*, vol. 11 (Malden, MA: Blackwell, Ridgeview, 1997), 375–399; Claus Emmeche, Simo Koppe, and Frederik Stjernfelt, "Levels, Emergence, and Three Versions of Downward Causation," in Downward Causation: Minds, Bodies and Matter, ed. Peter Bøgh Andersen et al. (Aarhus: Aarhus Univ. Press, 2000), 13–34; David J. Chalmers, "Strong and Weak Emergence," in The Re-Emergence of Emergence, ed. Philip Clayton and Paul Davies (Oxford: Oxford University Press, 2006), 244–256.

which I further characterize elsewhere in terms of Deacon's selection dynamics.[19] The interactions between ideas select a univocal meaning from the multivalent meanings of a symbol, which results in the apparent abstraction and universality.

In reductive idealism, although things other than ideas might exist, they are not generally knowable other than as ideas. For these other things to be experienced as phenomena and understood, they must have some intelligibility and without that ability to be experienced and interpreted in some way, they lack existence. Both reductive idealism and physicalism prioritize the phenomena of their levels as the foundation for existence. Xavier Zubiri identifies a tendency in philosophy to combine reductive idealism and reductive physicalism where reality consists of entities and where knowing (what he calls intellection) is reduced to pure reason (*logos*).[20]

Dualist Perspectives

While some monist perspectives emphasize one level's phenomena over others, dualist perspectives emphasize one transition over others. Three examples are vitalism, Cartesian dualism, and Platonic idealism.

Vitalism. Until the nineteenth century, many considered life as characterized by systems of the biological level as separate from the physical one, and the most refined theory of the distinction occurred in "vitalism," where organisms required a separate substance, the *élan vital*, to live.[21] Biologists now realize that life occurs not because of one big story of a vital force, but because of lots of little stories in metabolic pathways, genetic replication, and the like. The distinction between organic and inorganic was first bridged in the 1820s by the synthesis of urea from inorganic compounds, with further characterizations of the transition occurring with the discovery of DNA, its role in genetics, and the synthesis of genetics and evolutionary theory.

[19] Ted Peters, "Can We Hack the Religious Mind? The Interaction of Material Reality with Ultimate Reality in the Human Self," in *Interactive World, Interactive God: The Basic Reality of Creative Interaction*, ed. Carol Rausch Albright, John R. Albright, and Mladen Turk (Wipf and Stock Publishers, 2017), 207–239; Mark Graves, "Interaction in Emergent Human Systems," *Theology and Science* 21(2): 331–339 (2023).

[20] Xavier Zubiri, *Sentient Intelligence*, trans. Thomas Fowler (Washington, DC: The Xavier Zubiri Foundation of North America, 1999).

[21] Ian Coulter, Pamela Snider, and Amy Neil, "Vitalism–A Worldview Revisited: A Critique Of Vitalism And Its Implications For Integrative Medicine," *Integrative Medicine: A Clinician's Journal* 18, no. 3 (June 2019): 60–73, https://www.ncbi.nlm.nih.gov/pmc/articles/PMC7217401/

Cartesian Dualism. Descartes distinguished the realm of the mental (*res cogitans*) from the physical (*res extensa*), in part to reduce conflict between the powerful religious authorities' claim over study of the soul and emerging scientific examination of the human body. In Descartes' understanding of substances, the body and its extension in space exists separately from the mind (or rational soul) and its capacity to think and reflect, e.g., upon its own existence. By deemphasizing the distinction between physical and biological levels and not recognizing the interdependence between cognitive and social processes for self-reflection, Descartes overemphasized the transition between biological and psychological levels, deemphasized the physical-biological and psychological-cultural transitions, and left a legacy of mind-body dualism still affecting our understanding of both.

In its current incarnation as Cartesian dualism, the medieval understanding of substances has given way to an irreconcilable dualism between the reduction of matter to physical characteristics, such as extension and chemical properties, and the centrality of subjective thought in the socially constructed self. Social scientists since Mead have recognized the self as a result of complex social processes rather than a precursor to them.[22] An irreducible gulf between mind and body results from (i) understating the complex evolutionary and developmental processes giving rise to the human brain and physiology out of physical molecules and biochemical reactions and (ii) minimizing the complex social and development processes leading to a self capable of identifying itself as a social agent and reflecting upon its own existence. In both cases, the synchronic examination of a fully formed human body and a fully constructed mind reflecting upon itself ignores their mutual diachronic progressions over cosmological and evolutionary time and their individual human development.

Platonic Dualism. For Plato, the realm of ideas predominated over the material world, which he considered a reflected shadow of the essential ideas. Further development refined the essential idea as the form of living organisms. The essence of an organism became characterized by its existential patterns and relationships, and thinking about it became framed as concepts through a historical trajectory that includes Aristotle's monistic account of the soul, Cartesian dualism, and modern chemistry and

[22] George Herbert Mead, *Mind, Self & Society from the Standpoint of a Social Behaviorist* (Chicago: University of Chicago Press, 1934).

evolutionary biology.[23] Platonic dualism situates what we now consider as concepts (i.e., ideas) as fundamental to existence with sophisticated theories about their progression through Divine Mind to explain their origin and activity.

The return of souls toward the One in Neoplatonic thought was particularly influential on Western understanding of spirituality.[24] The realms of ideas through the lens of spirituality characterizes the essence (univocal meaning) of a person or other organism. This supports dualist perspectives on the soul and, when embodied, the Aristotelian and Thomistic monistic view of soul as form of the body. Elsewhere I characterize the emergence of the transcendentals Good, Truth, and Beauty and describe how Josiah Royce's religious philosophy of interpretative communities can result in traditional conceptions of spirituality.[25] The shared interpretive process within communities and traditions creates an emergent form with causal power and leads to recognizable characterizations of a range of spiritual phenomena when extending the interpretive process semiotically to organisms, societies, and nature as a whole. In the context of that emergent monist account, Platonic dualism divides the univocal ideas and essences from the cultural processes that generate and/or identify them. As a historically meaningful aside, considering the transcendent level as spiritual, the cultural and psychological levels as mental, and the prior levels as embodiment yields a tripartite understanding of human existence as spirit, mind, and body.

The monist and dualist perspectives on various level transitions influence scholarly discourse and scientific assumptions. Viewing human existence within nature from a systems perspective identifies levels that may be

[23] Paul S. MacDonald, *History of the Concept of Mind: Speculations about Soul, Mind, and Spirit from Homer to Hume* (Aldershot, Hants, England; Burlington, VT: Ashgate, 2003); Wladyslaw Tatarkiewicz, *A History of Six Ideas: An Essay in Aesthetics* (Hingham, MA: Kluwer Boston, 1980); Graves, *Mind, Brain, and the Elusive Soul*; Liane Gabora, Eleanor Rosch, and Diederik Aerts, "Toward an Ecological Theory of Concepts," *Ecological Psychology* 20, no. 1 (2008): 84–116, https://doi.org/10.1080/10407410701766676

[24] Pierre Hadot, *Plotinus, or, The Simplicity of Vision* (Chicago: University of Chicago Press, 1993); Bernard McGinn, *The Presence of God: A History of Western Christian Mysticism* (New York: Crossroad, 1991); Andrew Louth, *The Origins of the Christian Mystical Tradition from Plato to Denys* (Oxford, New York: Clarendon Press; Oxford University Press, 1981).

[25] Graves, "The Emergence of Transcendental Norms in Human Systems"; Josiah Royce, *The Problem of Christianity. Lectures Delivered at the Lowell Institute in Boston, and at Manchester College, Oxford* (New York: Macmillan, 1913); Graves, *Insight to Heal: Co-Creating Beauty amidst Human Suffering*, chap. 6.

over or underemphasized by various ontological positions. This systems perspective, and the transition between levels, also illuminates aspects of the metaphors chosen for metaphysical theories.

Metaphysical Metaphors

Metaphysics examines the fundamental nature of reality, and its first principles, such as being, causality, and modality.[26] In many metaphysical theories, to understand and explain what in many respects one cannot directly examine, various metaphors, models, and propositions are proposed and claimed.[27] In the prior section, I described five characterizations of being—two ontological monisms and three dualisms—and in this section, I outline three metaphysical theories that depend upon constructs from the three remaining (middle) levels—organisms, behaviors, and symbolic language, respectively. I also consider metaphysical implications of the two "boundary" levels associated with the beginning and transcendent chapters.[28] These positions are summarized in Table 6.1, which also indicates a "missing" historical dualism corresponding to human uniqueness.[29]

[26]This chapter focuses primarily on the nature of being, as it draws upon a systems approach to ontology. Elsewhere, I have briefly described how one can see new levels as leading to new types of causation drawing upon theories of strong emergence (Graves, *Mind, Brain, and the Elusive Soul*, 117–119) and how "real possibility" can underlie the emergence of new being drawing upon Peirce's understanding of real possibility (Graves, *Insight to Heal*, 265–267).

[27]Stephen C. Pepper, *Concept and Quality: A World Hypothesis* (La Salle, Ill.: Open Court, 1967); Ian G. Barbour, *Myths, Models and Paradigms: A Comparative Study in Science and Religion* (San Francisco: Harper, 1976); Thomas Leddy, "Metaphor and Metaphysics," *Metaphor and Symbolic Activity* 10, no. 3 (1995): 205–222, https://doi.org/10.1207/s15327868ms1003_4

[28]The metaphysics of physicalism and idealism have additional aspects beyond their ontological ones, such as causality and modality, which are not examined in this chapter.

[29]The additional "dualism" of human uniqueness does not appear to have a historical dualist position associated with the transition between psychological and cultural-linguistic phenomena, though sharp distinctions between humans and other animals have been frequently assumed. For historical reasons, the mental processing shared with animals as characterized by the psychological level has been ignored (e.g., by Descartes) or the ideas dependent upon human language has been ascribed to a Divine Mind (e.g., by Plato). The uniqueness resulting from symbolic language (Deacon, 1997) can be characterized as rationality for purpose of this chapter, which in turn may also conflate with human phenomenological consciousness. The transition also corresponds somewhat to the two aspects of a Thomistic dual-aspect monism derived from his distinctions between rational and sensitive powers of the soul.

Table 6.1 Systems classification of monism and dualisms

Level	*Key transition*	*Key metaphor for metaphysics*	*Example monism*	*Example dualism*
Transcendent		Idea/Spirit	Idealism	
	Idea			Platonic dualism
Cultural		Word	Peirce semiotic metaphysics	
	Language			Human uniqueness
Psychological		Behavior/act	Aristotelian/Thomistic substance	
	Synapse			Cartesian dualism
Biological		Organism	Organicism, process philosophy	
	DNA			Vitalism
Physical		Physical object	Physicalism	
	Elements			

Organicism. Biological metaphysical metaphors generally focus on the organism. Organicism is the doctrine that the universe is orderly and alive, like an organism. In the late eighteenth and early nineteenth centuries, Schelling responded to a mechanistic determination of nature common to his time through his emphasis on organism. The metaphor of living organism dates back at least to Plato's *Timaeus*, where he compares the cosmos to a living organism, crafted by the Demiurge. In developing his process philosophy, Alfred North Whitehead draws upon those metaphors of organism and organic process to help articulate his revised understanding of reality, i.e., "philosophy of organism," in light of quantum mechanics and its effect on philosophical understanding of reality.[30]

Behavior/Act. The use of psychological metaphors for metaphysics harkens back to Aristotle's original metaphysics and the use of the behavioral "act" to address the paradox of change. For Aristotle, the apparent contradiction was framed (i) by Heraclitus who observed that everything exists in flux like a river, in which one cannot step twice as the waters always continue to flow; and (ii) by Parmenides who claimed that something

[30] Alfred North Whitehead, *Process and Reality: An Essay in Cosmology*, Corrected (New York: Free Press, 1978).

either exists or does not exist and thus change is an illusion as there is no possibility between existence and non-existence. Aristotle argued that for something to change, it must actualize that which, in its nature, is possible. He acknowledged a certain immutability, but of the possibilities something has rather than of its existence, and change occurs in how it happens to exist. Thus, existence (or being) is an act where potency becomes actualized.

Semiotics. As a metaphysics metaphorically rooted in language, Charles Saunders Peirce's semiotic metaphysics extended the interpretation of words in language (and other "signs" used in communication) to a process in which everything is interpretation of signs. Peirce generalizes "sign" to any reference of something and examines the different ways in which interpretation could occur. For example in a biological interaction, one protein could interpret another molecule in terms of its shape, and cellular mechanisms interpret DNA to create a copy of the cell. By making interpretation of language the primary metaphor of his metaphysics, Peirce helps bridge the physical representation of things as signs and the univocal ideas toward which interpretation may occur.

Boundary Levels. These metaphors of organism, behavior, and symbolic language drive the formation of three respective metaphysical theories. Together with physicalism and idealism, they characterize philosophical perspectives on reality rooted in five chapters of nature and levels of systems. Reality is examined through the lens of the first chapter, too. Reality is considered through theories grounded in quantum electrodynamics and gravity, such as string theory and quantum gravity. These are also metaphors used to understand aspects of reality not obvious to humans outside of carefully conducted experiments and sophisticated technologies. The metaphors are also extended to scientific models and formalized in the language of mathematics.[31] However, as a metaphor, there remains an "is not" aspect of the scientific theories used to characterize the early, pre-element chapter of nature and subatomic level of systems.[32]

So far, I have considered idealism as a reductive monism and in the context of Platonic dualism as well as suggested nonreductive physicalism and emergent monism as alternatives to reductive physicalism. An additional metaphysical position considers the univocal ideas as having reality

[31] Barbour, *Myths, Models and Paradigms: A Comparative Study in Science and Religion.*

[32] The related view that nature *is* the mathematics theory would be a form of idealism—where the idea is formalized in the language of mathematics.

beyond the conceptualized constructs of Platonic dualism. Although mathematical, spiritual, and other transcendent-level phenomena may be situated within human systems, one can still postulate an additional, metaphoric "idea" as having metaphysical significance, just as one uses a biological process or behavioral act as a root metaphor for modeling fundamental reality. One could postulate a metaphoric univocal univocity, Monad, or "idea of God" as separate from the reality of the world, as discussed up to this point.[33]

Theological Metaphors

The level transitions also serve as metaphor for characterizing the difference between God and the world. This is explicit in panentheism (world-in-God) where some divide is identified between the world and aspects of God that exceed the world. The various level transitions identify aspects of nature where earlier levels are most associated with the world and subsequent levels are used metaphorically to characterize additional aspects of the Divine.

Process theology uses a physical-biological dualism to characterize a distinction between the physical world and an organismic (process) view of a dynamic God. Whitehead identifies reality with process stating "process is the becoming of experience."[34] For Whitehead, every entity has two poles: a concrete, physical pole and an abstract, conceptual pole; and experience reduces the abstract pole to concrete actuality. Much as biological processes direct the movement of molecules and their interactions, the abstract conceptual pole can lure the coming together (nexus) of experience into concrete actuality. Although Whitehead identified God non-religiously as the creative principle of that process, process theologians have furthered developed the role of God within process metaphysics.[35]

Sallie McFague makes metaphors explicit and methodological in her metaphorical theology and uses Cartesian mind-body dualism to describe the God-World relationship.[36] McFague identifies the "is" and "is not" aspects of metaphor and uses Paul Ricoeur's hermeneutic phenomenology

[33] John Edwin Smith, *Experience and God* (New York: Oxford University Press, 1968).

[34] Whitehead, *Process and Reality: An Essay in Cosmology*, 166.

[35] John B. Cobb and David Ray Griffin, *Process Theology: An Introductory Exposition* (Philadelphia: Westminster Press, 1976).

[36] Sallie McFague, *Metaphorical Theology: Models of God in Religious Language* (Philadelphia: Fortress Press, 1982).

and Ian Barbour's work on theology and science to capture the tensive relationship between the conceptual "is" and metaphoric narrative interpretation of what "is not."[37] The Cartesian mind-body dualism constitutes the source of the metaphor with God as mind and World as body, and with further narrative interpretation of what that means theologically. The metaphor also incorporates the unresolved need to articulate how the mental (psychological) and body (biological) interact in a dualist framework.

Josiah Royce's philosophy of community uses the transition from cultural (linguistic symbols) to transcendent (ideas) as a metaphor for understanding a World-Spirit relationship.[38] Royce builds upon Peirce's semiotic interpretation to describe the shared interpretation of a committed religious community, namely the Christian Church, as identifying the spirit of the church, i.e., the Holy Spirit. As mentioned above, the direct use of Peirce's theory of symbolic language interpretation helps clarify the transition between psychological (making sounds and movements) and cultural (linguistic) levels; and the generalization of that theory semiotically—where everything is a sign—forms a metaphysical foundation. Royce extends Peirce's linguistic theory to identify the importance of social processes in determining the meaning of words and symbols, and he adapts Peirce's semiotics to a religious context (besides others who developed semiotics to physical and biological contexts). For Royce, the symbol can refer to committed causes, such as a faith commitment, and the interpretation of those symbols has religious and spiritual significance. His communal semiotic theory clarifies the transition from cultural (linguistic) to transcendent levels, when those symbols refer to transcendent-level values and the interpretation has spiritual aspects (not merely linguistic ones). As a further theological development, Royce's position extends the symbol reference of the transcendent-level, shared, Christian idea of the Holy Spirit to the "real" Holy Spirit.[39]

[37] Paul Ricoeur, *Semeia 4: Biblical Hermeneutics*, ed. John Dominic Crossan (Society of Biblical Literature, 1975); Barbour, *Myths, Models and Paradigms: A Comparative Study in Science and Religion*.

[38] Royce, *The Problem of Christianity. Lectures Delivered at the Lowell Institute in Boston, and at Manchester College, Oxford*.

[39] John Smith further expands the semiotic approach to the "idea of God." Smith, *Experience and God*.

Conclusion

The Book of Nature and levels of human systems structure human existence and experience into six chapters (or levels) with five intervening transitions. The under and overemphasis of different transitions suffice to characterize several monist and dualist philosophical theories, especially ontological theories of human existence. The constructs of each chapter (or level) also serve as root metaphors for metaphysical theories. Theologically, the multilevel systems approach classifies natural, human, philosophical, and theological theories in a way that acknowledges their interdependence and provides an integrative perspective on human experience of reality.

The shift of focus from the realms divided to the division itself organizes and illuminates commonalities among dualisms. Attention to the transition between realms also provides a sufficient foundation for examining a range of dualist and monist positions on nature, human existence, metaphysics, and the God-world relation and clarifies the limitation and power in explicitly incorporating human perspectives. The systems framework and classified positions facilitate systematic study for integrative theological and scientific investigations.

Bibliography

Barbour, Ian G. 1976. *Myths, Models and Paradigms: A Comparative Study in Science and Religion*. San Francisco: Harper.

Bedau, Mark A. 1997. Weak Emergence. In *Philosophical perspectives*, vol. 11, 375–399. Malden, MA: Blackwell, Ridgeview.

Bellah, Robert Neelly. 2011. *Religion in human evolution: From the paleolithic to the axial age*. Cambridge, Mass: Belknap Press of Harvard University Press.

von Bertalanffy, Ludwig. 1975. *Perspectives on general system theory: Scientific-philosophical studies*. New York: G. Braziller.

Chalmers, David J. 2006. Strong and Weak Emergence. In *The Re-Emergence of Emergence*, ed. Philip Clayton and Paul Davies, 244–256. Oxford: Oxford University Press.

Clayton, Philip. 2004. *Mind and Emergence: From Quantum to Consciousness*. New York: Oxford University Press.

Cobb, John B., and David Ray Griffin. 1976. *Process Theology: An Introductory Exposition*. Philadelphia: Westminster Press.

Coulter, Ian, Pamela Snider, and Amy Neil. 2019. Vitalism–A Worldview Revisited: A Critique of Vitalism And Its Implications For Integrative Medicine. *Integrative Medicine: A Clinician's Journal* 18 (3): 60–73. https://www.ncbi.nlm.nih.gov/pmc/articles/PMC7217401/.

Deacon, Terrence W. 1997. *The Symbolic Species: The Co-Evolution of Language and the Brain*. New York: W.W. Norton.

———. 2006. Emergence: The Hole at the Wheel's Hub. In *The Re-Emergence of Emergence*, ed. Philip Clayton and Paul Davies, 111–150. Oxford: Oxford University Press.

———. 2011. *Incomplete Nature: How Mind Emerged from Matter*. New York: W.W. Norton.

Emmeche, Claus, Simo Koppe, and Frederik Stjernfelt. 2000. Levels, Emergence, and Three Versions of Downward Causation. In *Downward Causation: Minds, Bodies and Matter*, ed. Peter Bøgh Andersen et al., 13–34. Aarhus: Aarhus Univ. Press.

Gabora, Liane, Eleanor Rosch, and Diederik Aerts. 2008. Toward an Ecological Theory of Concepts. *Ecological Psychology* 20 (1): 84–116. https://doi.org/10.1080/10407410701766676.

Gelpi, Donald L. 2001. *The Gracing of Human Experience: Rethinking the Relationship between Nature and Grace*. Collegeville, Minn.: Liturgical Press.

Graves, Mark. 2008. *Mind, Brain, and the Elusive Soul: Human Systems of Cognitive Science and Religion*. Aldershot, Hants, England; Burlington, VT: Ashgate.

———. 2009. The Emergence of Transcendental Norms in Human Systems. *Zygon* 44 (3): 501–532.

———. 2013. *Insight to Heal: Co-Creating Beauty amidst Human Suffering*. Eugene, OR: Cascade Books.

———. 2016. Places of Information Generation: Bridging Pannenberg's Logos and Deacon's Emerging Semiosis. *Theology and Science* 14 (3): 305–324. https://doi.org/10.1080/14746700.2016.1191880.

———. 2021. Emergent models for Moral AI spirituality. *International Journal of Interactive Multimedia and Artificial Intelligence* 7 (1) Special Issue on AI, Spirituality, and Analogue Thinking: 7–15. https://doi.org/10.9781/ijimai.2021.08.002.

———. 2023. Interaction in Emergent Human Systems. *Theology and Science* 21 (2): 331–339.

Hadot, Pierre. 1993. *Plotinus, or, The Simplicity of Vision*. Chicago: University of Chicago Press.

Hefner, Philip J. 1993. *The Human Factor: Evolution, Culture, and Religion*. Minneapolis: Fortress Press.

Kim, Jaegwon. 1993. *Supervenience and Mind: Selected Philosophical Essays*. New York, NY, USA: Cambridge University Press.

Leddy, Thomas. 1995. Metaphor and Metaphysics. *Metaphor and Symbolic Activity* 10 (3): 205–222. https://doi.org/10.1207/s15327868ms1003_4.

Louth, Andrew. 1981. *The origins of the Christian Mystical Tradition from Plato to Denys*. Oxford, New York, Clarendon Press; Oxford University Press.

MacDonald, Paul S. 2003. *History of the concept of mind: Speculations about Soul, mind, and spirit from homer to Hume*. Aldershot, Hants, England; Burlington, VT, Ashgate.

Mead, George Herbert. 1934. *Mind, Self & Society from the Standpoint of a Social Behaviorist*. Chicago: University of Chicago Press.

McFague, Sallie. 1982. *Metaphorical Theology: Models of God in Religious Language*. Philadelphia: Fortress Press.

McGinn, Bernard. 1991. *The Presence of God: A History of Western Christian Mysticism*. New York: Crossroad.

Murphy, Nancey C. 1998. Nonreductive physicalism: Philosophical issues. In *Whatever Happened to the Soul?: Scientific and Theological Portraits of Human Nature*, ed. Warren Brown, Nancey Murphy, and H. Newton Malony. Minneapolis, Minn: Fortress Press.

Nagel, Thomas. 1974. What Is It Like to Be a Bat? *Philosophical Review* 83 (4): 435–450.

Parker, Kelly A. 1998. *The Continuity of Peirce's Thought*. Nashville: Vanderbilt University Press.

———. 2008. Josiah Royce: Idealism, Transcendentalism, Pragmatism. In *The Oxford Handbook of American Philosophy*. https://doi.org/10.1093/oxfordhb/9780199219315.003.0006.

Peacocke, Arthur Robert. 1993. *Theology for a Scientific Age: Being and Becoming–Natural, Divine, and Human*. Minneapolis: Fortress Press.

Pepper, Stephen C. 1967. *Concept and quality: A world hypothesis*. La Salle, Ill: Open Court.

Peters, Ted. 2017. Can We Hack the Religious Mind? The Interaction of Material Reality with Ultimate Reality in the Human Self. In *Interactive World, Interactive God: The Basic Reality of Creative Interaction*, ed. Carol Rausch Albright, John R. Albright, and Mladen Turk, 207–239. Eugene: Wipf and Stock Publishers.

Ricoeur, Paul. *Semeia 4: Biblical Hermeneutics*, edited by John Dominic Crossan. Society of Biblical Literature, 1975.

Royce, Josiah. 1913. *The Problem of Christianity. Lectures Delivered at the Lowell Institute in Boston, and at Manchester College, Oxford*. New York: Macmillan.

Skyttner, Lars. 2006. *General Systems Theory: Perspectives, Problems, Practice*. 2nd ed. Singapore; River Edge, N.J: World Scientific.

Smith, John Edwin. 1968. *Experience and God*. New York: Oxford University Press.

Jackson, M.G.S.A. 1986. "Formica Dei": Augustine's "Enarratio in psalmum" 66.3. *Vigiliae Christianae* 40 (2): 153–168.

Tatarkiewicz, Wladyslaw. 1980. *A History of Six Ideas: An Essay in Aesthetics.* Hingham, MA: Kluwer Boston.
Whitehead, Alfred North. 1978. *Process and Reality: An Essay in Cosmology, Corrected.* New York: Free Press.
Zubiri, Xavier. 1999. *Sentient Intelligence. Trans. Thomas Fowler.* Washington, DC: The Xavier Zubiri Foundation of North America.

CHAPTER 7

Reading the Signs of the Times: Nature-Culture Dualism and Human Feeling in the Anthropocene

Hannah Malcolm

Reports of distress in the face of climate change and ecological collapse have become increasingly frequent in the last decade, at the level of popular reporting and in psychological and sociological research.[1] These experiences—often referred to as climate or ecological grief, anxiety, or trauma—have even made it into international risk forecasting. In 2022, for the first time in its reporting history, the Intergovernmental Panel on Climate Change projected 'high confidence' in the widespread mental

[1] See Maria Ojala et al., "Anxiety, Worry, and Grief in a Time of Environmental and Climate Crisis: A Narrative Review," *Annual Review of Environment and Resources* 46, no. 35–38 (October 2021).

H. Malcolm (✉)
Department of Theology and Religion, Durham University, Durham, UK

T. J. Hastings, K.-W. Sæther (eds.), *Views of Nature and Dualism*,
https://doi.org/10.1007/978-3-031-42902-6_7

health and well-being impacts of climate change, both now and in the future.[2]

But these kinds of responses to anthropogenic loss are not a universal experience.[3] They are culturally conditioned, prompted, and shaped by narratives concerning human relation to the non-human and determined by local environment and context. The most clearly documented accounts of 'climate grief' belong either to those cultures with very close relation to their immediate environments (farmers and indigenous communities) or who have high levels of exposure to information about climate change and ecological collapse (scientists and environmental activists).[4] Climate grief is also temporally bound; those who experience profound sorrow over the human-caused extinction of other creatures do not, for example, tend to articulate those feelings in relation to former extinction events. Indeed, this latter point can be used as a way of downplaying the significance of human dread, anxiety, or guilt over *this* extinction event. How, then, do we account for its significance in directing our relationships with other creatures?

Objections to taking climate grief seriously express a kind of nature-culture dualism: in criticisms of climate grief as a purely cultural phenomenon, the critic implies that a culturally determined feeling cannot also be legitimately described as a 'real' response to something 'true'. In this reading, 'cultural' narratives which prompt climate grief are not necessarily related to the real fabric of an external 'nature', and so are fundamentally unreliable as a source of knowledge. By contrast, 'nature' is treated as that which is 'real', verifiable not through experience but through a supposedly separate realm of scientific data. On the other side, the speeches and campaigns of climate activists can also operate under an apparent nature-culture divide, claiming to tell wholly *natural* stories about human identity. The result, of course, is highly cultural; romanticising the non-human ('we need to learn from Mother Nature'), flattening the obvious differences between humans and other creatures in order to emphasise

[2] H.-O. Pörtner et al., "IPCC, 2022: Summary for Policymakers," Climate Change 2022: Impacts, Adaptation, and Vulnerability (Cambridge: Cambridge University Press, 2022), B.1.4, B.4.4.

[3] Many people *do not* claim to experience climate and ecological grief, despite living under the same changing material conditions as those who do. This can be compared to grief over the death of a loved one; its expression is highly culturally conditioned, but it can safely be read as a universal human experience.

[4] See, for example, Ashlee Cunsolo and Karen Landman (eds.), *Mourning Nature: Hope at the Heart of Ecological Loss and Grief* (Montreal/Kingston: McGill-Queen's University Press, 2019).

similarities ('we are nature defending itself'), or demonising all human activity ('humans are the virus (in an otherwise benign natural world)').[5] Such approaches share a common problem: they claim to transcend culturally determined interpretations of the relationship between humans and other creatures in order to give greater credence to their campaigns. Can a culturally determined set of emotions also reveal truth about the nature of those creatures or places which provoke such a movement? Or, to put it another way, if our responses to anthropogenic loss are humanly constructed feelings, can they be trusted to reveal truth which might bear moral authority?[6]

Using our creaturely identity as a basis for describing the relationship between culture and nature, I propose that the cultural conditioning of responses to anthropogenic loss is not necessarily a barrier to their role in revealing truth. I also seek to avoid collapsing the distinction between culturally shaped human feeling and the real 'nature' of non-human creatures. A nature-culture frame for responses to anthropogenic loss clarifies that they are both narratively shaped *and* narratively shaping; they connect the realm of sensory response to death to the realm of forming narratives which concern our spiritual and moral relation to the world. In this chapter I give an account of creation—including human creatures—as a world of signs which can be read. In a theological frame, encounters which bring about grief signify the consequences of sin and these signs can be received and interpreted by us. This is not to say that our reading of these signs is always equally effective. While the range of sorrows gathered under the umbrella of 'climate grief' can be more *or less* an approximation of what is real, they nevertheless communicate at least partial truth about us and our fellow creatures.[7] I begin with a critical overview of the modern relation between nature and culture before turning to a semiotic reading of human reception and interpretation of non-human creation.

[5] These are culturally determined phrases. Mother Nature is comfortably claimed as a teacher when we do not live in fear of smallpox. We might say we are 'nature defending itself' to decry littering while also promoting the nature-transcending ethics of veganism. And it is easy to claim 'humans are the virus' when no one is threatening to wipe out your particular people group.

[6] By 'culture', I mean those things which humans *make*. I draw here on John Milbank's description of 'human society'. See John Milbank, *Theology and Social Theory: Beyond Secular Reason* (Oxford: Blackwell Publishing, 2006), 272. I return to Milbank's interpretation of the relation between nature and culture later.

[7] The elasticity of the term 'climate grief' means that it can contain responses which are nihilistic, selfish, and actively violent towards other humans.

Nature-Culture

The climate and ecological crises have prompted a flurry of ideological scapegoating. Who or what is to blame for such an unprecedentedly dysfunctional relationship between human and non-human creatures?[8] Christianity, the reformation, colonialism, capitalism, anthropocentrism, and the Copernican and Cartesian revolutions have all variously come under fire, and while each critique certainly contains a varying degree of truth, they cannot really be successfully disentangled to identify a solo culprit.[9] What *does* seem to be a point of agreement is that the modern era ushered in a new and isolationist imagination concerning the human creature, both in removing divine power from interpretations of the movements of the cosmos and in an increasingly self-referential mode of human meaning-making. In *Passage to Modernity* Louis Dupré describes these related processes as a 'double breakup: the one between the transcendent constituent and its cosmic-human counterpart, and the one between the person and cosmos'.[10] He articulates the consequences of this new self-referential reading of humanity: humans, and humans alone, possess *inherent* meaning, and as such any meaning perceived outside the human subject is only the product of human meaning-making, rather than humans being creatures who can interpret meaning which exists independently of

[8] Humanity has a longer history of ecocidal behaviour. The arrival of the Norse in Greenland and the collapse of Mayan civilisation in Central America are two pertinent examples. But the distinctive scale and extent of this current collapse in the biosphere's stability is certainly unprecedented in human history.

[9] For example, see Lynn White Jr, "The Historical Roots of Our Ecologic Crisis," *Science*, 155.3767 (1967), 1203–1217, Peter Harrison, *The Bible, Protestantism, and the Rise of Natural Science* (Cambridge: Cambridge University Press, 1998), James H. Cone, "Whose Earth Is It Anyway?", *CrossCurrents*, 50.1/2 (2000), 36–46, Naomi Klein, *This Changes Everything: Capitalism* vs. *The Climate* (London: Penguin Books Ltd., 2014), David Keller, *Environmental Ethics: The Big Questions* (Oxford: Blackwell Publishing, 2010), and Michael Northcott, *A Political Theology of Climate Change* (Grand Rapids: W. B. Eerdmans, 2013).

[10] Louis Dupré, *Passage to Modernity: An Essay in the Hermeneutics of Nature and Culture* (New Haven and London: Yale University Press, 1993), 3. Bruno Latour similarly describes modernity as the simultaneous birth of 'humanity' and 'nonhumanity' and a 'crossed-out God', arising 'first from the conjoined creation of those three entities, and then from the masking of the conjoined birth and the separate treatment of the three communities while, underneath, hybrids continue to multiply as an effect of this separate treatment'. Bruno Latour, *We Have Never Been Modern*, trans. Catherine Porter (Cambridge, Massachusetts: Harvard University Press, 1993), 13.

human perception.[11] The loss of nature's transcendent *telos* left a vacuum of purpose which could only be filled by *human* purposes. The new 'nature' as distinct from human 'culture' therefore became characterised by utility—if God was there to be praised for creation, He was to be praised not out of wonder over the creatures themselves but because of their use in human advancement.

For examples of this attitude emerging out of modernity, Francis Bacon and Robert Boyle make regular appearances in literature concerning the modern reimagination of human and non-human nature, particularly concerning three interrelated ideas which remain fundamental to understanding the trust (or lack of it) which we place in our interpretations of the world: nature as knowable, as utility, and as manipulable.

Francis Bacon's development of 'true directions concerning the interpretation of nature' in *Novum Organum* argues that scientific knowledge (i.e., knowledge of the true nature of things) is pursued in accumulating observable data, freed from human preconception or tradition, and its purpose is the human use and control of creation, or the 'Empire' and 'Reign' of 'man'.[12] *Novum Organum* forms one part of his broader project—*Instauratio Magna*—concerned with the purpose of science, including partitioning knowledge into discrete categories.[13] It outlines Bacon's underlying theological anthropology: humanity is 'the servant [minister] and interpreter of nature' and 'human knowledge and power come to the same thing'.[14] Bacon reads scientific knowledge and the instruments it produces as re-establishing a pre-lapsarian order, in which humans once again wield control over creation. As Oliver Davies puts it, 'eliminative induction' becomes 'equated with the purification of mind' and a return to 'unity with the world'.[15] Bacon interprets Adam naming other animals as representing a pure form of knowledge, from which humanity has

[11] Dupré, *Passage to Modernity*, 24.

[12] Bacon identifies four 'idols'—the Tribe (projections of human nature), Cave (experience), Market (language), and Theatre (philosophers)—as deterring a truthful reading of nature. Francis Bacon, *The Instauratio Magna Part II: Novum Organum and Associated Texts*, ed. Graham Rees and Maria Wakely, Oxford Francis Bacon (Oxford: Clarendon Press, 2009), 79–83.

[13] This division of knowledge in *De Augmentis Scientiarum* is symptomatic of the broader 'purification' attempt as described by Bruno Latour in the division of nature and culture. I return to this later.

[14] Bacon, *Novum Organum*, 65.

[15] Oliver Davies, *The Creativity of God: World, Eucharist, Reason*, Cambridge Studies in Christian Doctrine (Cambridge: Cambridge University Press, 2004), 54.

fallen, and to which the study of nature helps us return.[16] While Bacon certainly characterises this work as for the glory of God, he nevertheless approaches nature as though it bears no intrinsic *telos*. He compares investigation of final causes to a consecrated virgin who has no children; there is no final causality to be discerned beyond the practical application of the uses of nature for humanity's advancement.[17] Dupré describes the consequences of negating nature's end as follows:

> Science for Bacon offered the most practical as well as the least expensive solutions to basic human problems. But without a common teleology that integrates humanity with nature, the mastery of nature becomes its own end, and the purposes originally pursued by it end up becoming secondary.[18]

Davies also picks up on this 'common teleology' gap in Bacon's thought. If, Davies argues, knowledge is only concerned with the 'fundamental material constitution of things' (understanding is knowing how something is made and being able to imitate that process), little room is left for acknowledging the *imaginative* work which humans do to 'comprehend the world *as a whole*' and to conceive of our place within it.[19] In driving a wedge between human imagination and intellectual enquiry into the nature of other creatures, a sense of human participation in God's cosmos is lost. With no common *telos*, the human is 'cut adrift or exiled from the world'.[20]

Like Bacon, mechanical philosopher Robert Boyle sought to protect the acquisition of knowledge from human discourse or prejudice. Against a Hobbesian appeal to natural law, Boyle argued that facts could be established through experimentation which would make hidden realities

[16] Davies, *The Creativity of* God, 54.

[17] Dupré, *Passage to Modernity*, 72. As Milbank puts it, 'in Francis Bacon, the 'truth of the made' begins to degenerate into the merely experimental confirmation of the utile working of a well-constructed machine… the indispensable disclosing *fictio*… is banished to the realm of a depoliticised rhetoric and poetics where it now discloses merely a decorous 'beauty' whose role tends to be reduced to subservience to either reason or utility or else to an integrating imaginative function that has a merely subjective import'. John Milbank, *Beyond Secular Order: The Representation of Being and the Representation of the People*, Illuminations: Theory and Religion Series (Hoboken: John Wiley & Sons, 2014), 219.

[18] Dupré, *Passage to Modernity*, 74.

[19] Davies, *The Creativity of God*, 54.

[20] Davies, *The Creativity of God*, 71.

apparent to the senses.[21] Bruno Latour describes Boyle's emphasis on laboratory experimentation as appealing to the 'inert' authority of the non-human, creatures who lack souls and are thus 'incapable of will and bias but capable of showing, signing, writing, and scribbling on laboratory instruments before trustworthy witnesses' (i.e., scientists). By contrast, the human possesses will, but is not trusted to 'indicate phenomena in a reliable way'.[22] Such an approach to the location of authority transforms both our perception of non-human creatures and our perception of scientists, who step in as 'authorized spokespersons', simultaneously arguing that natural forces cannot 'speak' to us, *and* that 'facts speak for themselves'.[23] While Hobbesian political power places the determination of destiny in the hands of human citizens, Boyle's legacy of a natural power which places scientific experimentation as the only reliable communicative bridge between the non-human object and the human agent promises that 'Nature' is simply there to be nakedly discovered, and upon discovery can be replicated by human artifice.[24] Latour describes this as the distinction between 'unifying but senseless nature, on the one hand, and on the other, cultures packed with meaning.'[25] What are the consequences of such a distinction? In transforming the location of meaning and meaning-making, the human response to encounters with non-human creation is also transformed.

The shift in nature's meaning changes human meaning-making. By way of example: Simon Oliver draws on Boyle's treatise on the final causes of natural things, in which *curiosity* is cited as that which ought to motivate the Christian's enquiries into nature:

> For certainly it becomes [Christian philosophers] to have curiosity enough to try at least, whether it can be discovered, that there are any knowable final causes, to be considered in the works of nature. Since, if we neglect this inquiry, we live in danger of being ungrateful, in overlooking the uses of

[21] Steven Shapin and Simon Shaeffer, *Leviathan and the Air-Pump: Hobbes, Boyle, and the Experimental Life* (Princeton, New Jersey: Princeton University Press, 2011).

[22] Latour, *We Have Never Been Modern*, 23.

[23] Latour, *We Have Never Been Modern*, 28–29. For more on this relationship, see Laura Ephraim, *Who Speaks for Nature? On the Politics of Science* (Philadelphia: University of Pennsylvania Press, 2017).

[24] Latour, *We Have Never Been Modern*, 30.

[25] Bruno Latour, *War of the Worlds: What about Peace?*, ed. John Tresch, trans. Charlotte Bigg (Chicago: Prickly Paradigm Press, 2002), 14.

> things, that may give us just cause of admiring and thanking the author of them, and of losing the benefits, relating as well to philosophy as piety, that the knowledge of them may afford us.[26]

This attentiveness is in the service of discerning utility, out of which it was assumed gratitude to God might flow.[27] Oliver describes the curiosity of early modern science as a 'prurient and quizzical gaze', unable to find *intrinsic* purpose or value in its subject, inimical to the kind of studious attention (*studiositas*) which encounters another creature for its own sake.[28] By contrast, early modernity's *curiositas* 'could not be studiously reverent because there was nothing to reverence'.[29] The operative theological assumption is that curiosity gives way to wonder (and as such worship), rather than the pursuit of knowledge working the other way round. For Augustine and Aquinas, fitting enquiry into the nature of creatures emerges out of wonder; Aquinas describes this curiosity as a 'vice', a product of an inappropriate desire for knowledge of sensible things.[30] Unlike *studiositas*, which pursues knowledge 'by reason of the necessity of sustaining nature' or for 'the study of intelligible truth', *curiositas* describes an unhelpful distraction, the pursuit of knowledge which leads to harm, or a desire to know the truth about another creature without due reverence to its *telos* in God.[31] The curiosity of early natural science lacks 'wonder and reverence towards the sheer givenness of being'.[32] Here, a particular narrative concerning the nature of creation (naked utility) is associated with a particular feeling (curiosity), and this feeling subsequently inhibits one's capacity to see the *telos* of the world.

By locating meaning-making in human activity and so treating non-human creation as a passive utility whose mechanisms can be revealed and mimicked, human capacity to know the world truthfully is limited to those investigations which claim to investigate what is presumed to be *simply*

[26] Robert Boyle, *A Disquisition about the Final Causes of Natural Things* (London, 1688), A2. Cited in Simon Oliver, "Life's Wonder," in *Astonishment and Science: Engagements with William Desmond*, ed. Paul Tyson (Eugene, OR: Cascade Books, 2023), 189.

[27] Boyle recommended that experiments be carried out on Sundays, participating in the Church's worshipping life. See Shapin and Shaeffer, *Leviathan and the Air-Pump*, 319.

[28] Oliver, "Life's Wonder," 196.

[29] Oliver, "Life's Wonder," 196.

[30] Oliver, "Life's Wonder," 187–189.

[31] Thomas Aquinas, *The Summa Theologiæ of St Thomas Aquinas*, trans. Fathers of the English Dominican Province, Online (New Advent, 2017), II-II, q. 167, a.2.

[32] Oliver, "Life's Wonder," 184.

there, as distinct from the meanings or desires humans project onto it.[33] Latour identifies two paradoxical practices emerging from this split: imaginative 'purification' (separating humans from non-humans into entirely separate 'ontological zones') and simultaneous 'translation' or 'mediation' (making networks of nature-culture hybrids).[34] This imagined divide between 'things-in-themselves' and 'humans-among-themselves'[35] permits us to operate as if there are separate categories of nature (with science as its intermediary) on the one hand and politics on the other. This collective self-deception, Latour argues, facilitates further translation work; 'the more we forbid ourselves to conceive of hybrids, the more possible their interbreeding becomes'.[36] Two obvious outcomes of these hybrids are climate change and ecological collapse, and Latour uses these as brief illustrative examples in *We Have Never Been Modern*.[37] He then treats them at length in *Facing Gaia*.[38] But here I will focus on Latour's broader

[33] Mechanistic descriptions of nature are often associated with losing intrinsic agency or meaning. But historian of science Jessica Riskin points out that mechanistic descriptions of life were not inevitably associated with passivity. The early modern period produced different approaches to describing life—one passive, one agential (if the world is a clock, is it full of inert or restless parts?). Riskin argues that the passive imagination won out. Her history of competing narratives recalls that a passive/inert description of 'natural things' is as much an act of faith as ascribing agency to living subjects. Riskin argues that the world can and *has been* imagined as full of agents; possessing 'an intrinsic capacity to act in the world, to do things in a way that is neither predetermined nor random... A thing with agency is a thing whose activity originates inside itself rather than outside'. If this is the case, the actions of those agents contain *intention*, which can potentially be read and interpreted by other actors—including us. Jessica Riskin, *The Restless Clock: A History of the Centuries-Long Argument over What Makes Living Things Tick* (Chicago: University of Chicago Press, 2016), 3.

[34] Latour, *We Have Never Been Modern*, 10–11.

[35] Latour, *We Have Never Been Modern*, 5.

[36] Latour, *We Have Never Been Modern*, 12.

[37] 'So long as Nature was remote and under control, it still vaguely resembled the constitutional pole of tradition, and science could still be seen as a mere intermediary to uncover it. Nature seemed to be held in reserve, transcendent, inexhaustible, distant enough. But where are we to classify the ozone hole story, or global warming, or deforestation?... Are they human? Human because they are our work. Are they natural? Natural because they are not our doing. Are they local or global? Both'. Latour, *We Have Never Been Modern*, 50.

[38] Latour describes climate and ecological breakdown (the Anthropocene) as a 'profound mutation in our relation to the world', an expression which in itself betrays our ongoing participation in the modern imagination of a nature-culture distinction. In response, Latour argues that the Anthropocene means nature-culture is one concept divided into two parts—one cannot be described without reference to the other. Latour proposes the word 'world' to hold together nature-culture. I return to a critical evaluation of this proposal later. Bruno Latour, *Facing Gaia: Eight Lectures on the New Climatic Regime* (Polity Press, 2017), 8.

description of attempts to cross this imagined nature-culture divide, the resistance they encounter, and how this illuminates suspicion concerning those claiming to experience climate grief.

Even as our networks of hybrids become larger and more complex, Latour observes that attempts to cross the nature-culture divide are treated as 'just… discourse, representation, language, texts, rhetorics'.[39] It is assumed that if we are not solely talking about some external 'nature of things' *or* solely talking about the 'pragmatic or social context' we are not talking about anything *real*:

> In the eyes of our critics the ozone hole above our heads, the moral law in our hearts, the autonomous text, may each be of interest, but only separately. That a delicate shuttle should have woven together the heavens, industry, texts, souls and moral law—this remains uncanny, unthinkable, unseemly… is it our fault if the networks are *simultaneously real, like nature, narrated, like discourse, and collective, like society*?[40]

If therefore we cannot describe an event or encounter in terms neatly framed by a distinct academic discipline, it cannot or should not be described at all. In *Facing Gaia* Latour outlines the consequences of such a move in climate communication. Successful attacks on climate experts by those profiting from inaction on climate change are rooted in the accusation that the expert has 'crossed the yellow line between facts and values'; they are inventing 'facts' to cover up a set of political desires.[41] The consequence of this manipulation is that those communicating climate science operate as though they cannot make moral or political recommendations and must only communicate pure data, even while this data is repeatedly undermined by an opposing political actor:

> Mr Spock's mechanical voice is not supposed to quaver before the measurements, the alarms, the warnings, and the imputations of responsibility. Yet the climatologists' voice never stopped quavering before discoveries that were all the more awkward in that the experts didn't know how to handle their moral and political charge, even though the implications were quite obvious.[42]

[39] Latour, *We Have Never Been Modern*, 5.
[40] Latour, *We Have Never Been Modern*, 5–6.
[41] Latour, *Facing Gaia*, 27.
[42] Latour, *Facing Gaia*, 28.

Such a hard distinction between data and its social implications is of course pure invention, and despite more recent efforts to overcome this divide, the gap persists. Between 2014 and 2020, Australian science communicator Joe Duggan approached climate scientists for a project called *Is This How You Feel*, asking them 'How does climate change make you feel'?[43] Despite being asked about personal feeling, many respondents turned to scientific data, either explicitly, via statistical information, or implicitly, by referring to 'the science' or 'scientists'. Many responses also reveal an implicit nature-culture divide, despite almost universally appealing for political action. For example, a research fellow at the University of Exeter distinguishes between the *data* itself as not producing an emotional reaction—it exists for its own sake—and the *politics* of climate change, which does:

> I don't really have feelings on the science of climate change, but I do have feelings about the cacophony of opinions and misdirection around the question of "what do we do about climate change?" On this topic, I feel discouraged, frustrated, and powerless.[44]

A senior researcher at the University of New South Wales (UNSW) makes a similar distinction, describing the 'science' as 'intriguing', though the 'debate' was getting 'tiresome' because its motivations are not 'purely science driven'.[45] The 'science' is assumed to be distinct from the messiness of human motivations, agency, and passion: scientific data is 'intriguing' (provoking curiosity) while human response to data is tiresome and discouraging. By the end of their letter, however, the UNSW respondent weakens the distinction by giving science a moral opinion, even if only indirectly. The science 'implies that we should somehow get our act together, the sooner the better, for the prosperity of future generations'.[46] A third respondent—a professor and theme leader on climate change at the University of Exeter—goes even further in differentiating his 'human' response from his response as a 'scientist':

[43] Joe Duggan, "This Is How Scientists Feel" Accessed 15 January 2023, https://www.isthishowyoufeel.com/this-is-how-scientists-feel.html

[44] Duggan, "This Is How Scientists Feel."

[45] Duggan, "This Is How Scientists Feel."

[46] Duggan, "This Is How Scientists Feel." (Emphasis mine).

> As a climate scientist I feel privileged to be alive when things are changing so fast. Humanity is currently carrying out an unintended experiment on the climate system, which means that new questions are being thrown-up continuously and our knowledge is advancing fast. As a research scientist that is exciting!
>
> As a human-being, and especially as a parent, I feel concerned that we are doing damage to the planet. I don't want to leave a mess for my children, or anyone else's children, to clear-up. We are currently creating a problem for them at an alarming rate—that is worrying.[47]

Rather than seeing excitement at doing ground-breaking work and alarm at the findings of that work as both manifestly *human* reactions, here the scientist transcends (or is at least unique amongst) the reactions of ordinary human beings. The appropriateness of any form of sorrow over climate change is restricted to the intra-human (political) realm of concern, while observations about changes in the non-human are restricted to curiosity.

Other contributors express their bewilderment that climate science (unlike other forms of data) is not accepted as truth. One research fellow at Australian National University's School of Earth Sciences writes:

> I feel perplexed at why many of our politicians, business leaders, and members of the public don't get that increased CO_2 in the Earth's atmosphere is a problem. The very premise that CO_2 traps heat is based on fundamental physics—the very same physics that underpins so much of modern society.[48]

This is a prime example of Latour's purification/translation paradox. While 'modern society' is clearly made up of a myriad of nature-culture hybrids, including a warming climate, climate science has been 'purified'—relegated to the realm of resourcing 'nature' as a utility, but not as an entity which can make demands on human decision-making. Even where it is accepted as truthful, it cannot be treated as innately politically relevant. A professor at the University of Lethbridge also appeals to the fundamental reality of 'nature', treating science as a mediating tool which helps us constrain the indifference of a non-human entity: 'listen to the science, *or* to nature. The latter will speak louder, with random and terrible viciousness—storms, heat waves, drought, floods, pollution—all

[47] Duggan, "This Is How Scientists Feel."
[48] Duggan, "This Is How Scientists Feel."

causing pain and suffering'.[49] One of the only responses which directly identifies climate change with ideology comes from an adjunct professor in an explicitly hybrid research context.[50] *Is This How You Feel* offers compelling examples of the ongoing challenge Latour identifies. Even amongst those dedicated to climate communication, something akin to climate grief must be treated in strict distinction from data, which communicates a 'nature' which is neither the product of human narrative nor makes any emergent narrative claims on us. Such an assumption does not, of course, make the hybrid disappear, despite our best efforts to conceal it.[51] To describe climate grief as cultural, then, is simply to acknowledge what was there all along: the context prompting such a response is a network of nature-culture hybrids whose story belongs as much to humans as to any other creature.

Agent

Another (related) way of examining the authoritative nature of climate grief is to consider whether we treat the world as populated by agents—capable of acting on us and thus capable of prompting us to feeling—or whether we assume that non-human existence lacks the agency to *move* us, and thus any sense of being *moved* is pure projection. For Latour, the Nature-Culture divide described above has precisely this latter effect: 'when we claim that there is, on one side, a natural world and, on the other, a human world, we are simply proposing... that an arbitrary portion of the actors will be stripped of all action'.[52] Rather than face awkward questions concerning who or what is speaking when, for example, a scientist communicates data, we make a linguistic move by which the material

[49] Duggan, "This Is How Scientists Feel." (Emphasis mine).

[50] 'Climate change is... ultimately a question of core values.' Adjunct Professor, Fenner School of Environment and Society. Duggan, "This Is How Scientists Feel."

[51] John Milbank writes similarly about denials concerning narrative in the sciences: 'As the phrase "natural history" suggests, natural science does not rid itself of narrative... it is just as possible to tell a story in which the characters are atoms, plants, animals, or quasars, as one where they are human beings. Moreover, these stories are always necessarily—however disguised this may become—stories of our human interrelationships, and our social relationships to the natural world... The gradual isolation of a more rigorous "natural science", in contrast to a vaguer, more speculative "natural philosophy", does not at all indicate success in prescinding from narrative and human relationship, to penetrate to an ontologically immutable level'. Milbank, *Theology and Social Theory*, 269.

[52] Latour, *Facing Gaia*, 58.

world is 'rendered mute', and thus deanimated, separate from the world of human subjects.[53]

Latour uses the Mississippi River as an example of a 'force of nature' which nevertheless possesses agency and imposes its agency on humans.[54] Beginning with the early eighteenth-century settlement of New Orleans, attempts to control the direction and height of the river using levees has created ongoing tension between the needs of human communities living along the river (and 41% of the population of the United States) and the river's own trajectory. As gravity encourages the Mississippi down the Atchafalaya riverbed to the Gulf of Mexico, humans divert water into the Mississippi using floodgates at the Old River Control Structure and upriver defences. If/when the river floods enough that these defences fail (i.e., if/when the river returns to its 'desired' course) the resulting destruction of human communities will be devastating. In his 1987 *New Yorker* essay on the Mississippi, John McPhee tracks the attempts made by humans to control the flow of the river, treating it as a water system to be managed rather than an 'individual' with direction:

As a result of settlement patterns (after Europeans settled by the river), this reach of the Mississippi had long been known as "the German coast," and now... industries were there because of the river. They had come for its navigational convenience and its fresh water. They would not, and could not, linger beside a tidal creek. For nature to take its course was simply unthinkable... Nature, in this place, had become an enemy of the state.[55]

McPhee describes the Atchafalaya—'this most apparently natural of natural worlds' as lying 'between walls, like a zoo. It is utterly dependent on the U.S. Army Corps of Engineers, whose decisions at Old River can cut it dry or fill it with water and silt'.[56] But if the river were not shaping and being shaped by its environment, one might imagine that one-off management would be sufficient: the river could be 'coded' to behave a certain way, and its behaviour would then be predictable. But the river persists in behaving like an agent. McPhee interviews Oliver Houck, a professor of law at Tulane University. Houck observes:

[53] Latour, *Facing Gaia*, 67.

[54] Latour, *Facing Gaia*, 51–54.

[55] John McPhee, "Atchafalaya", *The New Yorker*, 23 February 1987, https://www.newyorker.com/magazine/1987/02/23/atchafalaya

[56] McPhee, "Atchafalaya."

> "The greatest arrogance was the stealing of the sun," he said. "The second-greatest arrogance is running rivers backward. The third-greatest arrogance is trying to hold the Mississippi in place. The ancient channels of the river go almost to Texas. Human beings have tried to restrict the river to one course—that's where the arrogance began."[57]

The Mississippi continues pursuing its desired course, even as it is transformed by human intervention. Shifts in river management strategy also shift human relationship to it. Once it is treated as inert utility, expectations of its behaviour change too:

> In years gone by, when there were no control structures, naturally there were no complaints. The water went where it pleased. People took it as it came. The delta was in a state of nature. But now that Old River is valved and metered there are two million nine hundred thousand potential complainers, very few of whom are reluctant to present a grievance to the Corps. When farmers want less water, for example, fishermen want more, and they all complain to the Corps. In General Sands' words, "We're always walkin' around with, by and large, the black hat on. There's no place in the U.S. where there are so many competing interests relating to one water resource."[58]

However, language describing the river (or nature) as an agent even persists in the rhetoric of those trying to control the river. McPhee goes on to reference 'The Valley of the Giant', a 1940s documentary about the Mississippi River and Tributaries Project produced by the U.S. Army Corps of Engineers. The Mississippi becomes an opponent: 'we are fighting Mother Nature. . . It's a battle we have to fight day by day, year by year; the health of our economy depends on victory'.[59] It is easy to dismiss this as anthropomorphising reporting for dramatic effect—which we had best avoid if our description of the non-human is to be taken seriously. But whether we reject the language of 'Mother Nature' or not, it nevertheless appears that these engineers are describing a dynamic they have encountered: 'on the side of the subject, there is no mastery; on the side of the object, no possible deanimation'.[60]

[57] McPhee, "Atchafalaya."
[58] McPhee, "Atchafalaya."
[59] McPhee, "Atchafalaya."
[60] Latour, *Facing Gaia*, 54.

Latour responds to the accusation of anthropomorphism by proposing that we must see ourselves as inhabiting a *metamorphic* zone, characterised by a series of 'metamorphoses'; agents constantly exchanging properties with each other and being transformed in the process.[61] This zone of 'common exchange' *is* not simply a 'phenomenon of language *about* the world' but 'a property of the world itself'.[62] Rather than seeing human descriptions of agency in the non-human as mere metaphor, a cultural skin between human perception and what is 'really' there, Latour argues that speaking about the non-human is only possible because *signification* is a property of agents:

> As long as they are acting, agents signify. This is why their signification can be followed, pursued, captured, translated, formulated in language. Which does not mean that "every thing in the world is merely a matter of discourse" but, rather, that every possibility of discourse is due to the presence of agents in quest of their existence.[63]

We are not obliged to say that the Mississippi possesses the kind of free will we associate with humans, but we can say it is not inert—it *signifies,* which is another way of saying that it *acts*, and this action can change us, including changing our responses to encountering it.

Sign

A description of nature-culture being composed of agents who can signify to each other requires both a shared reference point which underpins the signs and (relatedly) a basis upon which to trust that the signs can be interpreted truthfully, at least in part.[64] We cannot simply 'reconcile' nature and culture or 'go beyond' the opposition to make peaceful relations

[61] Latour, *Facing Gaia*, 57–58.

[62] Latour, *Facing Gaia*, 69.

[63] Latour, *Facing Gaia*, 70.

[64] Otherwise, we might treat the world as full of agents who each speak a different language—seeking to communicate but unable to be understood. Such an approach may again solely rely upon scientific data to be the translator between agents, assuming that as soon as that data is given *meaning*, the translator has moved further away from truth.

between the two.[65] But we *can* identify a 'common core' which holds the parts together.[66] Latour's proposed shared reference is the term 'world'; 'which opens to the multiplicity of *existents*, on the one hand, and to the multiplicity of ways they have of existing, on the other'.[67] Latour's proposal is a helpful reminder of the particular material biosphere in which we find ourselves, whose present condition is as much the product of human cultural history as it is the product of non-human agents. These agencies are tangled up in each other. They cannot now be teased apart. And yet the entangled materiality of the world as a shared reference point still seems to bring us up short. As Latour points out, by moving from 'nature' to the multiplicity of the 'world', we may simply end up with a 'diversity of cultures', each presenting an unrelated story about the state of the world and human belonging to it.[68] And further: how do we know we can read our shared materiality *well*? How do we trust that data concerning the global average temperature over the last couple of centuries can approximate something true, and even something meaningful to our moral lives? In what follows I propose a theological approach to answering this question.

By framing nature-culture's shared reference point as being that which is created, the signifying capacity of creaturely agents—and our capacity to interpret the signs of other creatures—relies upon trust in God's faithful self-communication. We can read the world because the world is made to be read, and our status as cultural creatures does not alienate us from this condition because we are no less a creature than any other signifying agent. Another way of putting this is that the signs of the world are gifts. All that is created is freely given, and as creatures we therefore trust that we can give freely in communicating ourselves to the world and vice versa. The end of this self-communication is the revelation of truth—creaturely

[65] The problem of trying to 'reconcile' nature-culture can also emerge as a desire to 'return' to a premodern cosmological framework. As Davies notes—we cannot *return* to pre-modernity and must avoid simply indulging in 'cultural nostalgia' for a way of thinking which cannot be sincerely retrieved. However, with Davies, I think it is possible to retrieve something of the theological imagination concerning the world's *meaning* which has underpinned the Christian tradition for most of its existence. It is this retrieval with which I am concerned. Davies, *The Creativity of God*, 6.

[66] Latour, *Facing Gaia*, 19–20.

[67] Latour, *Facing Gaia*, 35.

[68] Latour, *Facing Gaia*, 36–37.

signs are not simply meaningful because we ascribe them meaning, but because they carry intrinsic meaning which we can at least partially discern.

John Milbank makes a very similar argument in responding to modern philosophy's desire to avoid metaphysical questions by concerning ourselves strictly with the 'inertly factual' (paradoxically called that which is 'given').[69] But of course we cannot treat any 'data' as 'inviolable… uncontaminated by synthesis or interpretation or evaluation'.[70] There is nothing, it seems, which is uncontestably given except for our existence. We take it as 'given' that we are beings reflecting on being, and we take it as 'given' that we occupy a 'specific existence in time and space'.[71] The same is true of other creatures, who also appear to us 'via spatial journeyings and temporal advents'—the sunrise and a budding tree arrive before us as though simply given, and that is how we respond to them.[72] We might say that the *givenness* of things—that things are *real*, just as we are, and we trust their continued existence—is distinct from a capacity to communicate or interpret meaning. But Milbank interprets the givenness of things as holding together our trust that other things are real and our trust that they are signifying agents:

> A sign has always a material vehicle, like the person speaking, the medium in which it is inscribed, the actions, place and time that accompany it. The vehicle itself supplements the import of the sign, and not just the next sign to which it gives rise. This ensures that *some* meaning is already realized… a sign proffered by a material someone deploying a material vehicle is not just a sign, it is also a gift. Inversely, a material thing handed over must be also a sign in order to be a gift. So, gift is the exact point of intersection between the real and the signifying.[73]

What, then, might this 'gift' framing make possible for how we perceive our responses to the world? Oliver emphasises the moral import of reading creation as a signifying gift. Nature-culture dualism treats culture as 'the domain of free and creative subjects… intention, purpose, and freedom' and nature as 'brute animal instinct and necessity… an objective domain

[69] John Milbank, "The Gift and the Given," *Theory, Culture and Society* 23, no. 2–3 (May 2006): 444.

[70] Milbank, "The Gift and the Given," 444.

[71] Milbank, "The Gift and the Given," 444.

[72] Milbank, "The Gift and the Given," 444–445.

[73] Milbank, "The Gift and the Given," 446–447.

governed by the laws of nature'.[74] By contrast, a shared creaturely 'gift' identity resists cutting off human culture from nature, and instead 'places moral demands on human agents' because we trust that a real relationship exists between 'giver' and 'recipient'.[75] By seeing the world as a product of gratuitous and divine love, relationships between creatures can be 'characterised as self-donation'; the essence of being a creature is being gift—and so creaturely agents communicate (i.e., *give*) themselves to other creatures. In a theological parallel to Latour's earlier observation concerning agent and signification, Oliver proposes that 'the act of being is the act of self-communication or self-donation'.[76] To be a creature is to signify, and this self-giving is in some sense trustworthy because it reflects God's faithful self-giving in creation. We trust that our encounters with loss—and the feeling they provoke—can reveal truth because our cultural creatureliness (that which humans make, whether meaning or physical structure) is not cut off from the creatureliness of the other agents around us. We also give and receive material signs, and we share an ontology which assures the possibility of relationship: we are given, and so can give to each other. This givenness has consequences for our response to receiving signs from other creatures. The reception of signs from another self-giving creature are not received merely as neutral data which resources curiosity or utility, but as the basis upon which a relationship between self-giving agents can flourish.[77]

Placing trust in the possibility of truthful self-communication is not at odds with acknowledging the partiality or incompleteness of this communication and its reception by other agents. Nathan Lyons addresses this problem in his use of mediaeval semiotics to propose an intimately related

[74] Simon Oliver, "Every Good and Perfect Gift Is from Above: Creation Ex Nihilo before Nature and Culture," in *Knowing Creation: Perspectives from Theology, Philosophy, and Science*, ed. Andrew B. Torrance and Thomas H. McCall, vol. 1 (Grand Rapids, Michigan: Zondervan, 2018), 28–29.

[75] Oliver, "Every Good and Perfect Gift Is from Above," 31.

[76] Oliver, "Every Good and Perfect Gift Is from Above," 35.

[77] Christoph Schwöbel makes a similar observation, reading creation as a divine 'speech-act' with an absolute unity of being and meaning. See Christoph Schwöbel, "'We Are All God's Vocabulary': The Idea of Creation as a Speech-Act of the Trinitarian God and Its Significance for the Dialogue between Theology and Sciences", in *Knowing Creation: Perspectives from Theology, Philosophy, and Science*, ed. Andrew B. Torrance and Thomas H. McCall, vol. 1 (Grand Rapids, Michigan: Zondervan, 2018), 57.

'natural culture' and 'cultural nature'.[78] Drawing on Nicholas of Cusa's distinction between *true* knowledge and *complete* knowledge, culture is understood as an elaboration of nature, in line with the longer tradition of understanding art as a *similitude* or *imitatio* of nature, while also acknowledging the 'determinative' attributes of art in subsequently constituting nature in a mutual fashion.[79] Speech and reasoning offer examples of this process: they are 'arts' which are learned, but each language also has 'a stable nature which becomes natural', and each creature who reasons is exercising a natural capacity of rational inference.[80] Culture, Lyons thus proposes, is a domain 'in which nature and art coincide and mutually create each other'.[81] This mutually constitutive framing of nature and culture challenges the notion that our responses to other creatures are arbitrary. Lyons puts it like this:

> The interpretation of environments that occurs in perception are not then an arbitrary imposition but rather a sort of active evocation of affordances that are in a sense already there in the things… Nature, then, to take up an old metaphor, is a book ready to be read, but its many meanings only come to the fore when the things of nature are set in an actual environmental context and in relation to a particular observer.[82]

We trust that our interpretations have some connection to the way things *are*—we can 'read' the world truthfully—and yet this does not mean that each individual instance of 'reading' is equally authoritative. Rather, the spatial and temporal relationships within which signs take place provide necessary interpretative context. The necessity of relational context is clear in relation to responses to climate change or ecological

[78] Nathan Lyons, *Signs in the Dust: A Theory of Natural Culture and Cultural Nature* (Oxford: Oxford University Press, 2019). He argues that moving away from treating human culture as 'uniquely meaningful' against an 'unmeaning material nature', requires cultural meaning to be understood as being 'at home in natural materiality' (p. 3).

[79] Lyons, *Signs in the Dust*, 75.

[80] Lyons, *Signs in the Dust*, 76. As Lyons later observes, culture is not limited to things like 'speech' or 'texts' whose production we often treat as the product of reason *as opposed to* the body, despite the necessarily embodied expression which both entail. Rather, the signs constituting culture 'includes the whole range of our biological being in the world', and as such 'every aspect of our natural bodily existence contributes to the meanings of our cultural life' (93).

[81] Lyons, *Signs in the Dust*, 76.

[82] Lyons, *Signs in the Dust*, 92.

collapse. We might enjoy a snowy winter's day or a temperate summer evening and receive these as signs that the world is not warming after all. Or we might admire the Lake District as an 'Area of Outstanding *Natural* Beauty' while being unaware that its sheep-denuded hills were, for most of their existence, covered in broadleaf woodland and their current condition is a cultural imposition. Of course, this does not mean that a snowy day, a temperate summer evening or the hills of Cumbria cannot truthfully inspire admiration or contentment, or that these responses bear no relation to what is 'really' there. Rather, we require discernment as to what kind of sign we encounter. We can distinguish between different kinds of signs by drawing on other signs available to us—our experience of the decline in snowy winters over our lifetimes, or instances of woodland returning to hills where sheep are excluded—to come closer to the possibility of more faithful sign-reception.

Sign action can and does offer truthful revelation, but 'there is always the possibility of ever greater precision through the addition of further signs'.[83] Cusa's creaturely semiosis, according to Lyons, treats both 'perceptual signs' (received by the senses) and 'stipulated signs' (those in a particular culture) as *partial* and yet *truthful*, in that it is a truth which "images' its full truth in God... this partial and culturally mediated truth is nonetheless *true*, for in the creaturely sign the infinite meaning of things 'shines forth' in a partial manner'.[84] This partiality is not, however, a static partiality, wherein all sign-making and sign-receiving is equally limited and thus equally limiting for the possibility of relationship. Rather, we can participate in sign-making and sign-receiving which is less like 'a circle of self-referring signs without purchase on the real' and more like 'an upward curve that endlessly approaches the "Preciseness of every nameable name"'.[85] The material nature of our sign-making and sign-receiving is vulnerable to both the corruption of sin and the limitations of physical finitude, but this does not mean that the material nature of our communication and culture is simply a frustration to which we are resigned.[86] To illustrate this point, Lyons compares spiritual and material creaturely communication: angels can 'beam' their thoughts to another spiritual creature

[83] Lyons, *Signs in the Dust*, 65.

[84] Lyons, *Signs in the Dust*, 67.

[85] Lyons, *Signs in the Dust*, 69.

[86] Not all aspects of finitude (e.g., the boundaries of time and space, being *this* kind of creature and not *that* kind) are products of sin, but part of the gift of being a material creature.

using 'perfect' immaterial signs, while we, as material creatures, cannot communicate externally unless we 'make something happen in the real world'.[87] But this 'detouring through the real' is not only a detour through corruption. It is also a detour through *creativity*:

> The material elements of our cultural productions and exchanges constantly mislead, disrupt, decay, and fail... The imperfection of materiality comes, however, with a concomitant blessing, because the constant encounter with the material real makes corporeal culture peculiarly open to creativity.[88]

Here Lyons touches upon similar ground to that articulated by Oliver. Our self-communication does not simply enable the passive reception of signs. Rather, our response to such signs both creatively shapes and is creatively shaped by relationships with other creatures. This creative potential—and responsibility—gives great significance to the narratives about the world and our place in it which we simultaneously encounter and create. The stories we tell about our material lives move us closer to or further away from the reception of truth, and even shape the fabric of reality itself.[89] Dupré articulates this concern in his closing reflection on the consequences of modernity's 'spiritual revolution':

> Ecological concerns as well as scientific theories have forced us to take seriously the idea that reality does not remain indifferent to modes of thinking and feeling. Their correlation appeared less obvious during the early modern period. Yet mental life is as essential a component of the real as neutrons and Milky Ways—and far more powerful in imposing its effect upon other forms of reality.[90]

In changing our understanding of the relationship between the human mind and the cosmos, we have, Dupré argues, changed 'the nature of the real'—'spiritual revolutions transform reality as much as physical changes do'.[91] In receiving the signs of other creatures, then, we not only passively

[87] Lyons, *Signs in the Dust*, 176.

[88] Lyons, *Signs in the Dust*, 177.

[89] For example, the ontological significance of *metaphor* in shaping human relation to other kinds of creatures, *and* the influence metaphor has on how we experience other creatures relating to us—recall, for example, the description of nature as enacting 'random violence' if we do not listen to scientists.

[90] Dupré, *Passage to Modernity*, 251.

[91] Dupré, *Passage to Modernity*, 252.

respond to naked data which we can act on (or not). The narrative that frames our reception of those signs is itself a kind of self-communication which we give back to other creatures. Climate grief is not only a (more-or-less) faithful response to a revelation of truth—we are not simply called to passively grieve, but to actively offer a narrative which more closely reflects the truth of the signs we have received. We are to tell a story which is as faithful as possible to our trust in God's self-revelation.

Story

In *Theology and Social Theory* Milbank argues that narrative is not a category limited to a 'special "human sphere"', but 'the mode in which the entirety of reality presents itself to us'.[92] Milbank proposes that the 'facts and motions' which we consider 'stable' or 'isolatable' are actually always presenting themselves to us as 'meanings'.[93] Rather than assuming nothing really exists apart from human perception or discourse, here nature and culture constantly make each other. The stories we tell are porous and plastic, open to the influence of other sign-making creatures. This does not mean that we cannot judge the truthfulness of these stories. As I have proposed, we receive a governing narrative—being a creature, and as such participating in gift—which guides our interpretations of the world and assures us that we can trust the sign-making agency of other self-giving creatures.

Milbank goes further in teasing out the implications of identifying specifically with the governing narrative of Christian theology. We not only receive meaning, but we also participate in *peace making*. It is out of peace that creation is freely given, and it is to peace that creation moves.[94] If the meaning-making of climate grief can be considered authoritative within this governing narrative of a peaceably offered creation, we are not merely passive observers of violence. We also receive signs of a particular relational obligation, respond in peace, and trust that this relational giving is itself *significant* of and for redemption. Along a similar vein, Rowan Williams describes sign-making as 'the action of hope'—specifically, 'the hope that this world may become other and that its experienced fragmentariness can

[92] Milbank, *Theology and Social Theory*, 362.

[93] Milbank, *Theology and Social Theory*, 362. Milbank describes natural scientific knowledge as a 'mode of narration', akin to narrated social knowledge (263).

[94] Milbank, *Theology and Social Theory*, 440.

be worked into sense'.[95] Our willingness to participate in grief over the loss of the world we have known can be an attentive receptivity to the significance of other creatures, and as such an action of hope.

Bibliography

Aquinas, Thomas. 2017. *The Summa Theologiæ of St Thomas Aquinas. Trans. Fathers of the English Dominican Province*, Online. New Advent.

Bacon, Francis. 2009. In *The Instauratio Magna Part II: Novum Organum and Associated Texts*, ed. Graham Rees, Maria Wakely, and Oxford Francis Bacon. Oxford: Clarendon Press.

Boyle, Robert. 1688. *A Disquisition about the Final Causes of Natural Things*. London.

Cone, James H. 2000. Whose Earth Is It Anyway? *CrossCurrents* 50 (1/2): 36–46.

Cunsolo, Ashlee, and Karen Landman, eds. 2019. *Mourning Nature: Hope at the Heart of Ecological Loss and Grief.* Montreal/Kingston: McGill-Queen's University Press.

Davies, Oliver. 2004. *The Creativity of God: World, Eucharist, Reason. Cambridge Studies in Christian Doctrine*. Cambridge: Cambridge University Press.

Duggan, Joe. *This Is How Scientists Feel*. Accessed 15 January 2023., https://www.isthishowyoufeel.com/this-is-how-scientists-feel.html

Dupré, Louis. 1993. *Passage to Modernity: An Essay in the Hermeneutics of Nature and Culture*. New Haven and London: Yale University Press.

Ephraim, Laura. 2017. *Who Speaks for Nature? On the Politics of Science*. Philadelphia: University of Pennsylvania Press.

Harrison, Peter. 1998. *The Bible, Protestantism, and the Rise of Natural Science*. Cambridge: Cambridge University Press.

Keller, David. 2010. *Environmental Ethics: The Big Questions*. Hoboken: Blackwell Publishing.

Klein, Naomi. 2014. *This Changes Everything: Capitalism vs. The Climate*. London: Penguin Books Ltd.

Latour, Bruno. 1993. *We Have Never Been Modern. Trans. Catherine Porter*. Cambridge, Massachusetts: Harvard University Press.

———. 2017. *Facing Gaia: Eight Lectures on the New Climatic Regime*. Cambridge: Polity Press.

Lyons, Nathan. 2019. *Signs in the Dust: A Theory of Natural Culture and Cultural Nature*. Oxford: Oxford University Press.

McPhee, John. 1987. Atchafalaya. *The New Yorker*. https://www.newyorker.com/magazine/1987/02/23/atchafalaya

[95] Rowan Williams, *On Christian Theology* (Malden, MA: Blackwell Publishing, 2000), 224.

Milbank, John. 2006. *Theology and Social Theory: Beyond Secular Reason*. Hoboken: Blackwell Publishing.

———. 2014. *Beyond Secular Order: The Representation of Being and the Representation of the People, Illuminations: Theory and Religion Series*. John Wiley & Sons.

Northcott, Michael. 2013. *A Political Theology of Climate Change*. Grand Rapids: W. B. Eerdmans.

Ojala, Maria (et al). "Anxiety, Worry, and Grief in a Time of Environmental and Climate Crisis: A Narrative Review." Annual Review of Environment and Resources 46, 35–38 (2021).

Oliver, Simon. 2018. Every Good and Perfect Gift Is from Above: Creation Ex Nihilo before Nature and Culture. In *Knowing Creation: Perspectives from Theology, Philosophy, and Science*, ed. Andrew B. Torrance and Thomas H. McCall, vol. 1, 28–29. Grand Rapids, Michigan: Zondervan.

———. 2023. Life's Wonder. In *Astonishment and Science: Engagements with William Desmond*, ed. Paul Tyson. Eugene, OR: Cascade Books.

Pörtner, H.-O., et al. 2022. IPCC, 2022: Summary for Policymakers. *Climate Change 2022: Impacts, Adaptation, and Vulnerability*. Cambridge: Cambridge University Press.

Riskin, Jessica. 2016. *The Restless Clock: A History of the Centuries-Long Argument over What Makes Living Things Tick*. Chicago: University of Chicago Press.

Schwöbel, Christoph. 2018. 'We Are All God's Vocabulary': The Idea of Creation as a Speech-Act of the Trinitarian God and Its Significance for the Dialogue between Theology and Sciences. In *Knowing Creation: Perspectives from Theology, Philosophy, and Science*, ed. Andrew B. Torrance and Thomas H. McCall, vol. 1. Grand Rapids, Michigan: Zondervan.

Shapin, Steven, and Simon Shaeffer. 2011. *Leviathan and the Air-Pump: Hobbes, Boyle, and the Experimental Life*. Princeton, NJ: Princeton University Press.

White, Lynn. 1967. The Historical Roots of Our Ecologic Crisis. *Science* 155 (3767): 1203–1217.

Williams, Rowan. 2000. *On Christian Theology*. Malden, MA: Blackwell Publishing.

CHAPTER 8

One Reality, Not Two: Bonhoeffer, Jesus Christ, and a Membraned World

Lisa E. Dahill

Introduction

For some time I have been exploring what happens when Christian prayer moves outdoors, into engagement with the more-than-human natural world. Living in a time of urgent ecological crisis, Christians need outdoor ritual experience of our faith: of what is wild, of the living Earth, stranger faces of the divine: taking eco-alienated people beyond the walls and into the river, the streets, the forest. We inhabit separation, alienation extending out into the world's desertification and bleaching, the unfolding climate catastrophe already killing vulnerable humans and species and entire biotic worlds, destabilizing or destroying all that we love ... as religious attention to what is holy still orients itself elsewhere.

I am grateful for a generous Sabbatical Grant for Researchers from the Louisville Institute, which made possible my work on this chapter.

L. E. Dahill (✉)
Hartford International University for Religion and Peace, Hartford, CT, USA
e-mail: ldahill@hartfordinternational.edu

T. J. Hastings, K.-W. Sæther (eds.), *Views of Nature and Dualism*,
https://doi.org/10.1007/978-3-031-42902-6_8

We need resources in every tradition that help knit back together this primal cognitive and interpretive splitting, the holy from the world. And so we need Christian resources too, not only Buddhist or Daoist or indigenous.[1] My work explores a non-dualistic reading of Dietrich Bonhoeffer to challenge dominant dualistic Christian readings of reality. I first situate this rejection of God-world dualism in Bonhoeffer's *Ethics* within his lifelong repudiation of what he considers abstractions. Next, I outline a methodology for reading Bonhoeffer ecologically based on his insistence in *Ethics* that God and world are one reality in Jesus Christ, not two. That is, I take this core conviction of the *Ethics* as a lens for interpreting Bonhoeffer himself, aiming to surface pieces of his thinking that contribute to a larger non-dualistic vision of God and world in Jesus Christ and to critique aspects of his thinking that fall into the dualism he will later reject. I enact this proposed methodology via an ecologically oriented reading of Bonhoeffer's *Creation and Fall* lectures of 1933, where he still takes for granted a strong God-world dualism, and of material from elsewhere in his writings that points to non-dual images of divine presence. In this way, I show how reading according to the criterion he himself lays out in his *Ethics* can generate an ecologically fruitful theology of God-world reality in Jesus Christ.

[1] This chapter does not engage the subject-object or self-other dualisms central in Buddhist, Daoist, and Hindu teaching. See David R. Loy, *Nonduality in Buddhism and Beyond* (Somerville, MA: Wisdom Publications, 1988). The terms "non-dualism" and "non-duality" are sometimes used interchangeably. Studies of Western thought tend to speak of "non-dualism," while those of Eastern philosophy or spiritual traditions often refer to "non-duality." I understand ***non-dualism*** to constitute a claim about what exists, namely just one [kind of] "thing," even as that may still include complex forms of differentiation. By ***non-duality***, I understand the rejection of binary thinking, often because the situation with regard to the terms identified (such as, e.g., gender) is plural, not binary. I use "non-dualism" in this project because I read Bonhoeffer allowing for differences in properties between "God" and "world" but asserting their shared reality. I also differentiate this dualism from indigenous forms of thought centering in complementary binary forms. The work of Seneca scholar Barbara Mann, in *Iroquoian Women: The Gantowisais* (New York: Peter Lang, 2000), lays out traditional Iroquoian worldviews regarding gender and other forms of binary complementarity. On this see also the work of George ("Tink") Tinker, "Why I Do Not Believe in a Creator," in *Buffalo Shout, Salmon Cry: Conversations on Creation, Land Justice, and Life Together*, ed. Steve Heinrichs (Waterloo: Herald Press, 2013).

Bonhoeffer

Ethics

Over the course of decades of scholarship and translation of Bonhoeffer, I have returned repeatedly to the vision he lays out at the end of the first chapter of his *Ethics*. From the heart of Nazi Germany, that shorthand for evil itself, Bonhoeffer insisted on the fundamental unreality of dualistic distinctions. A dualism of "realms," he writes, is the sickness infecting most of Christian history, a "Colossus obstructing our way."[2] The dominant theological conception after the New Testament has been the.

> basic conception [of] two realms [that] bump against each other: one divine, holy, supernatural...; the other worldly, profane, natural .... Reality as a whole splits into two parts, and [thus] .... the concern of Christ becomes [merely] a partial, provincial affair within the whole of reality. One reckons [that is] with realities outside the reality of Christ.[3]

That entire conception is false, he asserts, destroying the logic of the incarnation itself. There can exist nothing that is not ultimately contained already in the reality of Jesus Christ uniting God and world: "God and the world are enclosed in this name."[4] That is the point of the religion. For Bonhoeffer, "reality" (*Wirklichkeit*) is visible only when this fundamental unity of God and the world in Jesus Christ is perceived, beheld. "There are not two realities, but only *one reality*, and that is God's reality revealed in Christ in the reality of the world."[5] In fact, any attempt to think of "God" and "world" as separate, apart from this one reality of Christ, creates an abstraction, in his terms.[6] We can see the truth of his assertion as we notice how the accelerating perceptual separation of holiness from world in the West leads to the death of each: first the death of God in recent centuries, now the death of the world too unfolding before our eyes. Today's twin ecological and spiritual crises confirm Bonhoeffer's conviction that neither holiness ("God") nor world can exist apart from one another. He saw the

[2] Dietrich Bonhoeffer, *Ethics*, ed. Clifford J. Green, trans. Reinhard Krauss, et al., Dietrich Bonhoeffer Works (DBWE), volume 6 (Minneapolis: Fortress Press, 2005), 55.

[3] DBWE 6:56–57.

[4] DBWE 6:54.

[5] DBWE 6:58.

[6] DBWE 6:54.

deeply entrenched tendency to conceive of reality as split, "God" separate from "world," as a millennia-long perceptual error creating abstractions on both sides: a God increasingly un-credible to postmodern sensibilities in the "world come of age," and a Nazified, demonized, commodified, desecrated view of the human world and biological life stripped of their beauty, holiness, sacredness: a scarred world. A different way of articulating what I hear in Bonhoeffer's insight is that stripping transcendence from the world and locating it solely in a God conceived as separate from the world eventually extinguishes the human capacity to perceive and participate coherently in what David Abram refers to as the "more-than-human world"[7]—for such coherent participation in that larger world depends on the human perception of its ongoing transcendence of oneself, its mystery and sublimity: its holiness. The ultimate result of this process is that the holy dies—for this separative "God" doesn't exist—*and so does the world* robbed of its intrinsic holiness, made to be purely commodity. God-world dualism denies reality.

With this articulation of God-world dualism as a form of abstraction (in splitting apart aspects of reality that belong together) that is also hierarchically inflected (in that "God" is viewed as inherently superior to "world"), Bonhoeffer writes in a way congruent with the later work of Australian environmental philosopher Val Plumwood. In *Feminism and the Mastery of Nature,* Plumwood defines dualism as not just splitting or binary thinking in itself but a form of splitting that also includes value hierarchy: dualism is "an *alienated* form of differentiation, which construes and constructs difference in terms of an inferior and alien realm."[8] The problem is not the existence of distinction or difference themselves but division into fundamentally separate and differently valued realms of what belongs together as *one* reality—here, God as the holiness *of* the world, not a reality separate from it. She goes on:

> dualism is a relation of separation and domination inscribed and naturalized in culture and characterized by radical exclusion, distancing and opposition between orders constructed as systematically higher or lower, as inferior and superior, as ruler and ruled, which treats the division as part of the natures

[7] David Abram, *The Spell of the Sensuous: Perception and Language in a More-than-Human World* (New York: Vintage Books, 1996).

[8] Val Plumwood, *Feminism and the Mastery of Nature* (London and New York: Routledge, 1993), 42.

of beings construed not merely as different but as belonging to radically different orders or kinds, and hence as not open to change.[9]

Plumwood is speaking of the many forms of dualism creating value-laden and reified patterns of alienation between human beings around differences conceived as binary, such as gender, race, class, and the human relation to the natural world. But her definition holds as well for that theologically paradigmatic relation of "higher or lower, ... inferior and superior, ... ruler and ruled," namely that of a god imaged as the apotheosis of ruling reality and defined in absolute abstraction from the rest of reality, the "world."

By the time of his *Ethics*, Bonhoeffer has come to believe that the truly Christian worldview has been obscured via centuries of enmeshment in such dualism. Faith, he now asserts, is what makes possible Christian perception of and participation in the "Christ-reality": God and world reunited. Reading his work ecologically today, I hear his *Ethics* inviting us into a vision in which Jesus Christ—the bone-deep interweaving of divine and human biotic natural life in the incarnation—mends God-world dualism from both directions, both reclaiming the holiness of the natural world *and* opening to new forms of post-theism. Bonhoeffer's prison letters sketch fundamental shifts this insight demands, as he writes of learning to live in the world come of age *etsi deus non daretur*, without the illusion of God as working hypothesis or, in the terms of his *Ethics*, "God" understood as separate from the world, "God" as abstraction. Letters to his physicist brother Karl-Friedrich show him thinking in ways that anticipate science/theology work decades ahead and resonant with new materialist discourse, or Jane Bennett's proposal of "vibrant matter," *all that is* participating in emergent quantum weirdness, entangled and alive all the way down.[10] God and world, one reality.

[9] Plumwood, 47–48.

[10] Jane Bennett, *Vibrant Matter: A Political Ecology of Things* (Durham, NC: Duke University Press, 2010). Also cf. Larry Rasmussen, "The Brothers Bonhoeffer on Science, Morality, and Theology," *Zygon* 44/1 (March 2009): 97–113. For citations of selected New Materialist writings, see note 16.

Bonhoeffer's Lifelong Insistence on Concreteness

This ringing rejection of God-world dualism is the opening salvo of his *Ethics*, meant as his life's integrated theological statement and is of a piece with lifelong currents in Bonhoeffer's thinking repudiating what he also calls at various points ideals, a priori principles, or general truths, as invalid forms of theological thought. Only in the *Ethics* does he explicitly include theologies of radical distinction between God and world within this analysis. But if the subject matter of his analysis in the *Ethics* (God and world as "one reality, not two") is new, the logic he employs in the process is lifelong in him, visible from his first sustained academic project in *Sanctorum Communio*.[11] There he calls for a theology of human personhood rooted in the concrete givenness of every Thou, contra the traditions of German ethical idealism that ground personhood in ideas of universal human reason. Similarly, in *Life Together* he rejects any privileging of "ideals" over the concrete reality of the community and individuals God has given, along with attempts to know via a priori general rules or principles how a given person ought to be treated; he formulates his well-known critique of "cheap grace" in *Discipleship* precisely as an attack on "grace as doctrine, as principle, as system ... forgiveness of sins as a general truth"; and later in his *Ethics* he explores at length the question of how, without such a priori rules to guide one's decisions, one might in times of tremendous ethical ambiguity discern the concrete, ever newly emergent will of God.[12] Indeed, the very opening lines of his *Ethics* announce this paradigmatic refusal of general truths in favor of concrete, discerned reality as he asserts,

> Those who wish even to focus on the problem of a Christian ethic are faced with an outrageous demand—from the outset they must give up, as inappropriate to this topic, the very two questions that led them to deal with the ethical problem: "How can I be good?" and "How can I do something

[11] I trace this consistent repudiation of abstract principles and generalizations through key pieces of Bonhoeffer's writing (*Sanctorum Communio, Discipleship, Life Together,* and *Ethics*) in *Reading from the Underside of Selfhood: Bonhoeffer and Spiritual Formation,* Princeton Theological Monograph Series (Eugene: Wipf & Stock, 2007), 32–33, 83–92.

[12] See *Sanctorum Communio* (DBWE 1), 34–57; *Life Together,* ed. Gerhard Müller, Albrecht Schönherr, and Geffrey B. Kelly, trans. Daniel Bloesch and James H. Burtness, DBWE 5 (Minneapolis: Fortress Press, 1996), 34–36, 94–95; *Discipleship,* ed. Martin Kuske, Ilse Tödt, Geffrey B. Kelly, and John D. Godsey, trans. Barbara Green and Reinhard Krauss, DBWE 4 (Minneapolis: Fortress Press, 2001), 43, 53, 59, 81; and *Ethics* (DBWE 6):320–326.

good?" Instead they must ask the wholly other, completely different question: what is the will of God?[13]

Clearly Bonhoeffer rejects abstractions of all kinds throughout his life. What is new in the *Ethics* is his explicit inclusion of God-world dualism within this rejection of abstraction. His sustained denunciation of such dualism precisely here at its outset frames his *Ethics* as an attempt to take seriously and theologically this Christian insistence that in Jesus Christ God and world are revealed as one reality, and that the gift and task of Christian life is precisely to live in that reality. Even in the 1940s, his life at mortal risk in participating in the attempt to overthrow the Nazi government, his is not a good-versus-evil spirituality, the church a beleaguered citadel raging against the darkness—quite the contrary. What his *Ethics* offers and invites readers into is the radiant vision of the world "always *already* borne, accepted, and reconciled."[14] Within this utterly loved world, the church is those who perceive and live now in this already-accomplished redemption and who are inviting others also to live in reality. Two quotes from this section show Bonhoeffer's thinking:

> For it is [precisely] the "evil world" that is reconciled to Christ in God and has its ultimate and true reality not in the devil but, again, in Christ. The world is not divided between Christ and the devil; it is completely the world of Christ, whether it recognizes this or not. As this reality in Christ it is to be addressed (DBWE 8:65).

> In the body of Jesus Christ, God took on the sin of all the world and bore it .... Whoever perceives the body of Jesus Christ in faith can no longer speak of the world as if it were lost, as if it were separated from God.... [The role of the church-community is that of] calling the world into the community [Gemeinschaft] of the body of Christ to which the world in truth already belongs (DBWE 8:67).

His incarnational logic perceives the flesh of Jesus, that *sarx* (John 1:14) the church proclaims as fully matter and fully God, present in the world's own being today: encountered in and as the incarnate, crucified, and risen body of Jesus Christ himself. The vision faith makes possible is precisely this: "Whoever perceives the body of Jesus Christ in faith can no longer speak of the world as if it were ... separated from God."

[13] DBWE 6:47. On the centrality of discernment for Bonhoeffer and his spirituality, cf. *Reading from the Underside of Selfhood*, 87–92, as well as 205–207, 210–211, 217–222.

[14] DBWE 6:55.

Bonhoeffer's rejection of God-world dualism thus represents, I assert, his recognition that the logic of his lifelong suspicion of ideals and insistence on the concreteness of divine presence and revelation (as well as the logic of participation in a conspiracy demanding high levels of ethical humility, discernment, and courage) entails suspicion as well of traditional notions of "God" as separate from the world, the omnipotent puppet-master he will condemn in his prison letters not much later. As the essays in this volume and the deadly effects of divorcing God from world are making urgently clear today, an extension of this theology into the explicitly ecological realm is the move we need today. With the vision in his *Ethics*, I hear Bonhoeffer providing the foundation of a non-dualistic Christian ecospirituality.

Methodology

Therefore I propose taking seriously Bonhoeffer's assertion of God and world as "one reality" in Jesus Christ. That is, I propose developing it as a lens by which to critique inadequate theological assertions still reliant on theisms that divorce God from world and, more importantly, to surface and build on aspects of Bonhoeffer's spirituality—and, ultimately, other writings and practices within Christian spirituality—that ground a world-embracing, thoroughly entangled, indeed animist non-dual spirituality and thereby contribute to the necessary re-visioning today of what being Christian means. What emerges when reading Bonhoeffer himself through the lens of his rejection of dualism and with an eye to "God" and "world" being *one reality ecologically* in Jesus Christ throughout?

In this process I build on others' work. Environmental ethicist Lisa Sideris was an early voice in theological ethics to call for theology to reflect scientific insight much more rigorously, from biology in particular. With her I interpret Bonhoeffer's "one reality" as the conviction of fundamental coherence between theology and the biological sciences.[15] The New

[15] Lisa Sideris, *Environmental Ethics, Ecological Theology, and Natural Selection* (New York: Columbia University Press, 2003), and *Consecrating Science: Wonder, Knowledge, and the Natural World* (Berkeley: University of California Press, 2017). I rely on this way of thinking in recent work rethinking Jesus' resurrection in light of the bodily reality of decomposition and reincorporation into new creatures of all kinds: Lisa E. Dahill, "Eating and Being Eaten: Interspecies Vulnerability as Eucharist." *Religions* 204 (2020), 11 pages: https://doi.org/10.3390/rel11040204. Sideris's later book cautions against an overly deferential approach to science on the part of religion and the humanities and the crucial insight that the sciences too are interpretive projects that need their hermeneutical biases checked by theology and the humanities more broadly in turn.

Materialists (broadly) and thinkers like political philosopher Jane Bennett, eco-cultural theorist Donna Haraway, and anthropologists Tim Ingold and Eduardo Kohn are creating profoundly integrative forms of thought in diverse disciplines, each of which rejects a dualism of spirit from matter, creating powerful new forms of language and conception to try to bring into awareness the comprehensive entanglement and inseparability of all that is.[16] In so doing they approach and in various ways recognize the hermeneutical priority of the worldviews of indigenous scholars and practitioners, whose conceptions of spirits and beings, ancestors and persons, represent dense, rich, complex webs of relationships worlds removed from the stark God-world binary of Western thought.[17] Phenomenologist David Abram is perhaps the most thoroughgoing thinker I know in the depth and precision of his observations into the physical, sensory foundations of human perception and language. Regarding the intellectual process that gives rise to his writing, Abram notes:

> It is a way of thinking that strives for rigor without forfeiting our animal kinship with the world around us—an attempt to think in accordance with the senses, to ponder and reflect without severing our sensorial bond with the owls and the wind.[18]

[16] Cf., e.g., Catherine Keller and Mary-Jane Rubenstein, eds, *Entangled Worlds: Religion, Science, and New Materialisms,* Transdisciplinary Theological Colloquia (New York: Fordham University Press, 2017); Donna J. Haraway, *Staying with the Trouble: Making Kin in the Chthulucene* (Durham, NC: Duke University Press, 2016); Eduardo Kohn, *How Forests Think: Toward an Anthropology Beyond the Human* (Berkeley: University of California Press, 2013); Tim Ingold, *Imagining for Real: Essays on Creation, Attention and Correspondence* (London: Routledge, 2021).

[17] See, e.g., Workineh Kelbessa, "Indigenous Environmental Philosophy," in *The Oxford Handbook of World Philosophy,* ed. Jay L. Garfield and William Edelglass (New York: Oxford University Press, 2011); George ("Tink") Tinker, "Why I Do Not Believe in a Creator," in *Buffalo Shout, Salmon Cry: Conversations on Creation, Land Justice, and Life Together,* ed. Steve Heinrichs (Waterloo: Herald Press, 2013), 167–179; Robin Wall Kimmerer, *Braiding Sweetgrass: Indigenous Wisdom, Scientific Knowledge, and the Teachings of Plants* (Minneapolis, MN: Milkweed Editions, 2013); and chapters by Michael Onyebuchi Eze ("Humanitatis-Eco (Eco-Humanism): An African Environmental Theory") and Edwin Etieyibo ("Ubuntu and the Environment") in *The Palgrave Handbook of African Philosophy,* eds. Adeshina Afolayan and Toyin Falola (New York: Palgrave Macmillan, 2017).

[18] Abram, *Spell of the Sensuous,* 264.

To "think in accordance with the senses, to ponder and reflect without severing our sensorial bond with the owls and the wind"—this is the kind of thinking I hear an ecologically encompassing version of Bonhoeffer's vision calling for today and that am attempting here. This unitive worldview draws us not only out into expansive forms of ecological and intersectional thought and analysis but also more and more fully "down," into three-dimensioned multi-species interbeing with countless others right here in our given place.[19] This movement into place (as the ecological or interspecies contextual specification analogous to and expansive of Bonhoeffer's centering on particular concrete inter-human contexts), along with Abram's and Sideris's insistence above on theologies that illuminate and do not attempt to contradict biological truth-telling, describe the central features of how I understand Bonhoeffer's insistence on "one reality" ecologically.[20]

Finally, I experience this "one reality" approach also as an expression of the Lutheran insistence on the priority of the literal: that is, the assertion that symbolic or allegorical theological reflection (in Luther's case, on the biblical text) must cohere with, not contradict, and in many cases surrender to the priority of the literal meaning of a given text, that which is accessible to any reader.[21] So too here I assume a strong coherence between natural and revealed theology parallel to that between literal and symbolic scriptural meanings such that, in some sense (again, as noted earlier), biology and theology are equal partners. Thus, in my project's methodology—thinking with Bonhoeffer's insistence on God and world as one

[19] The language of *interbeing* comes from the work of Thich Nhat Hanh (see, e.g., his thinking collected in *Zen and the Art of Saving the Planet,* ed. and with commentary by Sister True Dedication (New York: HarperOne, 2021)), whose language and spirituality of interbeing is compatible with the vision I am attempting to generate from Bonhoeffer's insight.

[20] For more on the hermeneutical importance of place in reading Bonhoeffer ecologically, see Lisa E. Dahill, "Firm Grounding: Bonhoeffer, Place, and Coming Home to Reality," *Bonhoeffer and Eco-Theology. The Bonhoeffer Legacy: An International Journal* 8/2 (2022): 21–42; and "Rewilding Life Together: Bonhoeffer, Spirituality, and Interspecies Community," *Dialog: A Journal of Theology* 61/2 (2022): 1–11.

[21] Edward T. Oakes, "The Paradox of the Literal: The Voice of Canon Criticism in Reformation and Counterreformation Polemics," in *Reform and Counterreform: Dialectics of the Word in Western Christianity Since Luther,* ed. John C. Hawley (Berlin and New York: Morton de Gruyter, 1994), 15–29; Henry John Langknecht, *Giving Scripture Its Voice: The Tensive Impertinence of The Literal Sense of The Pericope, Metaphorical Meaning-Making, and Preaching the Word of God* (Th.D. diss., Emmanuel College of Victoria University and the University of Toronto, 2008).

reality, in particular thinking theologically inseparably from biological reality—I am attempting to enact Bonhoeffer's non-dualistic Lutheran vision.

This project includes a critical dimension of pointing to elements of Bonhoeffer's thinking that still perpetuate God-world dualism and thereby, along the lines of his analysis in *Ethics,* create abstractions of each. Its more creative dimension, however, is the second piece, namely highlighting and developing ecologically less-dualistic elements of Bonhoeffer's thinking. Here I develop ecological implications from these forms of Bonhoeffer's thinking in ways that honor his *Ethics'* non-dualism but go beyond what he himself articulated. Throughout, I do my best to be faithful to Bonhoeffer's thinking on its own terms and to do him justice; but my primary concern is not exploring his thinking itself. Rather, it is taking seriously his rejection of God-world dualism in favor of a theology of "one reality" in Jesus Christ and exploring how this commitment works as an eco-theological methodology. Ultimately this chapter's goal is to contribute toward the thinkability of a Christian non-dualism compatible with biological reality and useful for generating healthy human forms of life together with the divine goodness of the world itself, in all its complexity and particularity.

Critiquing Bonhoeffer's God-World Dualism: *Creation and Fall*

First, I use Bonhoeffer's non-dualism in his *Ethics* to critique his own God-world dualism elsewhere. Here I point to his winter semester 1932–1933 lectures for a course titled *Creation and Sin: A Theological Exposition of Genesis 1–3* at the University of Berlin, later published as *Creation and Fall.*[22] A longer, forthcoming version of this analysis works with these lectures to show also ecologically useful pieces of Bonhoeffer's thought in this material, such as his positive views of embodiment or his critique of "two-ness," *Entzweiung,* as the result of the fall—both important foundations of his later explicit non-dualism. But here I focus on places where Bonhoeffer exhibits the kind of thinking he will later critique as "abstraction."

[22] *Creation and Fall: A Theological Exposition of Genesis 1–3,* ed. Martin Rüter, Ilse Tödt, and John W. de Gruchy, trans. Douglas Stephen Bax, DBWE 3 (Minneapolis: Fortress Press, 1997). This was the period in which Hitler came to power, in January 1933.

Primal Splitting: God from World

In his exposition of Genesis 1, Bonhoeffer repeatedly points to the absolute and radical disjunction between God and the world. The following material is representative:

> In opposition to [extra-biblical] myths … the God of the Bible remains wholly God, wholly the Creator, wholly the Lord, and what God has created remains wholly subject and obedient, praising and worshiping God as Lord. God is never the creation but always the Creator. God is not the substance [Substanz] of nature …. God is never in the world in any other way than as one who is utterly beyond [jenseits] it.[23]

This quote and others like it through this section make clear that Bonhoeffer is assuming and upholding a perspective congruent with what he had been soaking up from his primary theological mentor through this period, namely Karl Barth, in which "God" and "world" name realities conceived as radically different from one another.[24]

[23] DBWE 3:40–41.

[24] Barth's strong God-world dualism is well known: it is starkest in his 1928 *Epistle to the Romans,* complexified in his *Church Dogmatics,* and softened slightly (in part from an appreciation of Bonhoeffer's increasingly incarnational theology) by the time of his death in 1968. On the first page of his treatment of creation in *Church Dogmatics* III/1, for example, Barth writes, "Though [the Apostles' Creed] speaks of God, [it does] not speak only of God, but also of *a reality which is distinct from God,* i.e., of heaven and earth as the two great distinctive but related spheres, intersecting in man [sic], of the whole being of the "world" as it exists apart from God. [It says] that He [sic] who alone is God the Father Almighty is not alone. And in order not to be alone and to have *this other quite different reality* before, with and near Him [sic], He [sic] deliberately gave it an existence and definite form" (Karl Barth, "Faith in God the Creator," §40, in Chapter IX, "The Work of Creation, *Church Dogmatics* III/1, ed. G. W. Bromiley and T. F. Torrance, trans. J. W. Edwards, et al. (London: T&T Clark [1945], 1958), 3, *italics mine*). Barth's language here of "a reality … distinct from God" and "this other quite different reality" shows the difference of his perspectives from Bonhoeffer's *Ethics,* written at around this same period, of God and world as "one reality," not two. Bonhoeffer had not read CD III/1, published in 1945; the final volume he was able to read was II/2 in 1942. Bonhoeffer's reliance on and differentiation from the work of Karl Barth is a much-analyzed topic. See, e.g., Andreas Pangritz, *Karl Barth in the Theology of Dietrich Bonhoeffer*, trans. Barbara Rumscheidt and Martin Rumscheidt (Grand Rapids: William P. Eerdmans Publishing Company, 2000), along with his essay, "Dietrich Bonhoeffer: 'Within not Outside the Barthian Movement'," in *Dietrich Bonhoeffer's Intellectual Formation,* ed. Peter Frick (Eugene: Wipf & Stock, 2017), 245–282; and Michael P. DeJonge, *Bonhoeffer's Theological Formation: Berlin, Barth, and Protestant Theology* (New York: Oxford University Press, 2012). Jens Zimmermann frames Bonhoeffer's non-dualistic perspective in the *Ethics* within a larger consideration of Barth and other twentieth-century Protestant and Catholic treatments of the nature/grace question in *Dietrich Bonhoeffer's Christian Humanism* (New York: Oxford University Press, 2019), especially 24–35 and chapter 8, "Recovering the Natural: Humanizing Politics," 291–330.

On the same page, Bonhoeffer does note a point of connection between these two otherwise utterly disjunctive spheres, namely the word: "There is no continuum that ties God to, or unites God with, God's work—except God's *word*.... [W]ere the word not there, the world would drop into a bottomless abyss."[25] This exception Bonhoeffer points to, namely the divine word as the sole locus of connection that "unites God with God's work," remains undeveloped in these lectures. Even in his Christology lectures of the same period he does not dwell long on the Johannine prologue and its framing both creation and incarnation around this metaphor of the word of God.[26] Nor does Bonhoeffer say much in

[25] DBWE 3:40. He rejects the idea of any "continuum" between God and world several times throughout this section of the lecture.

[26] The Christology lectures are found in *Berlin:1932–1933*, ed. Carsten Nicolaisen, Ernst-Albert Scharffenorth, and Larry L. Rasmussen, trans. Isabel Best and David Higgins, DBWE 12 (Minneapolis: Fortress Press, 2009). He treats the question of Christ as Word at DBWE 12:315–318, and begins this section with John 1:1 and 3, but its import for him has to do with the linking of the Word with truth and revelation, not with creation; the only place creation appears in this discussion is in his statement that "The Word carries within it the lightning that destroys and the rain that makes alive" (315). In fact, Bonhoeffer does not deal much with the natural world at all in his Christology lectures. The place he comes closest is in his discussion of the sacraments (DBWE 12:318–323) and his short section on Christ as the new creation, the center of nature (12:327), where again he locates the revelatory role of creation within the sacraments. In the earlier section he dismisses the possibility of natural revelation after the fall, because "the continuity between Word and creature has been lost. That is why the natural world is no longer a transparent world. That is why the whole creation is no longer sacrament. Sacrament exists only where God, in the midst of the world of creatures, names an element, speaks to it, and hallows it with the particular word God has for it by giving it its name. Through God's peaking to it, this element becomes what it is" (12:318–319). He does touch on Luther's doctrines of ubiquity (320) and of Christ's repletive presence (321), both of which provide potentially useful ways of articulating the incarnationally diffuse christological presence in creation with which my work is concerned. Of those, Bonhoeffer finds more traction in the repletive presence: "where something is everywhere and yet is not measurable in any place. *This* is the way in which Christ is present. He is everywhere....Christ is even in the rustling leaves, as Luther says, but his presence is not obvious; he is not there for *you*, not *pro-me*, so you cannot grope for him in the rustling leaves" (321, italics Bonhoeffer's, citing WA 20:382, 24–26. 29–30). Later he writes, "In the sacraments of the church, the old creation ... obtains its new freedom. Christ cannot be proved to be the redeeming creation within nature; he can only be preached as such" (327). In both sections, Bonhoeffer is highlighting the fact that perceiving nature as sacramental—filled with Christ—takes place only by means of the Word. He doesn't assert at this point the central role of faith in perceiving this sacramentality *as reality*, by the revelation of Christ in all that is—but that is the claim I hear him making ten years later in his *Ethics*.

those lectures either of how this "word" of Genesis 1 might or does indeed "[unite] God with God's world."[27] What becomes ten years later the profound uniting of the two abstracted "realms" of God and world, namely their becoming visible as "one reality ... not two" in Jesus Christ, hardly appears on Bonhoeffer's radar in these lectures. Here in his treatment of Genesis 1, this Christological reality that will later call into question all forms of Christianity marked by God-world dualism appears only to disappear again with vanishing speed, visible in only this glancing mention of the "word ... that ties God to, or unites God with, God's work."[28] His concern does not yet lie here but instead in making clear that "God is never in the world in any other way than as one who is utterly beyond [jenseits] it."[29]

Much later in prison Bonhoeffer will write very differently about this *jenseits* of God (DBWE 8:367). But at this point he clearly means a reality separate from and superior to the world. Thus later in *Creation and Fall*, reflecting on God's calling the world "good" (*tov*), Bonhoeffer takes pains to clarify that this joy of the Creator in the world does not mean that the world itself is a locus of goodness:

> That God looks upon God's work is the only thing that makes the work good. This really means, however, that the work is good only because the Creator alone is good. The work never has its goodness in itself, but only in the Creator. The goodness of the work consists precisely in its pointing

[27] The closest he comes is again in the material on the sacraments. There he writes, "The God-human Jesus Christ is wholly present in the sacrament. As God spoke at the creation, "'Let there be light'; and there was light," so the Word addressed to the sacrament becomes [reality]" (DBWE 12:320). Note 46 comments that the sentence is incomplete at this point in the original.

[28] DBWE 3:40. This way of speaking of the word of God is Barthian as well: both noting the word as the only point of connection between the "wholly other" God and the utterly separate world *and* the sense of the glancing nature of this connection. As Barth puts it, the word of God touches the world only barely, "as a tangent touches a circle" (Karl Barth, *The Epistle to the Romans*, sixth edition, trans. Edwin C. Hoskyns (London: Oxford University Press, 1968), 30). Dianne Rayson's study of ecological themes in Bonhoeffer (citation in note 37 below) helpfully traces other forms of christological presence in the Creation and Fall lectures, especially in the later chapters on Genesis 2 and 3.

[29] DBWE 3:41.

emphatically away from itself to the Creator and to the Creator's word alone as that which is good.[30]

He shows this essential separateness even more starkly in the subsequent section, on Genesis 1:26–27 ("The Image of God on Earth"), as he portrays with much care the extent to which the non-human creation is *not* of God, does not mirror God in any way, is alien to God and unfree, and is essentially dead:

What is decisive is that at the very moment when the Creator has brought it forth, the work is already torn away from, *and alien to*, the Creator; it is no longer the Creator. Even in its living nature the work is dead, because it is created, conditioned—because, though it arises out of freedom, it itself is not free but conditioned. Only that which is itself free would not be dead, would not, as a creature, be alien or torn away. Only in that which is itself free could the free Creator behold the Creator.[31]

As Bonhoeffer will go on to assert, only the human creature is "free" in this way. Only in the *human* creature, he believes, and decisively not in the rest of the creation, can "the free Creator behold the Creator," i.e., glimpse its own living image.

Human Distinctiveness

Here we see Bonhoeffer clearly marking human separation from the rest of creation as an essential part of what distinguishes our species in the sight of God. We also see his upholding of the *oben* (God)/*unten* (creation) schema that colors many other aspects of his theology and ethics, at least until later in his life when he begins to notice the irreconcilability of such schemas with the Gospel.[32] Operating within this up/down schema and God-world dualism, Bonhoeffer here inscribes such dualism *to the point of*

[30] DBWE 3:59.

[31] DBWE 3:60–61 (my italics).

[32] I have explored this topic at length as the larger framing of an analysis of gender within Bonhoeffer's nearly lifelong appreciation and upholding of such binary hierarchical thinking, an intellectual foundation of his thinking that begins to crack only in the conspiracy period and prison writings. See Lisa E. Dahill, "'There's Some Contradiction Here': Gender and the Relation of *Above* and *Below* in Bonhoeffer." In *Bonhoeffer and Interpretive Theory: Essays on Method and Understanding*, ed. Peter Frick, International Bonhoeffer Interpretation Series, volume 6 (Berne/Berlin: Peter Lang, 2013), 53–82.

creation's alienation and death into the very moment of creation.[33] He is attempting to set up the grandeur of *human* creation—precisely as free, and therefore in the image of the God who is free—in drawing this contrast with non-human creation, which he views as "not free but conditioned" and therefore ultimately just dead matter. His sketch of the primal God-world relationship of Genesis 1 thus exemplifies the definition of dualism (marked by both a clear binary and a hierarchy designating one pole of the binary as categorically superior to the other) that we saw from Val Plumwood earlier.

In his subsequent explications of the Yahwist creation story in Genesis 2 and 3, Bonhoeffer will dramatically shift his perspective, describing human inseparability from creation in beautiful and striking terms. Thus across the Genesis lectures as a whole, he does show both sides of the Christian tradition's paradoxical view of the human being as simultaneously uniquely close to God, i.e., radically separated from the rest of creation, *and* profoundly and essentially part of creation, inseparable from it. Indeed, one could assert that it is with this latter assertion, brilliantly developed in the later lectures of *Creation and Fall*, that Bonhoeffer's clear sympathies lie, given how comparatively unusual and original these interpretive moves were in his time and in the Christian tradition as a whole.[34]

[33] Interestingly, a few pages later in the same section Bonhoeffer seems to assert the opposite, denying that "nature [is]... alien to the spirit," as he writes, "It is my world, my earth, over which I rule. I am not free from it in any sense of my essential being, my spirit, having no need of nature, as though nature were something alien to the spirit. On the contrary, in my whole being, in my creatureliness, I belong wholly to this world; it bears me, nurtures me, holds me" (DBWE 3:66). This latter assertion foreshadows the much more clearly Earth-immersed creatureliness Bonhoeffer will develop in his explications of the Yahwistic creation story in Genesis 2–3, even as his language here of "ruling over" "my world, my earth," feels consonant with the dualistic flavor of his thinking here in the Genesis 1 material, a clear disjunction between humanity and the rest of creation. On Bonhoeffer and creatureliness, see Norman Wirzba, "Creaturely Humanity," in *This Sacred Life: Humanity's Place in a Wounded World* (New York: Cambridge University Press, 2021), 176–211, and Derek W. Taylor, "Can Bonhoeffer Speak Today? Resistance Beyond the Two Kingdoms"), paper delivered at the 2020 gathering of the American Academy of Religion.

[34] An important analysis in eco-theological anthropology by Adrian Langdon traces Bonhoeffer and Barth's views on the human being within creation (using *Creation and Fall* and CD III/1) in ways that contribute to the above discussion. Langdon finds a salutary and anti-anthropocentric window into human *embeddedness* within creation in Barth and a willingness also to go further than Bonhoeffer in highlighting human freedom *for* creation, while appreciating Bonhoeffer's ethical theology of I/Thou person-formation (see later section below) for inviting attention to how creation as *Thou* evokes human responsibility in relationship with creation. Adrian E.V. Langdon, "Embedded Existence: Bonhoeffer, Barth, and Ecological Anthropology," *Didaskalia* 25 (Fall 2015): 59–77.

But he equally clearly intends this creation-immersion to stand precisely *as the pole of a paradox*, held in tension with the equally decisive Genesis 1 insistence on the clear separation of humans from the rest of creation, a separation based in the underlying ontological split between God and the world. The fact that his portrayals of Christology (DBWE 12), Christian discipleship (DBWE 4), and community (DBWE 5) depict a world in which humans relate more or less exclusively with God and one another, with almost no mention of the creaturely entanglement with all the rest of creation celebrated in his lectures on Genesis 2, confirms that at this point in his thinking Bonhoeffer was not yet seeing that primal split between God and world/creation as the problem he would later perceive it as. For much of his life he, like most people of his time and context, considered the Christian life to take place within the "God" side of that split, in relationship spiritually only with those (humans and God) he here names as "free." The natural world with its embodied multiplicity of species is generally invisible on its own terms in these widely read works.[35] It is theologically inert to him, lifeless. And that was precisely how he portrayed non-human creation here in *Creation and Fall*: "Even in its living nature

[35] The indices to the German and English critical edition show nearly zero references to "Earth," "nature," and "creation" for *Life Together*; for *Discipleship* the few references point to things like Jesus or the disciples' lives on Earth. This does not mean that Bonhoeffer himself is heedless of creation, as other pieces of his writing demonstrate. As I note in a recent article on Bonhoeffer and place, "Bonhoeffer loves Germany as a larger natural home, and his letters contain many references to this love. His family's vacation home in Friedrichsbrunn and its forests is the site of many happy days and memories throughout his life; he takes pleasure visiting the sea and the expansive rural von Kleist-Retzow estate in the Finkenwalde and collective pastorate years; he enjoys hiking and skiing in the Bavarian mountains while writing his *Ethics*; and even in prison he finds profound solace in connection to creatures and elements of the natural world still accessible to him in brief or extended snatches of birdsong, in the wind on his body, in the sun through his prison window, the tree in the courtyard" (Lisa Dahill, "Firm Grounding" [cited earlier], 33). In note 21 on the same page I comment, "Derek Taylor, in the paper cited earlier, has compiled and listed all such references he found in reading the *Letters and Papers from Prison* (DBWE 8): the places Bonhoeffer mentions seeing or hearing birds (57, 110), birdsong (57, 407, 459), trees (110, 167), flowers (115, 119, 166, 459, 527), 'the full force of [the sun]...burning on one's skin' (448), as well as memories of gardens, mountains—even 'the bodily delight that comes with being present in the natural world' (Taylor, 19, drawing from Bonhoeffer's comments in DBWE 8:70, 81, 110, 165, 294, 407, 449)."

the work is dead, because it is created, conditioned—because, though it arises out of freedom, it itself is not free but conditioned. Only that which is itself free would not be dead, would not, as a creature, be alien or torn away."[36] This portrayal of the entire non-human creation as unfree dead matter is shocking and nonsensical to anyone who loves Earth's creatures and infinitely complex webs of beauty and life. In their essential separateness from one another, the God and world Bonhoeffer articulates in passages like these are precisely what he will later condemn as *abstractions.*

In other places in his writings as well, Bonhoeffer depicts a God-world relationship that seems to take for granted or (as here in *Creation and Fall*) actively to endorse the God-world dualism he later decries. To be fair, passages expressing this dualism as starkly as these are rare; and a reader of Bonhoeffer's entire corpus can hardly fail to be moved by his consistent and deepening orientation toward the world through the years of resistance and conspiracy, into his prison writings. Indeed, he provides plenty of material for a robust eco-theological-ethical perspective, as scholars Larry Rasmussen and Dianne Rayson have convincingly demonstrated.[37] My concern in this first step of my analysis is simply to note how God-world dualism nevertheless colors Bonhoeffer's thinking.

A Bonhoefferean Animist Ecology: Complexifying Views of the Sacred

While important, that initial piece of the project is not (to my mind) the most interesting. The constructive dimension of this project of reading Bonhoeffer non-dualistically surfaces pieces of his thinking that explicitly

[36] DBWE 3:60–61.

[37] Larry L. Rasmussen, "Bonhoeffer: Ecological Theologian," in *Bonhoeffer and Interpretive Theory: Essays on Method and Understanding*, ed. Peter Frick, International Bonhoeffer Interpretation Series, volume 6 (Berne/Berlin: Peter Lang, 2013), 251–268; Dianne Rayson, *Bonhoeffer and Climate Change: Theology and Ethics for the Anthropocene* (Lanham, MD: Lexington/Fortress Academic, 2021). Rayson's work, carefully tracing the theological foundations in Bonhoeffer for a robust ecological ethic, culminates in a beautiful vision: "If Christ is indeed the center of reality, and the center of the self,… at the center of every Earthling's being…. then Bonhoeffer's argument in *Sanctorum Communio* of the self-being freed of its boundary and responding to the Christ at the center of the other might extend to all our fellow Earthlings" (240). Rayson and Rasmussen are not attempting a non-dualistic reading (and I will use the "boundary" material in *Sanctorum Communio* differently in the next section, seeing it not as something the self needs to be freed of in Christ but precisely the means of the ultimately non-dualistic vision I hear in Bonhoeffer).

or implicitly ground a Christian non-dualism and make possible ecological spiritualities of much greater correspondence with reality.[38] In the next section of the chapter, I focus this constructive work on expanding Bonhoeffer's notion of "personhood" to include all species, revealing in the process a distinctively Bonhoefferean non-dual metaphor of the divine: the Christological boundary or, in my interspecies reading, the *membrane* dividing, connecting, and giving rise to all that is.[39]

Non-dual Divine Presence: Jesus Christ as a Boundary Between Persons

Already in his dissertation, *Sanctorum Communio,* Bonhoeffer insisted that human personhood is developed and experienced in the concrete encounter between selves. His description of this encounter with another focuses metaphorically on the experience of the barrier *(Schranke)* or boundary *(Grenze)* that separates us.[40] This barrier of another person's separate being confronts individuals with a reality alien to their own, drawing them into what Bonhoeffer calls the state of "responsibility," or ethical demand for some response.

Bonhoeffer expresses this insight with the language of "I" and "You." His use of this language echoes Martin Buber, although his focus is less on intimacy and more on that capacity for ethical responsibility generated by the encounter.[41] For Bonhoeffer, the "You" [German *Du*] is any other self who confronts the "I" as a barrier: "by recognizing a You, a being of alien consciousness, as ... distinct from myself, I recognize myself as a [separate]

[38] This criterion too emerges from Bonhoeffer's *Ethics.* In his chapter on "History and Good [1]," he writes, "Good is the action that is in accordance with the reality of Jesus Christ; *action in accordance with Christ is action in accord with reality*" (DBWE 6:228–229).

[39] I first developed these insights in my essay, "The View from *Way* Below: Inter-Species Encounter, Membranes, and the Reality of Christ," *Dialog: A Journal of Theology* 53/3 (Fall 2014): 250–258. Later material in this chapter draws directly on that essay.

[40] Bonhoeffer, *Sanctorum Communio: A Theological Study of the Sociology of the Church,* ed. Joachim von Soosten and Clifford J. Green, trans. Reinhard Krauss and Nancy Lukens, DBWE 1 (Minneapolis: Fortress Press, 1998). This paragraph and the following draw on material developed at more length in *Reading from the Underside of Selfhood,* pp. 32–35. See also David Ford's work on Bonhoeffer and alterity, in conversation with Levinas, in *Self and Salvation: Being Transformed* (Cambridge: Cambridge University Press, 2000).

[41] Buber's *I and Thou* (*Ich und Du*) was published in 1923, well prior to Bonhoeffer's completion of his dissertation, and it is possible that Bonhoeffer read this work though he nowhere cites it.

'I,' and so my self-consciousness awakens."[42] Such awakening by means of this encounter with an Other, a You, is what for Bonhoeffer makes possible the creation of a "person" [German: *Person*] in the fullest ethical sense; for him, a "Person" (vs. *Mensch* or other usage) is precisely the self drawn into relation and responsibility[43] by this encounter with the You's otherness.

As his faith develops, he comes to name this boundary *as Jesus Christ*, developing this christological insight in *Discipleship* and especially *Life Together*. That is, Christ for Bonhoeffer is the one who (a) protects each person from invasion by another, who (b) by divine power forms the personhood of each as they run up against Christ in encounter with one another, and who thereby (c) bridges these earlier separate selves into healthy, christologically mediated community.[44] He comes to experience the boundary that makes possible and gives rise to healthy, mature human selves as itself holy—as, indeed, the active living presence of Christ creating both responsibly bounded *Personen* and Christian community. In Bonhoeffer's non-dualism, therefore, boundaries and distinction are not the problem. Even in his *Ethics'* call for dissolving classical forms of Christian God-world dualism he never argues for monism, and he continues to use the language of "God" and "world" in ways that indicate he still experiences these words as having meaning in distinction from one another, even as his primary point is to argue for their being in fact "one reality" in Jesus Christ. He would, I suspect, appreciate the Scotian language of the "formal" distinction—especially given his lifelong exploration of language and images of the form (*Gestalt*, Gestaltung) of Jesus Christ.[45]

What all this looks like, however—what exactly it might mean to conceive of Jesus Christ as the "one reality" of both God and world, or how a spirituality along these lines might work—he never fully develops. If speaking of Jesus Christ (or the ongoing use of the language of *God*) implies a continuing presupposition of a being, a *You*, a *Person* with whom

[42] DBWE 1:71.

[43] The German term for "responsibility," *Verantwortung*, includes a similar wordplay to the English "response-ability." In German too, *Verantwortung* describes the capacity to make a response, an *Antwort*.

[44] Cf. Bonhoeffer, *Discipleship*, DBWE 4:92–99; and *Life Together*, DBWE 5:41–44, 95, 108–110.

[45] See the chapter in this volume by Jonas Gamborg Lillebø developing this conception of "formal distinction" using the philosophy of Duns Scotus.

one is in some form of inter-personal relationship, then that metaphor of divine personhood (implicit also in the Trinitarian language of three "Persons") would seem to uphold the divine distinction from the world Bonhoeffer is in fact attempting to repudiate. Yet to renounce all attempts to relate to this one reality, Jesus Christ—in favor perhaps of simply existing within an undifferentiated God-world "reality"—undercuts any capacity for inhabiting a *Christian* spirituality replete with such language and metaphors of relationship. It is not clear how Bonhoeffer himself envisioned a life of faith in the absence of the traditional God-world dualism on which most forms of Christian spirituality rely, although his late letters to Eberhard Bethge on "religionless Christianity" represent attempts in this direction.

I propose, however, that his own robust theology of Jesus Christ as the boundary between selves represents a key to such a spirituality. That is, I hear this image of the Christological boundary *as itself the divine presence* as a distinctively Bonhoefferean means by which to articulate a non-dualistic relationship of God and world in Jesus Christ. This image is far from that of God as a divine person—indeed, as the paradigmatic "Person"—to whom Christians relate, instead conceiving the divine as this nearly invisible but endlessly generative boundary formative of mature *human* persons. Rather than a vertically oriented relationship with a Person viewed as "higher" in the traditional Aristotelian cosmology of the Great Chain of Being, this image makes possible a view of the divine in Christ as the complex network of boundaries constituting, protecting, and connecting all ethically mature human selves in [Christian] community, a vast horizontal web making possible relationships *between* these human persons. This image does not imply a static conception of the divine; the boundary itself *is the formative agent* giving rise continually to new depths of selfhood and relationship. It opens a vision of persons relating to one another in all directions across and by means of this Christological boundary constantly forming and re-forming them. I now extend the image further into the bark and scales and fur, the womb linings and gullets and membranes of interspecies life itself.

The Boundary Becomes a Membrane

That is, I complexify Bonhoeffer's holy I/You boundary to point to its functioning not just as an *inter-human* boundary but as the membranes giving rise to *all of life*. I have described elsewhere what every cell wall does, as follows:

> …[Its] separating "in" from "out," mediating the transfer of energy and nourishment and waste across this boundary—has an analogue in multicellular animals like us. The *basement membrane* is the layer of epidermal tissue that not only anchors us together with skin (or other species' external analogues) but extends into the inner surfaces of our lungs, kidneys, digestive tract, and other organs. This basement membrane is what mediates the moment-by-moment exchange of oxygen and CO_2 in our lungs, flushes out wastes and maintains fluid balance in our kidneys, and houses the countless organisms in our guts—our microbiome—allowing elements of other creatures' flesh to cross over into us and become incorporated into our tissues. In short, this constitutive membrane performs precisely the same functions of protection and intimacy that Bonhoeffer *names as Christ* in his interpersonal ethics. We come to realize that our entire existence is comprised of divinely mediated I/Thou encounters as we negotiate the flow of life across the membranes and lungs and photosynthetic cells of countless microbes and mammals and plants, in every moment of our breathing lives.[46]

Thus as we begin to experience all kinds of creatures as *Yous* to us, relating across those countless christological membranes that bound and bridge all forms of life; as we see and honor their right to live and thrive as beings created as our kin in the fullness of the divine; as that seeing and honoring calls us into response and responsibility toward their fullness of life interwoven with ours, then these creatures simply are functioning as do human members of the communities Bonhoeffer describes. They call us into deeper and more ethically mature *humanity,* precisely as we encounter and honor their otherness and difference from us. In fact, they invite us to become *persons,* in Bonhoeffer's sense, precisely as we begin to perceive and acknowledge them too as persons, a profound reciprocity of mutual dependence and kinship. This perception of the personhood of countless non-human *Yous* with whom we are in christologically mediated biological and ethical inter-relationship creates a theological foundation for spiritualities of Christian animism. In the words of theorist Graham Harvey, animists are those who "recognize that the world is full of persons, only some of whom are human, and that life is always lived in

[46] Lisa Dahill, "The View from *Way* Below: Inter-Species Encounter, Membranes, and the Reality of Christ," *Dialog: A Journal of Theology* 53/3 (Fall 2014): 250–258.

relationship with others."[47] Conceiving of the divine as membrane—literally making possible every individual cell or being, mediating in some form all relationships, and giving rise to dizzying levels of complexity throughout all life on Earth—thus provides an utterly non-dual metaphor for the God-world relationship itself.

Reading Bonhoeffer through the lens of his own insistence on *one reality* in Christ both allows for critique of his own participation in theological abstraction via God-world dualism *and* makes possible a more fully interspecies Christian spirituality in which the divine is what permeates and mediates the flow of life across all the membranes of intercellular life, the emergence of new beings and species, the digesting and decomposition and renewal taking place on every level of life, in every moment. Reading in this way shifts the locus of divine presence from a God conceived as its own separate Person into a horizontal presence *forming* the experience of reciprocal personhood across all the membranes of our endlessly complex creaturely lives.

Conclusion

What is meant to be held together in Christ—the world and its holiness or mystery—have split into two. Bonhoeffer was, at least at times, a God-world dualist early on, articulating views that reinforce God's lofty separateness from and independence of creation; but his *Ethics* provides a hermeneutical lever to critique his own early conceptions of the all-transcendent god, toward images more befitting of the incarnational permeation of all that is in Christ. Bonhoeffer's repudiation of Christian dualism helps subvert the confusing and oppressive forms of alienation and splitting that have only continued to multiply in the decades since his time. This Bonhoefferean lens is useful in approaching other texts in the Christian tradition as well. What forms of God-world dualism might need critique? And what surprising openings to new images and metaphors of earthly-divine presence might we find or develop from the tradition's resources? What happens when we read other Christian texts and practices with place-based and/or animist eyes and hearts?

Ultimately these forms of thought complexify the language of non-dualism in ways that correspond to the entanglement and biological

[47] Graham Harvey, *Animism: Respecting the Living World* (New York: Columbia University Press, 2006), xi.

interpermeation of all that is. As noted earlier, I do not read Bonhoeffer's "one reality" as a form of monism that would flatten or subsume all this diversity into a divine oneness, another version of the dominant Western insistence on monocultures of all kinds (including monotheism). Rather, the one reality we share with all that is exists as endless multiplicity: the recognition that holiness is visible in countless strange and interwoven forms. It is not a mono-anything but the ceaseless divine plenitude of faces and bodies and forces and inter-digestions that make up life on Earth.[48] Taking seriously Bonhoeffer's "one reality" theology generates a Christian lens able to perceive divine presence and agency in all that is, into the very interstices of interspecies life: one world alive with holiness.

Bibliography

Abram, David. 1996. *The Spell of the Sensuous: Perception and Language in a More-than-Human World*. New York: Vintage Books.

Bonhoeffer, Dietrich. 1998. *Sanctorum Communio: A Theological Study of the Sociology of the Church*, ed. Joachim von Soosten and Clifford J. Green, trans. Reinhard Krauss and Nancy Lukens. Dietrich Bonhoeffer Works (DBWE), vol. 1. Minneapolis: Fortress Press.

———. 1997. *Creation and Fall: A Theological Exposition of Genesis 1-3*, ed. Martin Rüter, Ilse Tödt, and John W. de Gruchy, trans. Douglas Stephen Bax. Dietrich Bonhoeffer Works (DBWE), vol. 3. Minneapolis: Fortress Press.

———. 2001. *Discipleship*, ed. Martin Kuske, Ilse Tödt, Geffrey B. Kelly, and John D. Godsey, trans. Barbara Green and Reinhard Krauss. Dietrich Bonhoeffer Works (DBWE), vol. 4. Minneapolis: Fortress Press.

———. 1996. *Life Together and Prayerbook of the* Bible, ed. Gerhard Müller, Albrecht Schönherr, and Geffrey B. Kelly, trans. Daniel Bloesch and James H. Burtness. Dietrich Bonhoeffer Works (DBWE), vol. 5. Minneapolis: Fortress Press.

———. 2005. *Ethics*. Edited by Clifford J. Green, trans. Reinhard Krauss, et al. Dietrich Bonhoeffer Works (DBWE), vol. 6. Minneapolis: Fortress Press.

[48] A now classic text in this direction is Laurel Schneider, *Beyond Monotheism: A Theology of Multiplicity* (New York: Routledge, 2008), while the volume on which my own thinking in this area most resonates is Mary-Jane Rubenstein, *Pantheologies: Gods, Worlds, Monsters* (New York: Columbia University Press, 2018). See too the essays collected in *Polydoxy: Theologies of Multiplicity and Relation*, ed. Catherine Keller and Laurel Schneider (New York: Fordham University Press, 2010).

———. 2009. *Berlin:1932-1933*, ed. Carsten Nicolaisen, Ernst-Albert Scharffenorth, and Larry L. Rasmussen, trans. Isabel Best and David Higgins, Dietrich Bonhoeffer Works (DBWE), vol. 12. Minneapolis: Fortress Press.

———. 2010. *Letters and Papers from Prison*, ed. John W. de Gruchy, trans. Isabel Best, Lisa E. Dahill, Reinhard Krauss and Nancy Lukens. Dietrich Bonhoeffer Works (DBWE), vol. 8, Minneapolis: Fortress Press.

Barth, Karl. [1945], 1958. "Faith in God the Creator," §40. In "The Work of Creation", *Church Dogmatics* III/1. ed. Geoffrey W. Bromiley and Thomas F. Torrance, trans. J. W. Edwards, et al. London: T&T Clark.

———. 1968. *The Epistle to the Romans*, sixth edition, trans. Edwin C. Hoskyns. London: Oxford University Press.

Bennett, Jane. 2010. *Vibrant Matter: A Political Ecology of Things*. Durham, NC: Duke University Press.

Dahill, Lisa E. 2013. 'There's Some Contradiction Here': Gender and the Relation of Above and Below in Bonhoeffer. In *Bonhoeffer and Interpretive Theory: Essays on Method and Understanding*, International Bonhoeffer Interpretation Series, ed. Peter Frick, vol. 6, 53–82. Berne/Berlin: Peter Lang.

———. 2014. The View from *Way* Below: Inter-Species Encounter, Membranes, and the Reality of Christ. *Dialog: A Journal of Theology* 53/3: 250–258.

———. 2020. Eating and Being Eaten: Interspecies Vulnerability as Eucharist. *Religions* 204. https://doi.org/10.3390/rel11040204.

———. 2022a. Firm Grounding: Bonhoeffer, Place, and Coming Home to Reality. *Bonhoeffer and Eco-Theology. The Bonhoeffer Legacy: An International Journal* 8 (2): 21–42.

———. 2022b. Rewilding Life Together: Bonhoeffer, Spirituality, and Interspecies Community. *Dialog: A Journal of Theology* 61 (2): 1–11.

———. 2009. *Reading from the Underside of Selfhood: Bonhoeffer and Spiritual Formation*. Princeton Theological Monograph Series. Eugene: Pickwick Publications.

DeJonge, Michael P. 2012. *Bonhoeffer's Theological Formation: Berlin, Barth, and Protestant Theology*. New York: Oxford University Press.

Etieyibo, Edwin. 2017. Ubuntu and the Environment. In *The Palgrave Handbook of African Philosophy*, ed. Adeshina Afolayan and Toyin Falola. New York: Palgrave Macmillan.

Eze, Michael Onyebuchi. 2017. Humanitatis-Eco (Eco-Humanism): An African Environmental Theory. In *The Palgrave Handbook of African Philosophy*, ed. Adeshina Afolayan and Toyin Falola. New York: Palgrave Macmillan.

Ford, David F. 2000. *Self and Salvation: Being Transformed*. Cambridge: Cambridge University Press.

Frick, Peter. 2017. Dietrich Bonhoeffer: 'Within not Outside the Barthian Movement'. In *Dietrich Bonhoeffer's Intellectual Formation*, ed. Peter Frick, 245–282. Eugene: Wipf & Stock.

Haraway, Donna J. 2016. *Staying with the Trouble: Making Kin in the Chthulucene*. Durham, NC: Duke University Press.
Harvey, Graham. 2006. *Animism: Respecting the Living World*. New York: Columbia University Press.
Ingold, Tim. 2021. *Imagining for Real: Essays on Creation, Attention and Correspondence*. London: Routledge.
Kelbessa, Workineh. 2011. Indigenous Environmental Philosophy. In *The Oxford Handbook of World Philosophy*, ed. Jay L. Garfield and William Edelglass. New York: Oxford University Press.
Keller, Catherine, and Laurel Schneider, eds. 2010. *Polydoxy: Theologies of Multiplicity and Relation*. New York: Fordham University Press.
Keller, Catherine, and Mary-Jane Rubenstein, eds. 2017. *Entangled Worlds: Religion, Science, and New Materialisms, Transdisciplinary Theological Colloquia*. New York: Fordham University Press.
Kimmerer, Robin Wall. 2013. *Braiding Sweetgrass: Indigenous Wisdom, Scientific Knowledge, and the Teachings of Plants*. Minneapolis, MN: Milkweed Editions.
Kohn, Eduardo. 2013. *How Forests Think: Toward an Anthropology Beyond the Human*. Berkeley: University of California Press.
Langdon, Adrian E.V. 2015. Embedded Existence: Bonhoeffer, Barth, and Ecological Anthropology. *Didaskalia* 25: 59–77.
Langknecht, Henry John. 2008. *Giving Scripture Its Voice: The Tensive Impertinence of The Literal Sense of The Pericope, Metaphorical Meaning-Making, and Preaching the Word of God*. Th.D. diss., Emmanuel College of Victoria University and the University of Toronto.
Loy, David R. 1988. *Nonduality in Buddhism and Beyond*. Somerville, MA: Wisdom Publications.
Mann, Barbara. 2000. *Iroquoian Women: The Gantowisais*. New York: Peter Lang.
Nhat Hanh, Thich. 2021. *Zen and the Art of Saving the Planet*. Edited by and with commentary by Sister True Dedication. New York: HarperOne.
Oakes, Edward T. 1994. The Paradox of the Literal: The Voice of Canon Criticism in Reformation and Counterreformation Polemics. In *Reform and Counterreform: Dialectics of the Word in Western Christianity Since Luther*, ed. John C. Hawley, 15–29. Berlin and New York: Morton de Gruyter.
Pangritz, Andreas. 2000. *Karl Barth in the Theology of Dietrich Bonhoeffer*, trans. Barbara Rumscheidt and Martin Rumscheidt. Grand Rapids: William P. Eerdmans Publishing Company.
Plumwood, Val. 1993. *Feminism and the Mastery of Nature*. London and New York: Routledge.
Rayson, Dianne. 2021. *Bonhoeffer and Climate Change: Theology and Ethics for the Anthropocene*. Lanham, MD: Lexington/Fortress Academic.
Rasmussen, Larry. 2009. The Brothers Bonhoeffer on Science, Morality, and Theology. *Zygon* 44 (1): 97–113.

Rasmussen, Larry L. 2013. Bonhoeffer: Ecological Theologian. In Bonhoeffer and Interpretive Theory: Essays on Method and Understanding, ed. Peter Frick. International Bonhoeffer Interpretation Series, vol. 6, 251–268. Berne/Berlin: Peter Lang.

Rubenstein, Mary-Jane. 2018. *Pantheologies: Gods, Worlds, Monsters*. New York: Columbia University Press.

Schneider, Laurel. 2008. *Beyond Monotheism: A Theology of Multiplicity*. New York: Routledge.

Sideris, Lisa. 2003. *Environmental Ethics, Ecological Theology, and Natural Selection*. New York: Columbia University Press.

———. 2017. *Consecrating Science: Wonder, Knowledge, and the Natural World*. Berkeley: University of California Press.

Taylor, Derek W. *Can Bonhoeffer Speak Today? Resistance Beyond the Two Kingdoms*. Paper delivered at the 2020 gathering of the American Academy of Religion.

Tinker, George ("Tink"). 2013. Why I Do Not Believe in a Creator. In *Buffalo Shout, Salmon Cry: Conversations on Creation, Land Justice, and Life Together*, ed. Steve Heinrichs, 167–179. Waterloo: Herald Press.

Wirzba, Norman. 2021. *This Sacred Life: Humanity's Place in a Wounded World*. New York: Cambridge University Press.

Zimmermann, Jens. 2019. *Dietrich Bonhoeffer's Christian Humanism*. New York: Oxford University Press.

CHAPTER 9

Pentecostal Emotive (Non)Dualism: Pneumatology, Worship, and Context

Joel D. Daniels

Introduction

Pentecostalism is an extraordinarily diverse religious movement, comprised of believers from every branch of the many Christianities, from Catholic to Protestant, Orthodox, and innumerable indigenous forms. Research suggests that due to its extensive reach, Pentecostalism will likely exceed 800 million adherents by 2025.[1] The sheer size and breadth of Pentecostalism invites scholars and practitioners to examine the resources within the movement, discovering creative ideas and solutions to topics such as ecological dualism.

There are two important things to note from the start, however. First, I participate in this conversation as both a scholar and a practitioner. I was raised within Pentecostalism, and I remain in the movement as an American

[1] F. LeRon Shultz and Andrea Hollingsworth, *The Holy Spirit* (Grand Rapids: Eerdmans Publishing Company, 2008), 69.

J. D. Daniels (✉)
Claremont Graduate University, Claremont, CA, USA
e-mail: joel.daniels@claremont.edu

T. J. Hastings, K.-W. Sæther (eds.), *Views of Nature and Dualism*,
https://doi.org/10.1007/978-3-031-42902-6_9

Baptist Churches ordained minister. Thus, I focus primarily on positive elements in the movement, though there are numerous negative ways in which Pentecostalism interacts with the topic.[2] For example, Matthew Tallman claims that Pentecostals have embraced universal distinction and extended it to form an anthropocentric worldview, such as hierarchies and duality between humans and "nature" (nonhuman things). Consequently, as Tallman notes, there is dissonance between Pentecostals' concern for holy living and their tendency to ignore "the reality of unhealthy lifestyles and specific ecological sins that have led to disease and death."[3] These divisive ecological views by many Pentecostals, however, are not my focus here.

Second, narrowly defining Pentecostalism is inappropriate and exclusionary. There is not one characteristic, doctrine, or creed that unites Pentecostals across denominations and backgrounds. When definitions are proposed, they always exclude people, particularly those who do not neatly fit into "Western" religious ideals and proclivities. My admittedly broad description of the movement, then, is that Pentecostalism is defined by the Holy Spirit, moving how the Spirit wills, interacting with humanity through emotive experientialism.

With that brief introduction, my argument is that Pentecostalism's emotive non-dualism provides an important roadmap for our ecological reality, recognizing how people connect more deeply with emotions rather than "facts." The point is not to construct a debate between "reason" and "emotion," but instead to state that we first experience the world and then extract reason from those experiences. Although he is describing Daoist philosophy, Peng Feng's description of experience is instructive here. According to Peng, "Experience is directly given to us; fact can be proven only later, and even then such proof might still be faulty."[4] In other words,

[2] Articles continue to be written about Pentecostals' response, or at times the lack of response, to climate issues. For example, Graham Readfearn, "Ashes to Ashes: Pentecostalism, the PM and the Climate Crisis," *The Guardian*, May 8, 2021, https://www.theguardian.com/environment/2021/may/09/ashes-to-ashes-pentecostalism-the-pm-and-the-climate-crisis

[3] Matthew Tallman, "Pentecostal Ecology: A Theological Paradigm for Pentecostal Environmentalism," in *The Spirit Renews the Face of the Earth: Pentecostal Forays in Science and Theology of Creation*, ed. Amos Yong (Eugene, OR: Pickwick Publications, 2009), 143.

[4] Feng Peng, "Fact and Experience: A Look at the Root of Philosophy from the Happy Fish Debate," in *Zhuangzi and the Happy Fish*, ed. Roger T. Ames and Takahiro Nakajima (Honolulu: University of Hawai'i Press, 2015), 234.

experience happens, and human experience is true in that it occurs in reality. Fact, conversely, is an extraction from experience and is not inherently true or untrue. This is important because research suggests that most of our decisions are not rooted in fact-driven data points but rather in how we connect to topics as emotive people.[5]

Therefore, I begin with a very brief overview of Pentecostal cosmology and then move to the main section of the essay—emotions. What are emotions and how do emotions help us explain Pentecostalism's diversity? I conclude by looking at Genesis 1:2, noting the power of story. Pentecostals claim that it is the Spirit moving through the world, inviting people to experience and participate in creation, providing potential resources for our ecological reality.

Pentecostal Cosmology

Pentecostal spirituality and practice propose an interesting approach to the world. On the one hand, Pentecostals typically have a strong dualistic stance on good versus evil. The *Holy* Spirit is active, contending with unholy spirits that seemingly permeate the universe. There are variations of this position; however, by focusing on the Holy Spirit, there is an implied belief in evil spirits as well.

On the other hand, Pentecostals' emphasis on emotive experientialism removes all dualities between sacred and secular. In other words, there is never a time when or a place where the Spirit is not present. Although Pentecostal worship services are important, the Spirit can just as easily move in line at the grocery store. The space between extraordinary and mundane is transgressed, collapsing any-and-all dualities through immediate experience. After all, it is the Spirit that moves whenever, wherever, and on whomever the Spirit wills.

Without exhausting the topic here, many Pentecostal scholars and practitioners have entered the ecological conversation, suggesting interesting theological and biblical positions. Aimee Semple McPherson, the founder of the Four Square Church (1923), was a forerunner for Pentecostal eco-pneumatology. Although her doctrine primarily centered on people, she still acknowledged the goodness in "nature" (nonhuman world) and even

[5] For example, see Nadine Jung, et al., "How Emotions Affect Logical Reasoning: Evidence from Experiments with Mood-Manipulated Participants, Spider Phobics, and People with Exam Anxiety," *Frontiers in Psychology* 5 (2014): 1–12.

traded in her car for a buggy to conserve gasoline. She was also outspoken about "scarcity ideologies" believing the world had enough resources if people would simply stop hoarding based on fear of scarcity.[6] Additionally, John McConnell, Jr., the founder of Earth Day,[7] was Pentecostal, from a family of Pentecostals—his grandfather was "Spirit-baptized" at Azusa Street and his parents helped found the Assemblies of God.[8] McConnell believed that it is important for Christians to live out their faith through earth-care. This was visible not only in McConnell's "Earth Day" but also in his two previous endeavors: Meals for Millions (1961–1963) and Minute for Peace (1963-Present). Throughout his life, McConnell saw a connection between his Pentecostal faith and care for creation (all things).

Within theological scholarship, Brandon Hubbard-Heitz argues that Pentecostals have often constructed a duality between particular spaces as secular, or "dark"—in need of spiritual revival.[9] This dualistic notion informs ecological perspectives, spiritually dividing the Earth; Hubbard-Heitz explains, "The occupation of formerly 'immoral' or 'dark' spaces speaks to some early Pentecostals' belief that such locations needed to be transformed—a process best understood through the Pentecostal categories of salvation, healing, and the baptism in the Holy Spirit."[10] In other words, ecological issues are not reduced to fossil fuel emission or deforestation but rather spiritual conflict. Paul Ede, citing A. J. Swoboda and

[6] See, Brian K. Pipkin, "The Environment Theology of Aimee Semple McPherson," in *Blood Cries Out: Pentecostals, Ecology, and Groans of Creation*, ed. A. J. Swoboda (Eugene, OR: Pickwick Publications, 2014), 58–79.

[7] John McConnell, Jr. founded the original Earth Day (1970), which was "quickly eclipsed in prominence by a second Earth Day (celebrated on April 22)." Gaylord Nelson, a senator from Wisconsin, asked McConnell to merge Earth Day with his "Environmental Teach-In Day." Since McConnell had already established Earth Day on the March equinox, he was unwilling to move it to Nelson's Environmental Teach-In Day date, April 22. According to McConnell, the next thing he knew, Nelson stole the name "Earth Day." Nevertheless, McConnell's "Earth Day" is still celebrated on the March equinox by the United Nations. Darrin J. Rodgers and Nicole Sparks, "Pentecostal Pioneer of Earth Day: John McConnell, Jr.," in *Blood Cries Out: Pentecostals, Ecology, and the Groans of Creation*, ed. A. J. Swoboda (Eugene, OR: Pickwick Publications, 2014), 15–16.

[8] John McConnell, Sr. was also a popular Pentecostal evangelist. Rodgers and Sparks, "Pentecostal Pioneer of Earth Day: John McConnell, Jr.," 4–9.

[9] Brandon Hubbard-Heitz, "The Devil's Suicide: Early Pentecostal Hermeneutics of Space and Their Ecotheological Implications," in *Blood Cries Out: Pentecostals, Ecology, and the Groans of Creation*, ed. A. J. Swoboda (Eugene, OR: Pickwick Publications, 2014), 23.

[10] Hubbard-Heitz, "The Devil's Suicide: Early Pentecostal Hermeneutics of Space and Their Ecotheological Implications," 23–24.

Mark I. Wallace,[11] suggests that the Spirit has both a life-giving nature and a "dark side," or destructive nature.[12] Using biblical passages such as Genesis 19:24–25,[13] which describes the destruction of Sodom and Gomorrah, Ede claims that the Spirit judges sin through destruction: "The Holy Spirit is necessarily destructive in creation in order to ultimately cause its *ongoing sustenance*."[14] For Ede, the Spirit as infinite necessarily includes destruction, otherwise the Spirit would not be infinite.

According to Peter Althouse, Pentecostals must start by shifting the focus off individual salvation and instead move toward a "social and cosmic sense" that would better equip Pentecostals for a Spirit-led and informed ecology.[15] Daniela C. Augustine offers the analogy of a household to discuss this interrelationality:

> [The household] involves civil responsibility for the other that in turn demands 'internal restraints,' for example, a form of responsible and reverent consumption committed to economic and ecological justice. It cultivates the civic virtue of fasting from oneself on behalf of the fellow human and the rest of nature. In this spiritual discipline one is committed to sharing possessions with one's neighbor and redistributing wealth according to human needs rather than political and economic benefits.[16]

The world, according to Augustine, does not consist of isolated events but instead interrelated moments that must be recognized and then rectified. Extending interdependence and interrelationality, Michael J. Chan suggests that the events throughout history—past, present, and

[11] Ede uses Swoboda's book on Pentecostal ecology as the source for both Wallace and Swoboda's point on the "dark side" of the Spirit; A. J. Swoboda, *Tongues and Trees: Towards a Pentecostal Ecological Theology* (Boston: Brill), 136.

[12] Paul Ede, "River from the Temple: The Spirit, City Earthkeeping, and Healing Urban Land," in *Blood Cries Out: Pentecostals, Ecology, and the Groans of Creation*, ed. A. J. Swoboda (Eugene: Pickwick Publications, 2014), 208.

[13] Ede also cites Ezekiel 16:49.

[14] Ede, "River from the Temple: The Spirit, City Earthkeeping, and Healing Urban Land," 211.

[15] Peter Althouse, "Pentecostal Eco-Transformation: Possibilities for a Pentecostal Ecotheology in Light of Moltmann's Green Theology," in *Blood Cries Out: Pentecostals, Ecology, and Groans of Creation*, ed. A. J. Swoboda (Eugene, OR: Pickwick Publications, 2014), 131.

[16] Daniela C. Augustine, "Pentecostal Communal Economics and the Household of God," *Journal of Pentecostal Theology* 19 (2010): 224.

future—are interconnected as well. Using the idea of "generational curses,"[17] Chan argues for a "consequential nature of sin,"[18] particularly regarding the ecological crisis. In other words, generations to come will suffer from the ecological sins happening now.

The way the topic is discussed, it appears that Pentecostal scholars and practitioners view ecological discussions as a relatively recent development within the movement; however, I would argue that Pentecostalism, broadly understood, has been exploring this conversation for decades through the work of Charismatic Catholics like Leonardo Boff. In his book *Come Holy Spirit*, Boff states, "Spirit is the ability of all beings—even the most fundamental ones like Higgs bosons, hadrons, quarks, protons, and atoms—to relate to one another, to exchange information, and to create the networks of interconnection that make possible the complex unity of the whole."[19] Using Psalms 139:7–8 and Romans 8:10, Boff claims that the Spirit is life, permeating all things:

> We find ourselves plunged into a field of absolute Energy—the *Spiritus Creator*—who is manifested in the energies of the universe and in our own vital and spiritual energy. With and in the Spirit we make up a whole. The spirituality that arises from conviction sees itself as linked to natural and cosmic processes. To allow ourselves to be imbued and charged with them is to live according to the Spirit in a spontaneous and natural manner.[20]

Consequently, any conversation on the Spirit necessarily includes ecology. Moreover, because God desires to help the downtrodden, all Christians should include ecological ethics in their spirituality, because those most affected by climate change are the poor: "A spirituality that ignores the suffering of the poor is a false spirituality, deaf to the call of the Spirit."[21]

Pentecostal cosmology is non-dual because the Holy Spirit pervades all reality. Although some claim that there are "light" and "dark" spaces, the

[17] This idea is found in Bible verses like Exodus 20:5 and Deuteronomy 5:9, among others.

[18] Michael J. Chan, "Sins of the Ancestors: Generational Sin, Pentecostalism, and Ecological Crisis," in *Blood Cries Out: Pentecostals, Ecology, and the Groans of Creation*, ed. A. J. Swoboda (Eugene, OR: Pickwick Publications, 2014), 170–174.

[19] Leonardo Boff, *Come Holy Spirit: Inner Fire, Giver of Life and Comforter of the Poor*, translated by Margaret Wilde (Maryknoll, NY: Orbis Books, 2015), 58.

[20] Leonardo Boff, *Cry of the Earth, Cry of the Poor* (Maryknoll, NY: Orbis, 1997), 169.

[21] Boff, *Come Holy Spirit*, 163.

Spirit is present in and is actively moving throughout everything.[22] Because people experience reality, human affectivity and emotion is important for understanding and then responding to God's pervasive presence.

Emotive Experientialism

Pentecostal spirituality and practice is rooted in experiential emotion, making it is imperative to understand what emotion is. Interestingly, while there is no shortage of articles and books on the topic, no one really knows what emotions are. That is fascinating because everyone experiences emotions—emotions are ubiquitous to the human-*being*. The topic is immense, but for this section, I begin by exploring what Pentecostal pneumatology says about humanity. What is a person? Where does the Holy Spirit fit? I then briefly consider the convoluted topic of "emotion." Philosophers, theologians, and scientists alike have circled this idea for centuries, but the topic remains opaque. Depending on one's conception of humanity, however, there can be space to construct a theory of emotion, relating emotion to Pentecostal pneumatology. With that proposed, I conclude by examining the connection between emotion and the Holy Spirit.

What Are Humans?

In order to investigate emotion, one must advance innumerable theological, cosmological, and ontological positions before emotions even register. What is the human? From what are humans? How do humans fit within the universe? Without developing these notions, any discussion on emotion appears incomplete or at least open to refutation.

Christian cosmogony claims that God created the universe. Using the Torah's creation narratives, Christians believe that humans were God's final creation, coming after the incalculable elements of the vast universe. That version of the story is best represented in Genesis 1; however, Genesis 2 provides a slightly different story, giving more detail about humanity's construction: "Then the Lord God formed a man from the dust of the ground and breathed into his nostrils the breath of life, and the man

[22] Elsewhere, I explore the non-duality of "dark" and "light" spaces. See Joel Daniels, "What Lies Within: Uncovering the Holy Spirit with the Aid of Buddhist Philosophy," *Buddhist-Christian Studies* 40 (2020): 287–305.

became a living being."[23] In other words, humans are comprised of "universal" things—no part of the human body is uniquely "human." The universe is interconnected, swapping, and trading materials as things endlessly transform and change.

John 3:1–8 is another important text for understanding the body within Pentecostalism. After Nicodemus asks how someone can be born again, Jesus says that flesh gives birth to flesh and Spirit to spirit. For those alive in the Spirit, movement is fundamental. Jesus says that like the wind, the Spirit moves unpredictably and thus uncontrollably, and Jesus-followers are meant to do likewise. The message appears to address humanity within a Pentecostal framework. If one chooses to believe that the material flesh is reality—that is, the static body is *real*—then one will be drawn to "fleshly" desires such as power, prestige, and selfishness. When one's body, and consequently life, is ultimate, then achievement and advancement become central pursuits. Yet, when people experience life in the Spirit, they move and act however and wherever the Spirit leads.

The universe is movement and change, and humanity functions within that system. Problems occur when people believe that things inherently exist and therefore must be controlled, protected, or acquired. But Pentecostal pneumatology, which promulgates the Spirit's movement, refutes that construct, replacing static objects with fluidity and mutability. Perhaps the clearest example of this is in Pentecostal worship, where people unabashedly move, "verbally through speech, shouts, and singing and nonverbally through hand clapping, foot stomping, running, and the 'Holy Dance.'"[24] John's statement is embodied in Pentecostal nondualism, where those born of the Spirit move into and out of all space because they are not reduced to static flesh, only occupying determinant bodies and environments.[25]

What Are Emotions According to Theologians and Philosophers?

If humans are not strictly materiality but rather movement and change, then emotions must also exist within that arrangement, and that basic

[23] Genesis 2:7.

[24] Judith Casselberry, *Labor of Faith: Gender and Power in Black Apostolic Pentecostalism* (Durham: Duke University Press, 2017), 3.

[25] For a more detailed examination, see Joel Daniels, "From Normative to Nothing: A Pentecostal Theological Conception of the Body," *Pneuma* 43 (2021): 233–249.

premise shapes explorations into emotionality and experience. What makes things tricky is that no one really agrees on what an "emotion" is. For starters, "emotion" is a relatively new nomenclature, only getting used as a theoretical keyword since 1884.[26] Prior to that, human "emotion" was described as "passions" and "affections," where passions were tied to negative desires. For Augustine, feeling was connected to volition,[27] whereas later theologian Thomas Aquinas viewed it more in terms of movement between appetites of the soul and appetites of the intellect.[28] Friedrich Schleiermacher infused his theology with feeling (*Gefühl*), famously stating that religion was not about knowing or doing but rather feeling.[29] Everyone is religious, therefore, because everyone has a feeling of absolute dependence, which should be followed. Of course, not everyone was convinced by Schleiermacher. Addressing Schleiermacher, Karl Barth says, "Once again, the move found in *The Christian Faith* of granting supremacy to 'feeling,' in whose place of course one could then set 'faith' in order to move somewhat closer to the Bible or the Reformation; 'faith' on which was conferred sovereignty over everything which might be its ground, object, and content."[30] Although he disagreed with Schleiermacher, Barth proves Schleiermacher's influence by directly addressing his thought around 100 years later.

Within philosophy, Baruch Spinoza responded to Rene Descartes's dualist view of emotion and material, arguing that *émotion*—which Descartes proposed as an alternative to passion—was the byproduct of imperfect and defective thoughts and thus should be resisted.[31] Twentieth century philosopher Alfred North Whitehead constructs an intricate

[26] Thomas Dixon, "'Emotion': The History of a Keyword in Crisis," *Emotion Review* 4, no. 4 (2012): 338.

[27] John Corrigan, "Introduction: The Study of Religion and Emotion," *The Oxford Handbook of Religion and Emotion*, ed. John Corrigan (New York: Oxford University Press, 2007), 4.

[28] Dixon, "'Emotion': The History of a Keyword in Crisis," 339. This topic is taken up directly by Aquinas in Question 80, Article 2.

[29] Friedrich Schleiermacher, *The Christian Faith*, ed. H. R. Mackintosh and J. S. Stewart (New York: T&T Clark, 1999), section 3.3, 7.

[30] Karl Barth, *The Theology of Schleiermacher: Lectures at Göttingen, Winter Semester of 1923/24*, ed. Dietrich Ritchl, trans. Geoffrey W. Bromiley (Grand Rapids: William B. Eerdmans Publishing Company, 1982), 269.

[31] Corrigan, "Introduction: The Study of Religion and Emotion," 5.

system where emotions are called "subjective forms."[32] Subjective forms, according to Whitehead, "is determined by the subjective aim at further integration, so as to obtain the 'satisfaction' of the completed subject. In other words, final causation and atomism are interconnected philosophical principles."[33] Whitehead views the universe as interrelated and interconnected, and his philosophy attempts to explain how these connections function. For Whitehead, emotions shape our reality on the fundamental level, affecting how connections are made throughout the universe. To add more complexity, Whitehead uses "feeling" as a technical term,[34] which is comprised of three factors, one being subjective forms.[35]

More recently, Martha Nussbaum has suggested that emotions are value judgements and appraisals about others that aid in one's flourishing. She contends that there are three salient points to emotion, "the idea of a *cognitive appraisal* or *evaluation*; the idea of *one's own flourishing* or *one's important goals and projects*; and the idea of the *salience of external objects as elements in one's own scheme and goals*."[36] Nussbaum has written extensively about how anger as an emotion can function within society, explaining that most versions of anger have negative affect and effects; however, she argues that sometimes anger can create change through what she terms "Transition-Anger," which seeks meaningful change.[37] The feeling of changemaking anger is very close to payback pursuit, leading Nussbaum to say that those able to maintain the positive aspect of anger are "strange sorts of people, part Stoic and part creatures of love."[38]

What Are Emotions According to Scientists?

With the many advances in neuroscience have come new methods for exploring emotions. What happens in the brain when someone is afraid, for example? How are emotions cognitively constructed? Does it matter what someone *feels*, and if so, how would one measure that? Theoretically,

[32] Alfred North Whitehead, *Process and Reality: An Essay in Cosmology*, ed. David Ray Griffin and Donald W. Sherburne (New York: The Free Press, 1978), 24.

[33] Whitehead, *Process and Reality: An Essay in Cosmology*, 19.

[34] Whitehead, *Process and Reality: An Essay in Cosmology*, 164.

[35] Whitehead, *Process and Reality: An Essay in Cosmology*, 23.

[36] Martha C. Nussbaum, *Upheavals of Thought: The Intelligence of Emotions* (New York: Cambridge University Press, 2001), 3.

[37] Martha C. Nussbaum, *Anger and Forgiveness: Resentment, Generosity, Justice* (New York: Oxford University Press, 2016), 6.

[38] Nussbaum, *Anger and Forgiveness: Resentment, Generosity, Justice*, 8.

with the right equipment and perimeters, emotions could be located, described, and understood, resulting in the possibility to alter negative mental states.[39]

There is one crucial issue with all studies of emotion, though—what is an *emotion*? To an extent, theologians and philosophers have more freedom to avoid a strict definition; however, for studies to be conducted, scientists need a working definition in order to test hypotheses. In other words, "emotion" has to become a *thing* that can be measured against other things. For example, Nussbaum argues that compassion is an emotion whereas mercy is not.[40] Even if that were true, what differentiates mercy from compassion? Does it feel differently or is the function different? Nussbaum says that compassion is "the central emotion aroused by tragic spectatorship…an emotion that responds to the misfortunes of others."[41] Is that *compassion*? In other words, would all people label that as "compassion" or might that be love, empathy, or another "emotion"? Although she is writing widely on emotion, Nussbaum cautions scientific inquiry, arguing that there is no actual account of emotions, "not even the relatively simple emotion of fear, that identifies it with changes of a particular sort in a particular area of the brain."[42]

Scientists would likely disagree with Nussbaum's general claim, but they do agree with her point about "emotion" remaining undefined. Indeed, there is no shared definition or theory of what an emotion *is* within science.[43] The three proposals I explore here represent important

[39] I consider this issue as it relates to "happiness" in Joel Daniels, "Uncarved and Unconcerned: Zhuangzian Contentment in an Age of Happiness," *Dao* 18 (2019): 577–596.

[40] Nussbaum, *Anger and Forgiveness: Resentment, Generosity, Justice*, 206.

[41] Martha C. Nussbaum, *Political Emotion: Why Love Matter for Justice* (Cambridge, MA: The Belknap Press of Harvard University Press, 2013), 261.

[42] Nussbaum, *Anger and Forgiveness: Resentment, Generosity, Justice*, 253.

[43] For example, Ralph Adolphs, "Investigating Emotions as Functional States Distinct from Feelings," *Emotion Review* 10 no. 3 (2018): 191; Ralph Adolphs, "How Should Neuroscience Study Emotion? By Distinguishing Emotion States, Concepts, and Experiences," *Social Cognitive and Affective Neuroscience* 12, no. 1 (2017): 24; Ralph Adolphs and David J. Anderson, *The Neuroscience of Emotion: A New Synthesis* (Princeton: Princeton University Press, 2018): 260; Ralph Adolphs and Daniel Andler, "Author Reply: We Don't Yet Know What Emotions Are (But Need to Develop Methods to Find Out)," *Emotion Review* 10, no. 3 (2018): 233; Carroll E. Izard, "The Many Meanings/Aspects of Emotion: Definitions, Functions, Activations, and Regulations," *Emotion Review* 2, no. 4 (2010): 363; Caroll E. Izard, "Emotion Theory and Research: Highlights Unanswered Questions, and Emerging Issues," *Annual Review of Psychology* (2009): 4; Lisa Feldman Barrett, "Functionalism Cannot Save the Classical View of Emotion," *Social Cognitive and Affective Neuroscience* 12, no. 1 (2017): 35; Lisa Feldman Barrett, "Categories and Their Roles in the Science of Emotion," *Psychological Inquiry* 28, no. 1 (2017): 20.

theories that scientists implement in their studies: evolutionary psychology, functionalism, and constructionist. Each position contains excellent resources for Pentecostals attempting to understand how the Spirit is experienced non-dually; nonetheless, I adopt the constructionist theory because I argue that it best speaks to the complexity and diversity of Pentecostal emotionality.

Evolutionary psychology, like the other two theories, is extremely diverse, making any brief description inherently problematic and reductive. Consequently, I focus on Carroll E. Izard's work because he was a leading voice in the broad conversation on emotion, though not at all the only one. For Izard, emotion is a byproduct of evolution, helping humans survive and reproduce. Human emotion processes through data, organizing it in a way that maximizes basic existence and perpetuation. Izard argues that "emotion feeling should be viewed as a phase (not a consequence) of the neurobiological activity or body expression of emotion."[44] He adds that emotions are "always present" in human consciousness, meaning emotions "cannot be created, taught, or learned via cognitive processes."[45] Emotions, in short, are "evolved and unlearned neurobiological activity."[46] Humans, according to Izard's theory, are evolutionarily wired to involuntarily respond emotionally to neurological stimuli that naturally and universally exist in the human brain.[47] Consequently, emotions are *functional*, enhancing survival and reproduction.

The functional nature of Izard's position overlaps with "functionalism," though it is not identical. For functionalists like Ralph Adolphs, emotions are difficult to study because there is no way to know how another person feels. Evolutionary psychology has the benefit of

[44] Izard, "Emotion Theory and Research: Highlights Unanswered Questions, and Emerging Issues," 4.

[45] Izard, "Emotion Theory and Research: Highlights Unanswered Questions, and Emerging Issues," 4–5.

[46] Izard, "Emotion Theory and Research: Highlights Unanswered Questions, and Emerging Issues," 6.

[47] One of Izard's famous theories is Differential Emotions Theory (DET), which claims that emotions are universal and innate. See Jo Ann A. Abe and Carroll E. Izard, "The Developmental Functions of Emotions: An Analysis in Terms of Differential Emotions Theory," *Cognition and Emotion* 13, no. 5 (1999): 523–549.

universalizing human cognition and development, but those like Adolphs believe that is too difficult to determine. Instead, Adolphs considers emotion based solely on functionality, but he is not making a definitive claim or elevating functionalism above all others, stating, "a functional account is a promising *type* of account for a science of emotion in general (as an approach)."[48] He is not making an ontological claim about emotions but rather a methodological claim—one can only study the functionality of an emotion and not the ultimate source.[49] Moreover, Adolphs believes the only way the scientific study of emotion progresses is through interdisciplinary work, including "philosophy, ethology, psychology, neurobiology, and cognitive science."[50]

Adolphs seems to agree with Izard on "emotions" existing to some degree in human cognition when saying,

> (a) a science of emotion is constructed from observable data (whatever the aliens could measure), but not, at least in the first instance, by how emotions feel (the conscious experience of having one); (b) emotions are internal states (latent variables) of the same broad type of cognitions (i.e. functionally individuated states); (c) emotions are distinguished from the rest of cognition by specific operating characteristics or features.[51]

Adolphs uses the emotions of fear and disgust as examples, noting how both have a function. For disgust, a person will often involuntarily respond negatively, and fear produces a fight or flight action.

The challenge of the functionalism approach is twofold. First, can functionalism explain other emotions like love? Adolphs admits that emotions outside disgust and fear are difficult to account for with functionalism because there is no standard response.[52] Second, language is difficult to parse with functionalism, though this is also true for all theories of emotion. What is a "normal" emotional response? Even if one does not determine differentiated internal experiences, how is function categorized? For example, does a smile indicate happiness? What about when smiling is insincere? Does it help anything to say the function of smiling is based on

[48] Adolphs, "Investigating Emotions as Functional States Distinct from Feelings," 192.

[49] Adolphs and Andler, "Author Reply: We Don't Yet Know What Emotions Are (But Need to Develop Methods to Find Out)," 233.

[50] Adolphs, "Investigating Emotions as Functional States Distinct from Feelings," 200.

[51] Adolphs, "Investigating Emotions as Functional States Distinct from Feelings," 192.

[52] Adolphs, "Investigating Emotions as Functional States Distinct from Feelings," 198.

"happiness"? Without including feeling, which Adolphs correctly argues is difficult to determine, are "emotions" being studied? Anna Wierzbicka notes that the word "emotion" is not found in all languages, but "feel" is.[53] She concludes by saying, "A technical definition of 'emotion' which would not include the word *feel* would be so out of touch with ordinary language that it could not possibly ensure successful communication between scientists and 'ordinary people,' or indeed between scientists and themselves."[54] Adolphs is aware of the challenges, yet he maintains that the only way to determine emotions is through function.

Lisa Feldman Barrett's constructionist theory disagrees with both Izard and Adolphs, and with others who seem to reduce emotions through cognition or function:

> In constructionist theories, for example, it is no longer meaningful to ask what is, and what is not, an emotion. You don't 'have' emotions, 'display' emotions, or 'recognize' them. You construct emotions as experiences or perceptions—they emerge from complex dynamics within your nervous system, which is constantly in dynamic interaction with the surrounding context that usually includes other creatures, each of whom has a dynamically fluctuating nervous system.[55]

Emotions are not entities that exist somewhere in the brain, activated obligatorily based on survival requirements; rather, emotions are dynamic, interdependent expressions that are irreducible to any one thing. Scientific methodology entails categorization in order to properly test hypotheses, and Barrett argues that the categories that are used are arbitrary, subjective, and inaccurate. The problem Barrett observes is that scientists often fail to see how their own subjective goals influence what they pursue and focus on, assuming an objective perspective, particularly when categorizing emotions.[56] That scientists label and categorize emotions does not mean that those categorizations are accurate or even always useful; it only communicates how certain people organize perceptual data.

[53] Anna Wierzbicka, "On Emotions and on Definitions: A Response to Izard," *Emotion Review* 2, no. 4 (2010): 379.

[54] Wierzbicka, "On Emotions and on Definitions: A Response to Izard," 380.

[55] Lisa Feldman Barrett, "Categories and Their Roles in the Science of Emotion," *Psychological Inquiry* 28, no. 1 (2017): 25.

[56] Barrett, "Categories and Their Roles in the Science of Emotion," 20.

To present an alternative scientific proposal, Barrett and her colleagues developed the Conceptual Act Theory (CAT), which hypothesizes that physical changes in the world are received by an individual's sensorium and then are categorized as specific emotions that have been developed through language, socialization, and quotidian cultural artifacts. Barrett is convinced that an emotion like anger, for instance, is not a single thing that can be reduced to an evolutionary brain enactment or function—for some, anger functions as belligerence, but for others, it resembles depression or even glee. Barrett's point is that complex humans experience changing stimuli constantly and respond in multidimensional ways.[57] The category "anger," therefore, does not exist outside human determination and construction.[58]

Emotions are generated, Barrett suggests, by an intricate system that is interrelated and interdependent. She claims that "during every waking moment of life, mental states are constructed as interoceptive cues from the body and exteroceptive cues from the world are continually categorized and made meaningful with conceptual knowledge stored from past experience."[59] The interoceptive cues serve as a predictive device that "anticipate, rather than react to, sensory inputs from the world. Your brain is wired to be a generative model of your world by using past experience to actively create simulations that best fit the situation you are in."[60] The exteroceptive cues subsequently provide feedback for these simulations as correctives. The human emotional system is dependent on *and* irreducible to consciousness because stimuli on all levels are integrated into emotional expression.

Barrett's theory helps further and even strengthen Pentecostal emotionality. A typical trope proffered by sociologists, anthropologists, and religious scholars is that Pentecostal emotionality is a mechanism that charismatic leaders, either consciously or unconsciously, exploit. Cyril G. Williams, for example, argues that the rhythm and excitement of

[57] "The hypothesis is that emotion categories, like all abstract categories, do not have conceptual cores. Instead, the ideal of the emotion category might be those instances that best achieve the goal, relational theme, or prototypic situation." Lisa Feldman Barrett, "Psychological Construction: The Darwinian Approach to the Science of Emotion," *Emotion Review* 5, no. 4 (2013): 382.

[58] By explaining Barrett's point, I do not mean to suggest that Izard and Adolphs believe "anger" is ontological.

[59] Barrett, "Psychological Constructions, The Darwinian Approach to the Science of Emotion," 383.

[60] Barrett, "Categories and Their Roles in the Science of Emotion," 24.

Pentecostal worship lowers brain efficiency.[61] It is true that some Pentecostal leaders have inappropriately used the faith to promote personal wealth and power. That fact, nevertheless, does not represent the entire movement. As a curative response, Barrett's theory argues that emotional reality is constructed through *language*, *socialization*, and *culture*. In other words, although they dismiss it, critics who oppose Pentecostal emotionality exist in a linguistic, constructed, subjective cultural reality as well, and their specific emotional reality informs their research and theories in the same way emotions inform Pentecostal theology and praxis. There is, consequently, no elevated, objective perspective that is immune to emotionally constructed realities.

What Do Emotions Tell Us About the Spirit?

One primary reason why it is so important to examine emotion within Pentecostal non-duality is because the movement is regularly identified, both by insiders and outsiders, as "emotional."[62] Pentecostals claim to experience the Holy Spirit and then respond based on what they emotionally feel. Thus, the question remains, what is an emotion and how does it help inform Pentecostal non-duality? Barrett's theory, I contend, helps explain Pentecostal non-dual experiential emotionality, providing resources for addressing ecological challenges.

Barrett's theory can help Pentecostals defend their emotionality, and it also offers an interesting perspective on how and why Pentecostalism is so incredibly diverse. The Holy Spirit is emotionally experienced, as seen throughout the Bible, especially in the book of Acts. For example, Acts 2 explains how the people experienced the Spirit, embodying and emoting

[61] Cyril G. William, *Tongues of the Spirit: A Study of Pentecostal Glossolalia and related Phenomena* (Cardiff, Wales: University of Wales Press, 1981), 141–142.

[62] For example, Robert W. Hefner, "The Unexpected Modern--Gender, Piety, and Politics in the Global Pentecostal Surge," in *Global Pentecostalism in the twenty-first Century*, ed. Robert W. Hefner (Bloomington: Indiana University Press, 2013), 4; Martin Lindhardt, "Introduction," in *Practicing the Faith: The Ritual Life of Pentecostal-Charismatic Christians*, ed. Martin Lindhardt (New York: Berghahn Books, 2011), 3; Bernice Martin, "Tensions and Trends in Pentecostal Gender and Family Relations," in *Global Pentecostalism in the twenty-first Century*, ed. Robert W. Hefner (Bloomington: Indiana University Press, 2013), 118; Donald E. Miller, "Contextualizing the Contemporary Pentecostal Movement in China," in *Global Chinese Pentecostal and Charismatic Christianity*, ed. Fenggang Yang, Joy K. C. Tong, and Allan Anderson (Boston: Brill, 2017), 20; Amos Yong, *Spirit of Love: A Trinitarian Theology of Grace* (Waco, TX: Baylor University Press, 2012), 161–162.

Spirit. According to constructivists, human emotion is not uniform or universal, meaning the Spirit's presence and embodied experience *must be different in each context*. Consequently, it should come as no surprise that Pentecostals experience the Spirit differently, leading to different practices and even beliefs.

Another extraordinary outcome emerges here. Because the Spirit is experienced differently across the various emotional responses, that means there is not *one* Spirit, which would create an immediate duality between "Spirit" and all "non-Spirit" entities. The Holy Spirit is no-one-thing, meaning the Spirit can be and is expected to be experienced in anyway and at any-time. Pentecostal emotive experientialism is adaptive to every and all contexts through each Pentecostal community's *pneumatological claims* (language), *worship communities* (socialization), and *location* (culture). In other words, emotion translates experience across boundaries limited by language, society, and culture alone. Regarding ecological solutions, this theory removes the idea that giving facts and figures will always work. What people connect to is experienced emotions, whether through direct contact or through stories.

Non-dualism in Genesis 1:2

These explorations into emotion through scientific inquiry are vital; however, Pentecostals are Biblicists, relying on scripture to guide and direct. The Genesis 1 creation story connects many of these themes together, and this text, at times combined with Genesis 2–3, inform Christian positions on ecology. After Genesis 1:1 explains the separation of heaven and earth, Genesis 1:2 brings in the Spirit, saying, "The Spirit of God moved upon the face of the waters" (KJV). The verse appears to be dualistic, where the Spirit is contrasted with the chaotic waters; however, the original language reveals something closer to Pentecostal non-dualism, explaining first that the term for Spirit is multifarious, meaning wind, breath, and mind. Here, God is not described as singular but rather plural (*ṣelem*). Furthermore, the Spirit's movement is not meant to convey distinction, where the Spirit forcefully acts on creation; rather, the Hebrew word for moves can also be translated as relaxes and even flutters. The term "upon" also seems to express division, but the word is fluid, able to translate as between, beyond, among, because of, and accordingly. Even the word "face" is more non-dual, meaning before and behind, in front of, and before time.

Amos Yong argues for a similar pneumatology focused on interrelationality. Yong examines the creation stories in Genesis 1–2 to establish the Spirit's presence throughout history; one of his objectives for this is to remove the duality often found in Christianity. Yong argues that there is no duality between the Spirit and the "chaotic void" because the Spirit hovered above the void and churned the waters—there was no separation.[63] For Yong, that means there's no duality between transcendence and immanence, saying, "Hence the Spirit can be understood to be both the universality of togetherness and interrelatedness of things on the one hand, and the principle of particularity that makes things what they are and distinct from other things on the other."[64] In short, the Spirit binds the entire universe together.[65]

Thus, the Spirit moves in people and in Creation through non-dualism, relaxingly moving between, among, before, behind, in front of, and beyond all.

Conclusion: Stories of Healing

How can Pentecostal emotive experientialism aid conversations on ecology? First, understanding emotion helps Pentecostals—a global community growing rapidly—engage in the discussion. Rather than being politicized, Pentecostals can focus on stories. Candy Gunther Brown argues that the main Pentecostal characteristic is not speaking in tongues or prophecy but instead healing.[66] In fact, converts regularly cite healing, whether their own or someone close to them, as the main reason that they adopt a Pentecostal spirituality. Conversion and confession tend to follow healing within Pentecostalism. Thus, Pentecostals can seek out "evil" places in the world creating disease and destruction. These human stories invite emotional connection, seeking healing through the Spirit. Although there remains the potential for emotions to be misused or

[63] Amos Yong, *Pneumatology and the Christian-Buddhist Dialogue: Does the Spirit Blow through the Middle Way?* (Boston: Brill, 2012), 39.

[64] Yong, *Pneumatology and the Christian-Buddhist Dialogue: Does the Spirit Blow through the Middle Way?* 39.

[65] Yong, *Pneumatology and the Christian-Buddhist Dialogue: Does the Spirit Blow through the Middle Way?* 43.

[66] Candy Gundy Brown, "Introduction: Pentecostalism and the Globalization of Illness and Healing," in *Global Pentecostal and Charismatic Healing*, ed. Candy Gunther Brown (New York: Oxford University Press, 2011), 8.

misguided—and that same problem surrounds determinations made through "reason"—rooting Pentecostal emotionality in healing provides a framework to mitigate those concerns. Healing, like emotions, is not one thing but rather innumerous things, each unique to the individuals and communities. This encourages human contact and interaction, which is needed for strong emotional attachment, rather than nameless people "out there" who seemingly threaten one's way of life.

Second, Pentecostal non-dualism can extend beyond Pentecostalism, providing resources for understanding human emotion in general. Studies have shown that mountains of data do not change public opinion. Pentecostal non-dualism takes emotions seriously, valuing the ability for change to occur through the affective. Rather than translating facts and figures into different languages and then disseminating, the goal should be to translate stories that connect humans emotionally. Change happens when people emotionally experience the world, and Pentecostal non-dualism can provide that needed resource.

Bibliography

Abe, Jo Ann, and A. and Carroll E. Izard. 1999. The Developmental Functions of Emotions: An Analysis in Terms of Differential Emotions Theory. *Cognition and Emotion* 13 (5): 523–549.

Adolphs, Ralph. 2017. How Should Neuroscience Study Emotion? By Distinguishing Emotion States, Concepts, and Experiences. *Social Cognitive and Affective Neuroscience* 12 (1): 24–31.

———. 2018. Investigating Emotions as Functional States Distinct from Feelings. *Emotion Review* 10 (3): 191–201.

Adolphs, Ralph, and David J. Anderson. 2018. *The Neuroscience of Emotion: A New Synthesis*. Princeton: Princeton University Press.

Adolphs, Ralph, and Daniel Andler. 2018. Author Reply: We Don't Yet Know What Emotions Are (But Need to Develop Methods to Find Out). *Emotion Review* 10 (3): 233–236.

Althouse, Peter. 2014. Pentecostal Eco-Transformation: Possibilities for a Pentecostal Ecotheology in Light of Moltmann's Green Theology. In *Blood Cries Out: Pentecostals, Ecology, and Groans of Creation*, ed. A.J. Swoboda, 116–133. Eugene, OR: Pickwick Publications.

Augustine, Daniela C. 2010. Pentecostal Communal Economics and the Household of God. *Journal of Pentecostal Theology* 19: 219–242.

Barrett, Lisa Feldman. 2017a. Functionalism Cannot Save the Classical View of Emotion. *Social Cognitive and Affective Neuroscience* 12 (1): 34–36.

———. 2017b. Categories and Their Roles in the Science of Emotion. *Psychological Inquiry* 28 (1): 20–26.
———. 2013. Psychological Construction: The Darwinian Approach to the Science of Emotion. *Emotion Review* 5 (4): 379–389.
Barth, Karl. 1982. *The Theology of Schleiermacher: Lectures at Göttingen, Winter Semester of 1923/24*, ed. Dietrich Ritchl, trans. Geoffrey W. Bromiley. Grand Rapids: William B. Eerdmans Publishing Company.
Boff, Leonardo. 1997. *Cry of the Earth, Cry of the Poor*. Maryknoll, NY: Orbis.
———. 2015. *Come Holy Spirit: Inner Fire, Giver of Life and Comforter of the Poor*, trans. Margaret Wilde. Maryknoll, NY: Orbis Books.
Brown, Candy Gunther. 2011. Introduction: Pentecostalism and the Globalization of Illness and Healing. In *Global Pentecostal and Charismatic Healing*, ed. Candy Gunther Brown, 3–26. New York: Oxford University Press.
Casselberry, Judith. 2017. *Labor of Faith: Gender and Power in Black Apostolic Pentecostalism*. Durham: Duke University Press.
Chan, Michael J. 2014. Sins of the Ancestors: Generational Sin, Pentecostalism, and Ecological Crisis. In *Blood Cries Out: Pentecostals, Ecology, and the Groans of Creation*, ed. A.J. Swoboda, 169–184. Eugene, OR: Pickwick Publications.
Corrigan, John. 2007. Introduction: The Study of Religion and Emotion. In *The Oxford Handbook of Religion and Emotion*, ed. John Corrigan. New York: Oxford University Press.
Joel, Daniels. 2019. Uncarved and Unconcerned: Zhuangzian Contentment in an Age of Happiness. *Dao* 18: 577–596.
Daniels, Joel. 2020. What Lies Within: Uncovering the Holy Spirit with the Aid of Buddhist Philosophy. *Buddhist-Christian Studies* 40: 287–305.
———. 2021. From Normative to Nothing: A Pentecostal Theological Conception of the Body. *Pneuma* 43: 233–249.
Dixon, Thomas. 2012. 'Emotion': The History of a Keyword in Crisis. *Emotion Review* 4 (4): 338–344.
Ede, Paul. 2014. River from the Temple: The Spirit, City Earthkeeping, and Healing Urban Land. In *Blood Cries Out: Pentecostals, Ecology, and the Groans of Creation*, ed. A.J. Swoboda, 205–224. Eugene: Pickwick Publications.
Hefner, Robert W. 2013. The Unexpected Modern--Gender, Piety, and Politics in the Global Pentecostal Surge. In *Global Pentecostalism in the 21st Century*, ed. Robert W. Hefner, 1–36. Bloomington: Indiana University Press.
Hubbard-Heitz, Brandon. 2014. The Devil's Suicide: Early Pentecostal Hermeneutics of Space and Their Ecotheological Implications. In *Blood Cries Out: Pentecostals, Ecology, and the Groans of Creation*, ed. A.J. Swoboda, 22–40. Eugene, OR: Pickwick Publications.
Izard, Carroll E. 2009. Emotion Theory and Research: Highlights Unanswered Questions, and Emerging Issues. *Annual Review of Psychology* 60: 1–25.

———. 2010. The Many Meanings/Aspects of Emotion: Definitions, Functions, Activations, and Regulations. *Emotion Review* 2 (4): 363–370.

Jung, Nadine (et al.). "How Emotions Affect Logical Reasoning: Evidence from Experiments with Mood-Manipulated Participants, Spider Phobics, and People with Exam Anxiety," Frontiers in Psychology 5 (2014): 1–12.

Lindhardt, Martin. 2011. Introduction. In *Practicing the Faith: The Ritual Life of Pentecostal-Charismatic Christians*, ed. Martin Lindhardt, 1–48. New York: Berghahn Books.

Martin, Bernice. 2013. Tensions and Trends in Pentecostal Gender and Family Relations. In *Global Pentecostalism in the 21st Century*, ed. Robert W. Hefner, 115–148. Bloomington: Indiana University Press.

Miller, Donald E. 2017. Contextualizing the Contemporary Pentecostal Movement in China. In *Global Chinese Pentecostal and Charismatic Christianity*, ed. Fenggang Yang, Joy K.C. Tong, and Allan Anderson, 15–32. Boston: Brill.

Nussbaum, Martha C. 2001. *Upheavals of Thought: The Intelligence of Emotions.* New York: Cambridge University Press.

Nussbaum, Martha C., and Political Emotion. 2013. *Why Love Matter for Justice.* Cambridge, MA: The Belknap Press of Harvard University Press.

Nussbaum, Martha C. 2016. *Anger and Forgiveness: Resentment, Generosity, Justice.* New York: Oxford University Press.

Peng, Feng. 2015. Fact and Experience: A Look at the Root of Philosophy from the Happy Fish Debate. In *Zhuangzi and the Happy Fish*, ed. Roger T. Ames and Takahiro Nakajima, 229–247. Honolulu: University of Hawai'i Press.

Pipkin, Brian K. 2014. The Environment Theology of Aimee Semple McPherson. In *Blood Cries Out: Pentecostals, Ecology, and Groans of Creation*, ed. A.J. Swoboda, 58–79. Eugene, OR: Pickwick Publications.

Readfearn, Graham. 2021. Ashes to Ashes: Pentecostalism, the PM and the Climate Crisis. *The Guardian*, May 8. https://www.theguardian.com/environment/2021/may/09/ashes-to-ashes-pentecostalism-the-pm-and-the-climate-crisis

Rodgers, Darrin J., and Nicole Sparks. 2014. Pentecostal Pioneer of Earth Day: John McConnell, Jr. In *Blood Cries Out: Pentecostals, Ecology, and the Groans of Creation*, ed. A.J. Swoboda, 3–21. Eugene, OR: Pickwick Publications.

Schleiermacher, Friedrich. 1999. In *The Christian Faith*, ed. H.R. Mackintosh and J.S. Stewart. New York: T&T Clark.

Shultz, F. 2008. *LeRon and Andrea Hollingsworth. The Holy Spirit.* Grand Rapids: Eerdmans Publishing Company.

Swoboda, A.J. 2013. *Tongues and Trees: Towards a Pentecostal Ecological Theology.* Boston Brill.

Tallman, Matthew. 2009. Pentecostal Ecology: A Theological Paradigm for Pentecostal Environmentalism. In *The Spirit Renews the Face of the Earth:*

Pentecostal Forays in Science and Theology of Creation, ed. Amos Yong, 135–154. Eugene, OR: Pickwick Publications.
William, Cyril G. 1981. *Tongues of the Spirit: A Study of Pentecostal Glossolalia and related Phenomena*. Cardiff, Wales: University of Wales Press.
Whitehead, Alfred North. 1978. In *Process and Reality: An Essay in Cosmology*, ed. David Ray Griffin and Donald W. Sherburne. New York: The Free Press.
Wierzbicka, Anna. 2010. On Emotions and on Definitions: A Response to Izard. *Emotion Review* 2 (4): 379–380.
Yong, Amos. 2012a. *Spirit of Love: A Trinitarian Theology of Grace*. Waco, TX: Baylor University Press.
———. 2012b. *Pneumatology and the Christian-Buddhist Dialogue: Does the Spirit Blow through the Middle Way?* Boston: Brill.

CHAPTER 10

The Role of Formal Distinction in the Articulation of *Univocity of Being*: "Neutral" and "Expressive" *Univocity of Being* in the Thinking of Duns Scotus and Spinoza

Jonas Gamborg Lillebø

Introduction

This paper examines the role distinctions play in philosophical reasoning when considering *univocal being* or the idea that the fundamental reality consists of *only one substance*. The problem for those who wish to think about *univocity of being* or hold a so-called "monist" view on substance is how to account for differences and variations without introducing a division into being. A monist philosophy still needs to deal with substantial differences and variations. As the paper aims to show, distinctions play a role in a rationalist approach to univocity.

J. G. Lillebø (✉)
Department of Religious Studies, Volda University College, Volda, Norway
e-mail: jonas.gamborg.lillebo@hivolda.no

T. J. Hastings, K.-W. Sæther (eds.), *Views of Nature and Dualism*,
https://doi.org/10.1007/978-3-031-42902-6_10

The German philosopher Christian Wolf (1679–1754) used the term monist for the first time to characterize philosophers who admit to only *one kind* of substance.[1] Today the term is connected to a variety of topics or fields of research (like the relation between mind and body, subjective internal experience and objective external reality, between the human being and God, between human being and nature) and covers various kinds of monist positions (materialist, spiritual, neutral). Even though there are several ways of thinking about univocity of being or monism of substance, one of the first—and to some perhaps the only—names that comes to mind is the seventeenth-century Dutch philosopher Baruch Spinoza. Even though Spinoza inherited much of the vocabulary from René Descartes, he became famous for claiming against Descartes that *thinking* and *extension* are not different substances (*res cogitans* and *res extensa*), but rather different attributes of the same substance.

However, as the French philosopher Gilles Deleuze writes, there is a philosophical tradition and other philosophers with whom it is fruitful to compare Spinoza. Deleuze mentions two philosophers in this tradition, Friedrich Nietzsche, and John Duns Scotus, both philosophers Spinoza himself never read.[2] However, when it comes to thinking about *being*, they have similar or common thoughts. This is particularly interesting in the case of Duns Scotus, to whom we will turn later in this text. While Spinoza was expelled from the Jewish congregation in the Netherlands due to his heretical thoughts and was regarded as a pantheist and even an atheist, and Nietzsche 200 years later launched an attack on Christian civilization itself, the name of the thirteenth-century Christian theologian Duns Scotus may seem more out of place.

Deleuze, however, in the book *Difference and Repetition* (2004), makes the following statement: "There has only ever been one ontological proposition: Being is univocal. There has only ever been one ontology, that of Duns Scotus, which gave being a single voice."[3] In the late sixties, Deleuze published two books that addressed the connection between

[1] Horst Hillermann and Anton Hügli, "Monismus," in *Historisches Wörterbuch der Philosophie*, edited by Joachim Ritter (Darmstadt: Wissenschaftliche Buchgesellschaft, 2019); Carlo Mazzantini and Dario Sacchi, "Attribut," in *Enciclopedia filosofica* (Milano: Bompiani, 2010b).

[2] Deleuze writes that Spinoza probably never read Scotus's own writings, but that he might have been influenced by Juan of Prado, who "definitely knew Duns Scotus." Gilles Deleuze, *Expressionism in Philosophy: Spinoza* (New York: Zone Books, 2013), 359ff.

[3] Gilles Deleuze, *Difference and Repetition* (London/New York: Continuum), 2004, 44.

Duns Scotus and Spinoza: *Expressionism in Philosophy: Spinoza* (*Spinoza et le problème de l'expression*), which is an interpretation of Spinoza, and subsequently *Différence et Répétition* (*Difference and Repetition*) on the ontology of difference, identity, and repetition. For those who know some of Deleuze's other writings, which emphasize *immanence*, the above statement might come as a surprise. Duns Scotus was, however, Deleuze continues, "the one who elevated univocal being to the highest point of subtlety." He immediately adds, however, that Scotus thought this was "at the price of abstraction"[4] by rendering univocal being *indifferent*. Spinoza, on the other hand, makes univocal being into "an object of pure affirmation." In other words, we are confronted with two traditions for thinking about the univocity of being with significant differences among them.

The task in this text is not to discuss whether Deleuze's reading of either Scotus or Spinoza holds. I aim to examine and demonstrate how important distinctions—particularly the so-called *formal distinction*—are a rationalist argument for univocity of being. The question is thus: *What kind of role does formal distinction play in the univocity of being?* Without a formal distinction, it would, for instance, be difficult to account for observable differences in nature and the ontological univocity of being at the same time. Scotus and Spinoza represent different philosophies simultaneously giving them radically different takes on formal distinction.

Before comparing the two thinkers, I want to examine them separately by looking at the background of their thinking and some of their texts. Doing so requires us to take a tour through metaphysics, logical inferences, and distinctions in medieval and seventeenth-century logical and metaphysical thinking. This is rather complex, but I will make it as clear and brief as possible.

Univocity, Analogy, and Equivocity

John Duns Scotus (1266–1308) was a Scottish-born Franciscan monk and theologian. He was one of the so-called "Oxford theologians" alongside, amongst others, William of Ockham (1285–1347), but he also taught for many years in Paris and Cologne. So where can we place Scotus philosophically and theologically? As the theologian and historian of philosophy Frederick Charles Copleston writes in his *A History of Philosophy* (1993), the interpretation of Scotus has ranged from being a direct

[4] Deleuze, *Difference and Repetition*, 44.

precursor to Ockham and Martin Luther (1483–1546) to those who try to soften the differences between Scotus and the thinking of Thomas Aquinas (1225–1274). However, though the analogical thinking of Thomas is certainly central, the work of Henry of Ghent (1217–1293) is perhaps at least as important when we want to understand Scotus's conception of *being*. What kind of problem is the question of being connected to in Scotus's philosophy?

First, we must clarify the relation between theology and philosophy. The question of being has been a problem at least since Aristotle's *Metaphysics*. While we do not have the time to go into this here, there is a problem finding a univocal term for being in nature, since being has an irreducible plurality of meanings. Moreover, the law of contradiction prohibits that a substance, quality, or being can be and not be under the same relation simultaneously.[5] An example discussed by Aristotle is whether "health" harbors an identical meaning in all contexts (the health department, being healthy, healthy nutrition, etc.). Since it does not, Aristotle's solution is to see its meaning across individual contexts as analogous. This means univocity and analogy offer rival understandings of being. As Nathan Widder explains, this philosophical problem became a way of framing theological problems in the medieval period. The choice of univocity or analogy to resolve the problem of the categories shapes the answers offered, not only for the philosophical problem of individual diversity but also "a later Christian theological difficulty of the relationship within irreducible disparity between God and His creatures."[6] Against this backdrop, one of the main questions in medieval Christian theology was whether we can have knowledge of God, and if so, what kind of knowledge we can have of God. There were many answers, and the two main contributors were Thomas Aquinas and Henry of Ghent.

Thomas Aquinas was trying to find a middle ground between a negative theology (God is good, but his goodness is not human) and (an impossible) positive theology. Thomas proposed that we can acquire knowledge of God in an *analogical* way to knowledge of material things and creatures. Without going into too much detail here, we limit ourselves to mentioning that Thomas distinguishes between two kinds of

[5] Nathan Widder, "The Univocity of Substance and the Formal Distinction of Attributes: The Role of Duns Scotus in Deleuze's Reading of Spinoza," *Parresia* 33 (2020), 157.

[6] Widder, "The Univocity of Substance and the Formal Distinction of Attributes: The Role of Duns Scotus in Deleuze's Reading of Spinoza."

resemblance found in two kinds of analogical predication, the analogy of proportion (*analogia secundum convenientiam proportionis*) and the analogy of proportionality (*analogia secundum convenientiam proportionalitatis*).[7] This is the difference between univocity and analogy of meaning.[8] Thomas admits that the meaning of the words we use to describe God, like "wise," is the same as when applied to creatures. But he is negative when it comes to finding an analogy on a referential level.

Given the context of this article as part of the discussion on different cosmologies and views on nature, it is interesting to observe how the analogical solution of Thomas is part of a Christian holism, a holism that could also be found in a pre-Christian version in Aristotle. This holism takes the form of a hierarchy in the sense that, for instance, human goodness both pre-exists and, in a higher modality, is in the divine substance. While I do not pretend to have completely understood Thomas, given that the cosmos is a hierarchical holistic structure, it follows that every part is always already connected to the whole. Or put differently, lower and higher modalities are connected within the same holistic structure.

Henry, for his part, suggested that the concept of being as applied to God and to creatures is *not univocal*. Following Frederick C. Copleston in his *A History of Philosophy* (1993), Henry refused to admit to a univocal concept of being: "There is no univocal concept of being common to God and creatures: There are two concepts, that of a necessary being and that of a contingent being, and our concept of being must be one or the other."[9] Scotus is critical of both these positions. On the one hand, while Thomas's analogical method is in a nowhere land between positive and negative theology, Henry's thinking ends, according to Scotus, on the other hand, in agnosticism regarding knowledge of God.

To go further than *analogical* thinking and certainly to avoid the dead end that the understanding of *being as equivocal* leads to, Scotus proposes seeing *being as univocal*. We can read this in § 26 of Book 1 distinction 3 in his work *Ordinatio*. Duns Scotus here writes:

[7] Frederick Charles Copleston, *A History of Philosophy Volume II. Medieval Philosophy. From Augustine to Duns Scotus* (New York: Image Books, 1993), 356.

[8] James F. Ross and Todd Bates, "Duns Scotus on Natural Theology," in *The Cambridge Companion to Duns Scotus*, edited by Thomas Williams (Cambridge: Cambridge University Press, 2013), 213.

[9] Copleston, *A History of Philosophy Volume II. Medieval Philosophy. From Augustine to Duns Scotus*, 471–472.

> Secondly, I am saying that God is not merely conceived in a concept analogous to the concept of the creature, that is to say, which is entirely different from what is said of the creature, but in some concept univocal to Him and creatures.[10]
>
> (Secundo dico quod non tantum in conceptu analogo conceptui creaturae concipitur Deus, scilicet qui omnino sit alius ab illo qui de creatura dicitur, sed in conceptu aliquo univoco sibi et creaturae.)

In this paragraph a univocal concept is contrasted to both analogical and equivocal concepts. While an analogical concept accepts that there is something "entirely different" (*omnino sit alibus*) between two phenomena, a univocal concept is the same for both. In other words, not just purely equivocal concepts, as Henry of Ghent holds, but also analogical concepts contain the fallacy of equivocal concepts. Scotus's reasoning seems sound. The concept of *being* has a quality that an analogical thinking cannot express since analogies assume a difference or an otherness. The problem is not that there are differences between beings, but that these differences are taken as constituting the most fundamental conceptual level. Analogical thinking does not distinguish between different ontological levels. But if we, on the contrary, accept that there is a <ontological> difference between *being* (singular and general) and *specific beings* (plural), *being* must be expressed through a univocal concept. For instance, even though a mouse and a frog are different beings, they share, on a more fundamental level, the fact that they both *are*. Put differently; their differences do not express their most fundamental level. As far as I can understand, Scotus's criticism of analogical thinking is because it confuses these two levels. So, according to Scotus, there is a problem or lack of depth in analogical thinking. This critique could be understood as either a refutation of analogical thinking or a condition for analogical thinking. As Widder writes, "Any analogy of predicates must therefore be grounded in a prior univocity between infinite and finite substances."[11] Due to limited space, we will not discuss this matter further.

The main point is that though there are perceivable differences in nature, univocity of being is more fundamental and "prior" to the disjunction created/not created. The French expert on Duns Scotus, Olivier

[10] John Duns Scotus in *Ordinatio* I, 3, 2 no. 5. Thank you to colleague Timothy Saunders for commenting on my translation of this quote from Latin to English.

[11] Widder, "The Univocity of Substance and the Formal Distinction of Attributes: The Role of Duns Scotus in Deleuze's Reading of Spinoza," 158.

Boulnois, writes in the lexicon article in *Historisches Wörtebuch der Philosophie*: "Even the concept of being, applied to God, is prior to the disjunction created/uncreated, consequently it is univocal."[12]

Maybe I am too close to Deleuze's reading of Duns Scotus, but it seems to me that the novelty of Duns Scotus is that he does not make the mistake of scrutinizing ontological categories from an epistemological perspective or being from the perspective of knowledge of attributes. While negative and analogical theology remains on the secondary/epistemological level of attributes (e.g., comparing the goodness of humans with that of God), to determine ontologically God's being, Duns Scotus discovers a more fundamental ontological level that can be expressed through the logical study of concepts and inferences. Nevertheless, there is a commonality behind the differences (of being and/or degree of supremacy) between God and creation. As Scotus expresses in § 39 Book 1 distinction 8, "God is formally being, but being states a concept said of God in the 'what'—and this concept is not proper to God but is common to him and creatures (Deus formaliter est ens, ens autem dicit conceptum dictum de Deo in 'quid'—et iste conceptus non est proprius Deo, sed communis sibi et creaturae)."[13]

This commonality of concepts is however not an identity and the univocity of being is not generic. On the contrary, the univocity of being is transcendental and, as Widder writes, "characterized by its indifference to the difference between the finite and infinite. Being's transcendental univocity thus applies to God and creatures without gathering them under an identity."[14] As will be discussed later, this is linked to Scotus's understanding of the neutrality of the univocity of being.

Even though the notion of univocity of being is perhaps what Duns Scotus is most famous and celebrated for, this is closely connected to another aspect that Deleuze claims is "one of his most original concepts, which complements that of univocity: the idea of formal distinction."[15] The formal distinction is also an interpretative key to understanding Spinoza and ultimately the discussion of monism. So let us have a look at this.

[12] Sven. K. Knebel and Olivier Boulnois, "Univozität; univok," in *Historisches Wörterbuch der Philosophie,* edited by Joachim Ritter (Darmstadt: Wissenschaftliche Buchgesellschaft, 2019), 227. Quotes originally in German and Italian are translated by the author.

[13] As Scotus expresses in § 39 Book 1 distinction 8.

[14] Widder, "The Univocity of Substance and the Formal Distinction of Attributes: The Role of Duns Scotus in Deleuze's Reading of Spinoza," 159.

[15] Deleuze, *Expressionism in Philosophy: Spinoza*, 63.

Real Distinction, Conceptual Distinction, Modal Distinction, and Formal Distinction

We saw in the previous section the differences between the various ways of understanding *being* in medieval thinking. To understand what the *formal distinction* is, we must take a similar tour and give a relatively brief account of the main kinds of distinctions in medieval thinking. Defining and locating different kinds of logical inferences had been an important part of the philosophical endeavor for a long time. In medieval philosophy, this reached a climax in the art of distinction to which Duns Scotus contributed. This might seem very abstract to those without special training or interest in logic, philosophical arguments, and inferences. Still, the reader must remember that the aim is to clarify different aspects and levels of our thinking. The first two distinctions are (a) *real distinction* and (b) *conceptual distinction/distinction of reason.*

(A) *Real distinction*: Two items are distinct if and only if they are separable (if and only if one can exist without the other).[16] A *real distinction* is *not* the product of some sort of mental activity. To say that the distinction exists *in the things* themselves is probably anachronistic since, the Kantian distinction between the thing in itself (*Ding an sich* and *Ding für mich*) was of course unknown to Scotus. An example, though, is the distinction between *cat* and *dog*. There is a real distinction between them, and not "only" in the concepts. For reasons that become clearer later when we discuss the *formal distinction*, I think it might be less ambiguous to call the *real distinction* a *real separation*.

(B) *Conceptual distinction*: A distinction that is the product of some sort of mental activity.[17] A well-used example is the "morning star" and the "evening star." Both concepts have the same reference, the planet Venus. However, the distinction between the different concepts cannot be found in Venus but exists only conceptually.

Scotus discusses two other distinctions: (c) *modal distinction* and (d) *formal distinction*. The second is the most important for our purposes. Even though the formal distinction went back to

[16] Peter King, "Scotus on Metaphysics," in *The Cambridge Companion to Duns Scotus*, edited by Thomas Williams (Cambridge: Cambridge University Press, 2003), 21.

[17] King, "Scotus on Metaphysics," 22.

Bonaventura (1221–1274) and was also discussed before Duns Scotus, the latter developed it systematically.[18] The problem that the formal distinction is a response to is accounting for phenomena that are real (*ex natura rei*) and independent of mental activity (*differentia quae ponitur praecedere omnem actum intellectus*),[19] but it still does not entail separation between two or more entities like a *real distinction* does. It is sometimes explained as a distinction that is less than a real distinction but more than a conceptual distinction. But this does not, as far as I can see, adequately articulate the difference between distinction and separation that seems to be at the core of Scotus's reasoning because how can we account for phenomena that are, on the one hand, distinct but not separated, and on the other hand, real and not conceptual?

(C) *Modal distinction*: As far as I have understood it, *modal distinctions* are distinctions of degrees or nuances. There are, for instance, distinctions of the color red but it is still the same color. The distinction is real and does not stem from our concepts. The modal distinction was later developed by Francisco Suárez (1548–1617) who had an influence on Descartes.[20] Deleuze claims that both Suárez and Descartes rejected the *formal distinction*, which we shall soon deal with, by confusing it with the *modal distinction*. Furthermore, it was the introduction of modality into the understanding of substance that, according to Spinoza, is the fundamental error of Descartes.[21] But for now, we will leave this aside and try to understand the *formal distinction*.

(D) *Formal distinction*: A distinction that is real and does not entail separation. An example from Scotus is the distinction between will and intellect and their relation to the soul. It is a real distinction and not the product of mental activity, but the distinction does not entail separation since will and intellect are both parts of the soul. At the risk of oversimplifying, the formal distinction could be characterized as aspects of the same thing. An aspect is not separated from that which it is an aspect of. On the one hand, soul and

[18] Otto Muck, "Distinctio formalis," in *Historisches Wörterbuch der Philosophie*, edited by Joachim Ritter (Darmstadt: Wissenschaftliche Buchgesellschaft, 2019a).

[19] Scotus in Muck, "Distinctio formalis," 269.

[20] Otto Muck, "Distinktion," in *Historisches Wörterbuch der Philosophie*, edited by Joachim Ritter (Darmstadt: Wissenschaftliche Buchgesellschaft, 2019b)

[21] Deleuze, *Expressionism in Philosophy: Spinoza*, 30.

> will are distinct but not separated. On the other hand, will and intellect are also distinct but not separated. According to Deleuze, however, there is a potential pitfall (which my example might fall into) in seeing the formal distinction as an aspect if we want to think about monism or the univocity of being. In thinking that formal distinction means aspects of the same thing, the question is whether we have (unconsciously?) reintroduced a duality or separation between being and an aspect of being. If we are going to consider monism or univocity of being, then being and an aspect of being must be ontologically identical.

The *formal distinction* is, on the one hand, comprehensible from known examples. A coin, for instance, is one but has two sides (heads and tails) that cannot be separated from the coin itself. There is no coin if we take away either of the sides. And yet there is a distinction between the two sides. On the other hand, our thinking tends to draw this in the direction of either real distinction or conceptual distinction. But as Peter King writes, "[t]he core intuition behind Scotus's formal distinction is, roughly, that existential inseparability does not entail identity of definition."[22]

To anticipate how the *formal distinction* is useful to us vis a vis Spinoza's concept of substance, we see that the relation between the substance and its attributes "needs" a formal distinction to be reasonable. As is well known, Spinoza's claim against Descartes is that substance is one, and not two, but expresses itself in an infinity of attributes. The attributes of the substance are ontologically inseparable from the substance but still distinct from the substance and each other. In other words, despite an infinite number of distinct and mutually heterogeneous attributes, this heterogeneity does not entail a division of the one substance into many. However, before examining Spinoza, we must also look at Deleuze's main objection to Duns Scotus in comparison to Spinoza: the neutralization of being.

Neutrality or Pantheism?

Despite the potential Deleuze sees in Scotus's concept of *univocity of being*, which he articulates through the *formal distinction*, the problem, according to Deleuze, is that Scotus neutralizes the univocity of being:

[22] King, "Scotus on Metaphysics," 22.

> In the greatest book of pure ontology, the *Opus Oxiniense*, being is understood as univocal, but univocal being is understood as neutral, neuter, indifferent to the distinction between the finite and the infinite, the singular and the universal, the created and the uncreated.[23]

And rightly, in Scotus's text we can read that "every common concept is neutral with respect to the things to which it is common; no concept is neutral with respect to contradictories, because it is one or other of them" (*omnis conceptus communis est neuter respectu illorum quibus est communis; nullus est conceptus neuter respectu contradictoriorum, quia est alter eorum*). There seems to be both a logical and a theological reason for calling univocal being transcendental[24] and neutral. The question is whether the theological guides the logical.

The logical reason is, for instance, that a predicate such as materiality only applies to finite (creation), not to infinite (God) creatures. Another example is that the difference that constitutes an individual is not common to anyone other than that individual. For example, the "Socraticity" of Socrates is not common to others, only to Socrates. For such predicates, Duns Scotus claims, transcendental or neutral terms do not hold. It then seems logical to say that only transcendentals are univocal.

According to Widder and Deleuze, the theological reason for thinking of univocal being as neutral is obvious. If every predicate was univocal, not just transcendental, then "every predicate could be considered univocal, making it impossible to limit what can be affirmed of God and resulting in precisely the pantheism univocity and analogy both seek to avoid."[25] In other words, if universal being is not neutralized but affirmed in all predicates, there is no difference between God and creation. Deleuze writes that Scotus neutralized univocity to avoid pantheism, and accordingly did not take the step into an affirmative philosophy as did Spinoza:

[23] Deleuze, *Difference and Repetition*, 48.

[24] Transcendental in Scotus's sense "is characterized by its indifference to the difference between the finite and infinite (Widder, "The Univocity of Substance and the Formal Distinction of Attributes: The Role of Duns Scotus in Deleuze's Reading of Spinoza," 159). In other words, being as transcendental is indifferent to the difference between the finite and the infinite. However, being has two modes and is accordingly "quantified into infinite and finite manifestations" (King, "Scotus on Metaphysics," 27).

[25] Widder, "The Univocity of Substance and the Formal Distinction of Attributes: The Role of Duns Scotus in Deleuze's Reading of Spinoza," 160.

> Furthermore, univocity in Scotus seems compromised by a concern to avoid pantheism. For his theological, that is to say 'creationist,' perspective forced him to conceive univocal being as a neutralized, indifferent concept.[26]

Deleuze and Widder might be quite right in reading Scotus this way. Furthermore, this objection is highly relevant to the question of monism since it provides a valuable nuance for evaluating "monistic philosophical positions."

Could not Scotus also be understood consistently with the meaning of the notion "neutral"? According to Latin dictionaries, "neuter" means "neither of the two," "which does not belong to either the one or the other side of a class, group, etc."[27] If I understand this correctly, Scotus contrasts common concepts (*conceptus communis*) that are neutral (*neither* this *nor* that) with contradictory concepts that are *either* this *or* that. This stems from a discussion of the difference between univocal and equivocal concepts in medieval logical thinking. The difference between equivocal and univocal terms Scotus took from Simplicius (ca. 480–560 CE), who stated that "the univocal does not allow contradiction" (*Univocum autem non suscipit contradictionem*).[28] Having said that, Scotus seems unclear on the matter since there are no equivocal concepts according to Boulnois's reading of Scotus. If a concept has opposing meanings, it is not one concept but two. In other words, if it is a concept, it does not have opposing meanings (*si sum certus de aliquo uno conceptu, ille univocum est*).[29] In the article on Duns Scotus in *Enciclopedia filosófica*, Poppi writes:

> Even though being indetermined and neutral regarding any concrete determination of beings or categories, the concept of being possesses equally its own semantical identity capable of excluding contradictions and guaranteeing the validity of the demonstrated conclusion.[30]

To sum up: On the one hand, the reason for neutralizing the univocity of being is to avoid pantheism; on the other hand, from a logical point of view, one could argue that the univocity of being must be neutral to avoid contradiction. So even though Deleuze might be right that Scotus tries to

[26] Deleuze, *Expressionism in Philosophy: Spinoza*, 67.
[27] *Latinsk ordbok* (Oslo: Cappelen Damm, 2007).
[28] Simplicius in Knebel and Boulnois, "Univozität; univok," 226.
[29] Scotus in Knebel and Boulnois, "Univozität; univok," 228.
[30] Antonio Poppi, "Duns Scoto," in *Enciclopedia filosofica* (Milano: Bompiani, 2010), 3141.

avoid pantheism, it still seems that Scotus's reasoning could also be understood as genuinely logical and that such an inference is tenable from a certain perspective in logical reasoning.

Another question, though, is whether univocal being *necessarily* means that univocity of being is neutral. If we leave Scotus's intentions of avoiding pantheism out of it, the question remains whether univocity is necessarily neutral or if this is just one way of interpreting univocity. As far as I understand Deleuze, or perhaps my simplification of what he says, univocity does not have to be neutral (in the sense of *neither* this *nor* that) but, as we will see in Spinoza, can be affirmative (*both* this *and* that).

This is a bit technical but still relevant to the difference between dualist and monist thinking and between different kinds of monist thinking. To anticipate matters that will be discussed later, I would say that whereas dualism separates being (or substance) into *either* this *or* that (like in Descartes, substance is either *res cogitans* or *res extensa*), Scotus and Spinoza represent alternatives to dualism in two different ways. To Scotus, being is more a fundamental category than, for example, Descartes's *res cogitans* or *res extensa*. To Scotus, being is *neither res cogitans nor res extensa*, so to speak. According to Spinoza, however, substance is *both res cogitans and res extensa* (as well as all other attributes we do not know). And it is in articulating how *res cogitans* and *res extensa* can be distinct but not separated, real, and not conceptual that Deleuze "needs" Scotus's formal distinction. But before returning to this, we will have a look at Spinoza.

Spinoza and the Distinction Between Substance and Attributes

Baruch Spinoza (1632–1677) was a Dutch philosopher with a Jewish background who had emigrated from Portugal. Spinoza did not publish much in his lifetime due to the radicality of his thinking. Like many other philosophers of his time, Spinoza's main opponent was René Descartes. Central is the discussion above of Descartes's claim that there are two substances (*res cogitans* and *res extensa*). This idea that there are two kinds of substances is one version, or rather a broad category, of what we call "dualism." In the case of Descartes, his dualism could be characterized as an "instantiated-attributive dualism," since he not only identifies substances from their most fundamental attributes (thought and extension) but also holds that "there actually are substances of each of these kinds."[31]

Today, dualism has fallen into disfavor, and it seems almost unnecessary to argue against it. However, with Spinoza's critique of Descartes, it becomes clear that the problem of dualism and the argument in favor of monism is more intricate than we think. Furthermore, it becomes clear that monism and dualism need clarification through reason, arguments, and distinctions. After all, Spinoza rejects Descartes on rational and not empirical grounds. And it is through making a distinction of his own that Spinoza both criticizes Descartes and builds his own thinking. Central here is how distinction and connection are considered, particularly between the two concepts of *substance* and *attribute* in Descartes's and Spinoza's reasoning. Let us begin with how Spinoza himself understands substance and attribute:

> By substance, I understand that which is in itself and is conceived through itself; in other words, that, the conception of which does not need the conception of another thing from which it must be formed.[32]
>
> (Per substantiam intelligo id, quod in se est, & per se concipitur: hoc est, id, cujus conceptus non indiget conceptu alterius rei, à quo formari debeat).[33]

Firstly, we must be aware that there is some confusion related to vocabulary when facing Spinoza's text. *Substance* is sometimes used in the sense of *the one and only substance* and sometimes used in a plural form as the different *attributes* at times are called "substances." For instance, part I, Proposition II: "Two substances having different attributes have nothing in common with one another."[34] (Duae substantiae, diversa attributa habentes, nihil inter se commune habent).[35] Furthermore, the concept of substance only occurs in the two first parts of his work and is, throughout the work, substituted by *God* or *Nature*.[36] From reading some of the primary texts of Spinoza and some secondary sources, it becomes clear that

[31] Roger. S. Woolhouse, "Spinoza and substance," in *Descartes, Spinoza, Leibniz. The concept of substance in the seventeenth-century metaphysics* (Milton Park: Routledge, 1993), 35.

[32] Benedict Spinoza, *Ethics* (Ware: Wordsworth Editions, 2001), 3.

[33] Baruch Spinoza, *Éthique*. Bilingue Latin-Français (Paris: Points, 2010), 14.

[34] Spinoza, *Ethics*, 4.

[35] Spinoza, *Éthique*, 18.

[36] Charles Ramond, *Dictionnaire Spinoza* (Ellipses, 2007), 177.

he is notoriously difficult to understand. But to make it easier, let us just for now accept that to Spinoza, there is only one substance, and that following Deleuze, we can say that Spinoza is thinking about being as *univocal*.

A second problem is the intimate relation between the one and only *substance* and the plurality of *attributes*. Even though there is only one substance, this does not mean that everything is the same or identical. According to Spinoza, we only know of two attributes: thinking and extension. Furthermore, these attributes have *modes*. On the one hand, we have a unity of substance; on the other, we have a plurality of attributes and modes. This calls for distinctions, but which ones? Deleuze presents Spinoza's challenge to understand what kind of distinction can be introduced into the absolute eternal. How can we understand difference and variation without dividing the univocity of being into parts? As we have seen, in Scotus's thinking, this was the problem with equivocal and analogical concepts: They both assume that *being is equivocal*.

The relation between the (many) attributes and the (one) substance is at play here. Traditionally we use the notion of attribute as an attribution to a subject, as a characteristic through which we can identify something or someone, but where there is a distinction between the essence and the attributes of something or someone.[37] For example, we distinguish between the essence of a human being and the properties of this human being (gender, nationality, temperament, etc.). Spinoza admits to a *distinction* between substance and attribute but not to an ontological separation. He defines attribute thus: "By attribute, I understand that which the intellect perceives of substance, as if constituting its essence."[38] (Per attributum intelligo id, quod intellectus de substantiâ percipit, tanquam ejusdem essentiam constituens).[39]

A problem with interpreting this passage is that we are led to believe that by attributes, he only refers to *how the intellect sees it* and that attributes "in reality," so to speak, could be something other than the one and only substance. We will not go into this debate here but only mention that there is a debate on how to interpret the Latin meaning of *tanquam* in the quote from Spinoza as either "as if" or "as." This has led to a long debate

[37] Ramond, *Dictionnaire Spinoza*, 27.

[38] Spinoza, *Ethics*, 3.

[39] Spinoza, *Éthique*, 14.

between a "subjective" and an "objective" interpretation of Spinoza.[40] That the attributes are something that exists as a perspective for the intellect, "but in reality it might be something else," is a "subjective" interpretation. As highlighted by many, in the "objective" strand, this is a false conception,[41] since there is no "outside" from where the intellect can grasp the substance.

I neither want to go into this debate here nor claim to have sufficient knowledge of Spinoza to evaluate this. However, in the light of the formal distinction, we are given the opportunity to see the role of the intellect as distinct but not separated from the substance. We encountered a similar problem in Scotus's teaching of formal distinction. A formal distinction accounts for a distinction that (1) is real and not the product of some mental activity[42] and (2) does not entail separation. The above passage from *Ethics* could be understood in a similar way here. If attribute is inseparable but distinguished from substance, then the distinction between substance and attribute and between intellect and substance are formal. As stated by Deleuze:

> All qualified substances form only one substance from the point of view of quantity. So that attributes themselves have at once identity of being and distinction of formality. Ontologically one, formally diverse, such is their status (ontologiquement un, formellement divers, tel est le statut des attributs).[43]

This means, however, that there is no room for thinking of attributes as an *aspect of* or *perspective on* substance. As Charles Ramond writes, an attribute is not a fragment of the essence of the substance but "expresses

[40] Jason Waller, "Benedict de Spinoza: Metaphysics." In *Internet Encyclopedia of Philosophy*.

[41] Gilles Deleuze, *Spinoza et le probléme de* l'expression (Paris: Éditions de minuit, 1968); Charles Ramond, *Le vocabulaire de Spinoza* (Ellipses, 1999); Hanne White, "Attributtets ekspressivitet," *Agora. Journal for metafysisk spekulasjon* 2–3 (2003); "The Univocity of Substance and the Formal Distinction of Attributes: The Role of Duns Scotus in Deleuze's Reading of Spinoza"; Woolhouse, "Spinoza and substance." According to Widder, it was precisely the need to avoid a subjective idealist philosophy (in particularly Hegel) that led Deleuze to emphasize the idea of expression in Spinoza's philosophy (Widder, "The Univocity of Substance and the Formal Distinction of Attributes: The Role of Duns Scotus in Deleuze's Reading of Spinoza," 153).

[42] King, "Scotus on Metaphysics."

[43] Deleuze, *Spinoza et le probléme de l'expression*, 56; *Expressionism in Philosophy: Spinoza*, 66.

the essence of God in its totality."[44] For Deleuze, the formal distinction seems to be a "game changer" for articulating an immanent and objective interpretation of the relation between attributes and substance.

There are three kinds of distinctions Deleuze discusses in *Expressionism in Philosophy*[45]: real distinction, numerical distinction, and formal distinction. If we concentrate for the moment on the first two, Deleuze's point is that Spinoza tries to avoid Descartes's error of confusing numerical and real distinctions. As Deleuze highlights, Descartes was confronted with Scotus's notion of formal distinction by the Dutch theologian Johannes Caterus (1590–1657). As a reaction to Descartes's book *Meditations on First Philosophy*, Caterus writes: "In order for something to be conceived distinctly and separately from another it is sufficient that there is what he [Duns Scotus] calls a formal or objective distinction."[46] Descartes dismisses this, saying that "it is no different from the modal" (ibid.).

Descartes wrote in the later work *Principia philosophiae* that we recognize substance from its attribute since there are no attributes that come from nothing (*substantia…agnosimus ex quodlibet eius attributo, per communem illam notionem, quod nihili nulla sint attributa*).[47] The conceptual distinction or distinction of reason plays a role here. An attribute is considered a constitutive element of a substance to which it is really identical and only distinguished by a *distinction of reason*:

> Finally, a conceptual distinction [or 'rational distinction'] is a distinction between a substance and some attribute of that substance without which the substance is unintelligible; alternatively, it is a distinction between two such attributes of a single substance. Such a distinction is recognized by our inability to form a clear and distinct idea of the substance if we exclude from

[44] Ramond, *Dictionnaire Spinoza*, 28.

[45] What is confusing in entering this domain of distinctions is that sometimes different notions refer to the same phenomena (conceptual distinction and distinction of reason are the same), and sometimes the same notions refer to different phenomena depending on the philosophers (Spinoza and Descartes have a different understanding of attributes and substance). In other words, it is, at times, relatively difficult to understand whether two thinkers are in agreement or not, and to what extent.

[46] Caterus in Deleuze, *Expressionism in Philosophy: Spinoza*, 359.

[47] Descartes in Helmut K. Kohlenberger and Ludger Oeing-Hanhoff, Attribut," in *Historisches Wörterbuch der Philosophie*, Band 1, edited by Joachim Ritter (Darmstadt: Wissenschaftliche Buchgesellschaft, 2019).

> it the attribute in question, or, alternatively, by our inability to perceive clearly the idea of one of the two attributes if we separate it from the other.[48]

If we recall the example of the *conceptual distinction/distinction of reason* in the second section on Duns Scotus above, we saw that the distinction between the "evening star" and the "morning star" was only conceptual. There is no corresponding *real distinction* on the planet Venus. The difference between the "evening star" and the "morning star" is only nominal and exists only in our minds or in language. So, when in Descartes's reasoning the relation between substance and attribute is conceptual, this is the same as saying that attribute and substance have no real distinction between them after all: In Descartes's view, they are really identical, only conceptually distinct.

But is this not the same thing as what Spinoza is saying? Even though Spinoza takes up many similar thoughts and, in part inherits Descartes's vocabulary, there are subtle but pivotal differences between them. The most central work of Spinoza is his *Ethics*, where he crafts his main ideas. In this book, Spinoza introduces some principal distinctions (though he does not use the notion of "distinction" so much) both to problematize Descartes and to find a new path for thinking. In fact, as Deleuze emphasizes, distinction is one of the main keys to understanding Spinoza's philosophy: "One of the sources of Spinoza's anti-Cartesianism is to be found in the theory of distinctions."[49] I do not claim to offer an exhaustive list here, but there seem to me to be several critical points in Spinoza's theory of distinctions in comparison to Descartes.

(1) Contrary to Descartes, there are no *conceptual distinctions/distinction of reason* in Spinoza's philosophy at all.[50] I could be mistaken, but this seems to be because, in contrast to Descartes, there is no thinking "outside" the substance itself in Spinoza. Thinking does not have a privileged position from where to see the reality of the substance. As Deleuze writes in his second book on Spinoza, *Spinoza: Philosophie pratique* (1981), there can be "neither emi-

[48] René Descartes, *Principles*, I, 63, CSM, I, 214 AT 30. Descartes in Noa Shein, "Spinoza's Theory of Attributes," in *The Stanford Encyclopedia of Philosophy, edited by* Edward N. Zalta, (Spring 2018 edition).

[49] Deleuze, *Expressionism in Philosophy: Spinoza*, 37.

[50] White, "Attributtets ekspressivitet," 183.

nence of the substance in relation to an attribute, nor of one attribute in relation to another" (*nulle éminence de la substance par rapport aux attributs, ni d'un attribute sur un autre*).[51]

(2) As far as I understand it, there are no *real distinctions* in Spinoza's thinking either. By this, I mean that *only formal distinctions are real.* My somewhat paradoxical statement here is deduced from a rejection of what Descartes says in the *Principles of Philosophy*:

> Strictly speaking, a *real distinction* exists only between two or more substances; and we can perceive that two substances are really distinct simply from the fact that we can clearly and distinctly understand one apart from the other.[52]

As far as I can see, what Descartes is describing here with real distinction is a separation on the level of substance. For his part, Spinoza would accept the relation between thinking and extension as a distinction between attributes of the same substance (i.e., *formal distinction*), but *not* as a *real distinction* between substances. The substance has real distinctions, since the attributes of thinking and extension are distinct, but these *real distinctions* in Descartes's reasoning mean, rather, separation. In other words, there are some challenges in comparing Descartes and Spinoza since they not only disagree but also do not understand the terms and concepts in exactly the same way.

(3) Descartes confuses numerical and real distinction. A numerical distinction is between individuals or singular things of the same category. For instance, the distinction between the dogs "Rex," "Fido," and "Goethe" is numerical. The difference between Descartes and Spinoza is that numerical distinctions (the dogs are numerically distinct) for Spinoza do not refer to different substances, only to modes. Furthermore, if numerical distinctions or modes referred to different versions of a substance, then this would mean that the substance was made of parts that these dogs expressed each in their own way.

[51] Gilles Deleuze, *Spinoza: Philosophie pratique* (Paris: Éditions de minuit, 1981), 72.
[52] Descartes in Shein, "Spinoza's Theory of Attributes."

And the main reason that leads Descartes to his "instantiated-attributive dualism" (that substances are deduced from fundamental attributes) is his confusion of *real distinction* and *numerical distinction* and the fact that he calls the difference between attribute and substance a *conceptual distinction* (which he calls *distinction of reason*). Returning to Deleuze, Spinoza's rejection of this confusion is one of the principal themes of *Ethics*.[53] Since Descartes thinks that numerical distinction (from the two fundamental attributes) is real, he introduces not just a division of being into two substances, but also a plurality of individual substances that share the same attribute: "In other words, *there are numerical distinctions that are at the same time real or substantial*" (En d'autres termes, il y a des distinctions numériques qui sont en même temps réelles ou susbtantielles).[54]

We see that the understanding of distinctions is central in Spinoza's criticism of Descartes, but also in the former's philosophical kinship with Duns Scotus. Deleuze writes:

> We still find in Descartes the same repugnance toward conceiving a real distinction between things that does not lie in different subjects, that is, which isn't attended by a division of being (qui ne s'accompagneraient pas d'une division dans l'être) or a numerical distinction. The same is not true of Spinoza: In his conception of a nonnumerical real distinction, it is not hard to discern Scotus's formal distinction.[55]

If Spinoza and Scotus are so close, what is it that nevertheless separates them? It has already been indicated above in the discussion on the reasons why Scotus neutralizes being. But if Scotus neutralizes being or renders it indifferent to the distinction between the finite and the infinite, how is this considered in Spinoza?

Substance and Attribute as Expression and as Formal Distinction

One of the conceptual keys to understanding Spinoza and his contrast both with Descartes and with Duns Scotus is the understanding of the "attribute," "expression," and how they articulate a formal distinction. Again, different kinds of distinctions become important.

[53] Deleuze, *Expressionism in Philosophy: Spinoza*.

[54] Deleuze, *Expressionism in Philosophy: Spinoza*, 30; *Spinoza et le probléme de l'expression*, 24.

[55] Deleuze, *Expressionism in Philosophy: Spinoza*, 65; *Spinoza et le probléme de l'expression*, 55.

If we look at how Descartes sees it, he assumes: (1) that attributes can be used to distinguish between different substances—thought and extension (as fundamental attributes) are distinguished from each other "insofar as one constitutes the nature of the body, and the other of the soul"[56] he also distinguishes the attribute from the substance (thought and extension) by seeing them as modes or dependents. (This is a bit confusing since attribute is the only way to identify a substance, but at the same time attribute is not identical to substance.)

The notion of "expression" becomes central here. In definition VI in the first part of *Ethics,* we read: "By God, I understand Being absolutely infinite, that is to say, substance consisting of infinite attributes, each one of which expresses eternal and infinite essence"[57] (*Per Deum intelligo ens absolute infinitum, hoc est, substantiam constantem infinitis attributis, quorum unumquodque æternam & infinitam essentiam exprimit*). Not only Deleuze, but other researchers highlight how the notion of "expression" becomes important in Spinoza's explanation of how attributes (thought and extension) that have nothing in common could be the essence of the same indivisible substance.[58] However, as Widder points out, it is due to Deleuze that what we could call the "expressive definition" of God and substance is given privilege over other definitions and formulations Spinoza gives.[59]

The Latin verb *exprimere* means "to press, force out, or squeeze out."[60] But in Spinoza, this takes a specific meaning. In the light of the formal distinction, "expression" or the verb "to express" accounts for a reality where what expresses, the expression, and what is expressed are distinct but not separated. The substance is hence a *triad*.[61] This distinction, which, as I have repeatedly stressed, is *not* a separation and is explicitly connected to the notion of expression by Spinoza himself many times in this work. In Scholium 7, in the second part of *Ethics* we read:

[56] Descartes in Deleuze, *Expressionism in Philosophy: Spinoza*, 30.

[57] Spinoza, *Ethics*, 4.

[58] Ramond, *Dictionnaire Spinoza*, 28.

[59] Widder, "The Univocity of Substance and the Formal Distinction of Attributes: The Role of Duns Scotus in Deleuze's Reading of Spinoza," 153.

[60] Charlton T. Lewis and Charles Short, *A Latin Dictionary*, 1879, https://www.perseus.tufts.edu/hopper/text?doc=Perseus%3Atext%3A1999.04.0059%3Aalphabetic+letter%3DE%3Aentry+group%3D42

[61] Part one of Deleuze's book *Expressionism in Philosophy: Spinoza* is called "The triads of substance."

> Everything that can be perceived by the infinite intellect as constituting the essence of substance pertains entirely to the one sole substance only, and consequently that substance thinking and substance extended are one and the same substance, which is now comprehended under this attribute and now under that (quae jam sub hoc, jam sub illo attributo comprehenditur). Thus, also, a mode of extension and the idea of that mode are one and the same thing expressed in two different ways.[62]

Even though Spinoza is conscious of the confusing use of language here (the attributes are called "substance thinking" and "substance extended"), it sometimes remains difficult for the reader to understand that "substance" has a double meaning throughout the text. Having said that, by calling the attributes (thinking and extension) substances, he at the same time manages to stress that the attributes are not separated or a fragment of the substance but express the essence of the substance in its totality.[63] Put differently, if the attributes and the substance are not ontologically separated but only distinct, then it is, as far as I see it, consistent also to call an attribute a substance. But what does the notion of "expression" add of analytical value to the formal distinction?

While expression in the sense of "to press, force out, or squeeze out" might easily give us the idea that there is a separation of what is inside (trying to get out) and what is outside, in Spinoza's understanding, it means two things at the same time. On the one hand, expression means explication or evolvement, and on the other, expression means implication or involvement. As Spinoza writes, "the true definition of any one thing neither involves nor expresses anything except the nature of the thing defined."[64] (veram uniuscujusque rei definitionem nihil involvere, neque exprimere præter rei definitæ naturam).[65] Deleuze explains this double movement of explicating and implicating further:

> To explicate is to evolve, to involve is to implicate. Yet the two terms are not opposites: They simply mark two aspects of expression. Expression is, on the one hand, an explication, an unfolding of what expresses itself, the One manifesting itself in the Many (substance manifesting itself in its attributes, and these attributes manifest themselves in their modes). Its multiple expres-

[62] Spinoza, *Ethics*, 50.
[63] Ramond, *Dictionnaire Spinoza*, 28.
[64] Spinoza, *Ethics*, 8.
[65] E I, Pr.VIII, S II. Spinoza, *Éthique*, 27.

sion, on the other hand, involves Unity. The One remains involved in what expresses it, imprinted in what unfolds it, immanent in whatever manifests it: Expression is in this respect an involvement.[66]

Hence, it becomes clear that the univocity of being in Scotus and Spinoza is different and that expression marks the difference in the understanding of formal distinction. The formal distinction as a way of accounting for the univocity of being is, in Scotus's thinking, never affirmed but withdraws itself. In Spinoza's reasoning, the formal distinction means that being is always affirmed and never withdraws itself. What will be a question for another article is whether both neutral and expressive positions count as monist philosophies and whether pantheism is a more authentic monist position than neutrality.

Conclusion

I have highlighted *how* the formal distinction is at the core of the argument for the univocity of being in two different thinkers, Scotus and Spinoza. For Scotus, the formal distinction is explicitly part of his reasoning against the philosophers and theologians of his time. For Spinoza, the formal distinction is more implicit in his critique of Descartes. In both cases, a discussion about distinctions as such constitutes a part of their reasoning. However, even though the formal distinction is pivotal for considering being as univocal, the two thinkers have very different versions of univocity. While Scotus neutralizes the univocity of being for both logical and theological reasons, Spinoza affirms it.

Bibliography

Bouaniche, Arnaud. 2007. *Gilles Deleuze, une introduction*. Paris: La découverte.

Busa, Roberto. 2010. Distinzione. In *Enciclopedia filosofica*. Milano: Bompiani.

Copleston, Frederick Charles. 1993. *A History of Philosophy Volume II. Medieval Philosophy. From Augustine to Duns Scotus*. New York: Image Books.

Deleuze, Gilles. 1968. *Spinoza et le probléme de l'expression*. Paris: Éditions de minuit.

———. 1981. *Spinoza: Philosophie pratique*. Paris: Éditions de minuit.

———. 2004. *Difference and Repetition*. London/New York: Continuum.

———. 2013. *Expressionism in Philosophy: Spinoza*. New York: Zone Books.

[66] Deleuze, *Expressionism in Philosophy: Spinoza*, 16.

Hause, Jeffrey. John Duns Scotus. In *Internet Encyclopedia of Philosophy* [Downloaded 16.02.2022 at:] https://iep.utm.edu/scotus/#H7

Hillermann, Horst, and Anton Hügli. 2019. Monismus. In *Historisches Wörterbuch der Philosophie*, ed. Joachim Ritter. Darmstadt: Wissenschaftliche Buchgesellschaft.

Jacquet, Chantal. 2004. *L'unité du corps et de l'esprit. Affects, actions et passions chez Spinoza*. Paris: PUF.

King, Peter. 2003. Scotus on Metaphysics. In *The Cambridge Companion to Duns Scotus*, ed. Thomas Williams. Cambridge: Cambridge University Press.

Knebel, Sven K., and Olivier Boulnois. 2019. Univozität; univok. In *Historisches Wörterbuch der Philosophie*, ed. Joachim Ritter. Darmstadt: Wissenschaftliche Buchgesellschaft.

Kohlenberger, Helmut K., and Ludger Oeing-Hanhoff. 2019. Attribut. In *Historisches Wörterbuch der Philosophie* Band 1, ed. Joachim Ritter. Darmstadt: Wissenschaftliche Buchgesellschaft.

Latinsk ordbok. 2007. Oslo: Cappelen Damm.

Lewis, Charlton T., and Charles Short. 1879. A Latin Dictionary. https://www.perseus.tufts.edu/hopper/text?doc=Perseus%3Atext%3A1999.04.0059%3Aalphabetic+letter%3DE%3Aentry+group%3D42

Mazzantini, Carlo, and Dario Sacchi. 2010a. Attributo. In *Enciclopedia filosofica*. Milano: Bompiani.

———. 2010b. Monismo. In *Enciclopedia filosofica*. Milano: Bompiani.

Muck, Otto. 2019a. Distinctio formalis. In *Historisches Wörterbuch der Philosophie*, ed. Joachim Ritter. Darmstadt: Wissenschaftliche Buchgesellschaft.

———. 2019b. Distinktion. In *Historisches Wörterbuch der Philosophie*, ed. Joachim Ritter. Darmstadt: Wissenschaftliche Buchgesellschaft.

Nodé-Langlois, Michel. 1999. *Le vocabulaire de saint Thomas d'Aquin*. Ellipses.

Poppi, Antonio. 2010. Duns Scoto. In *Enciclopedia filosofica*. Milano: Bompiani.

Ramond, Charles. 1999. *Le vocabulaire de Spinoza*. Ellipses.

———. 2007. *Dictionnaire Spinoza*. Ellipses.

Ross, James F., and Todd Bates. 2003. Duns Scotus on Natural Theology. In *The Cambridge Companion to Duns Scotus*, ed. Thomas Williams. Cambridge: Cambridge University Press.

Scotus, John Duns. 2011. *Sur la connaissance de Dieu et l'univocité de l'étant*. Paris: PUF.

———. Authors/Duns Scotus/Ordinatio/Ordinatio I/D8/Q3 - The Logic Museum. http://www.logicmuseum.com/wiki/Authors/Duns_Scotus/Ordinatio/Ordinatio_I/D8/Q3

Shein, Noa. Spinoza's Theory of Attributes. In *The Stanford Encyclopedia of Philosophy*, ed. Edward N. Zalta, (Spring 2018 edition). https://plato.stanford.edu/archives/spr2018/entries/spinoza-attributes/

Spinoza, Benedict. 2001. *Ethics*. Ware: Wordsworth Editions.

Spinoza, Baruch. 2010. *Éthique. Bilingue Latin-Francais.* Paris: Points.

Waller, Jason. Benedict de Spinoza: Metaphysics. In *Internet Encyclopedia of Philosophy*. https://iep.utm.edu/spinoz-m/. Accessed September 12, 2022.

White, Hanne. 2003. Attributtets ekspressivitet. In *Agora. Journal for metafysisk spekulasjon* 2–3.

Widder, Nathan. 2020. The Univocity of Substance and the Formal Distinction of Attributes: The Role of Duns Scotus in Deleuze's Reading of Spinoza. *Parresia* 33: 150–176.

Woolhouse, Roger S. 1993. Spinoza and substance. In *Descartes, Spinoza, Leibniz. The Concept of Substance in the Seventeenth-Century Metaphysics.* Milton Park: Routledge.

CHAPTER 11

You Are What You See: Environmental Ethics from Aesthetic Experience Via David Bentley Hart and Gilles Deleuze

Caleb Gordon

Introduction

I should probably start off by noting that a more *accurate* title for this piece might be 'You are what you *sense*,' but I couldn't resist riffing off the well-known refrain 'You are what you *eat*.' This isn't just for my own amusement, however, as the idea of ingestion and the self-making reincorporation of the world provides the basis for the following discussion. Just as food is ingested and in a distinct way *becomes* us, so too do our sensory experiences; what I refer to as the 'aesthetic' is the process whereby our various sensory inputs are interpreted or combined with each other to produce sensory knowledge. The idea that our sensory experiences *become* us, of course, is not as simple as saying we *are* our sensory experiences, at least not any more than we become a falafel if our diet consists of them. Yet just as we require food to persist, we also understand that a poor diet often results in a failure to thrive; low-quality food may leave us feeling

C. Gordon (✉)
Lindisfarne College of Theology, North Shields, UK

T. J. Hastings, K.-W. Sæther (eds.), *Views of Nature and Dualism*,
https://doi.org/10.1007/978-3-031-42902-6_11

lethargic, uncomfortable, or simply in need of another meal, and over the course of this chapter I suggest something similar about sensory environments. Many discourses within environmental ethics proceed from the recognition that poor air quality, polluted water, and other environmental factors are of ethical concern precisely because they pertain to the continuing existence of humans and other creatures, and the degradation of our constitutional requirements is a moral problem inasmuch as we or any other creatures have moral worth. I propose that the quality of our sensory inputs is likewise of moral concern—the worker must have bread, but she must have roses too.

It is this 'must'—this sense of *need*—which is at the center of this inquiry, and which I am presenting as a necessary consideration when evaluating aesthetic frameworks. This is largely to do with the way that aesthetic experience has been characterized within the history of philosophy. Where it builds from Kant's *Critique of Judgement*—which is often—philosophical aesthetics asserts the requirement of 'disinterestedness' for aesthetic appreciation. For Kant, this hinges on the assumption that the sensory and the rational operate *separately*.[1] And it is very *important* that their separation be clarified and sustained as a categorical division across which no overlap is permitted. The requirement for disinterestedness is essentially just that aesthetic appreciation is that which satisfies the reason without any complicating reference or entanglement to one's baser—which is to say, sensory or *material*—interests.[2] I don't have the space in this chapter to fully flesh out the debates around disinterestedness, but Emily Brady does an admirable job of setting up a contemporary version of Kant's natural aesthetics in *The Aesthetics of the Natural Environment*, while much of Arnold Berleant's work extends and defends a rejection of the same dualist Kantian paradigm on the basis that modes of cognition cannot be separated from each other.[3] However, the prevalence of the former position is one of the primary reasons that environmental *aesthetics* has not been effectively developed as a legitimate sphere of environmental

[1] Immanuel Kant, *Critique of the Power of Judgement*, ed. Paul Guyer, trans. Paul Guyer and Eric Matthews (Cambridge University Press, 2000), §2 (5:205), 90–91.

[2] Kant insists that our ability to differentiate between the good and the merely pleasurable rests in the distinction between our reason and sense; 'That is good which pleases by means of reason alone,' as compared to 'The agreeable which as such represents the object solely in relation to sense.' Kant, *Critique of the Power of Judgement*, §4 (5:208), 92–93.

[3] Emily Brady, *Aesthetics of the Natural Environment* (Tuscaloosa: University of Alabama Press, 2003); Arnold Berleant, *The Aesthetics of Environment* (Philadelphia: Temple University Press, 1992).

ethics. My own position is closer to the latter, but rather than wrestling with disinterestedness head-on, I will offer a constructive comparison between ontological frameworks and then return to the idea of disinterestedness near the end of the essay. The purpose of this chapter is to provide an account of aesthetic self-making which puts sensory experience on the same level as other kinds of ingestion, and in doing so I undermine the most persistent barrier to the use of aesthetics in environmental ethics: the kind of dualism between self and world which conceives of aesthetic experience as something that a self *does*, instead of also something that a self is *made of*. I do this through the positions of two philosophers, David Bentley Hart and Gilles Deleuze, and a comparison of the ways their respective ontologies situate creatures relative to both the world and God. Based on their recognition that sensory experience is self-making, I will show how the overlaps between their positions generate significant implications for environmental ethics.

I develop this claim through an analysis of David Bentley Hart's engagement with Giles Deleuze in Hart's book *The Beauty of the Infinite*. This will be a two-step process: the first concerns the significant overlaps between Hart and Deleuze, which together present a case for the consideration of aesthetic experience as an important issue for environmental ethics. The second step is distinguishing Hart's theological premises from Deleuze, and in so doing, articulating theological reinforcement for the philosophical position identified from their overlap. The effect of this is the framing of aesthetic experience as a critical area of concern for environmental ethics generally, but also for Christian environmental ethics as a specific field.

Hart and Deleuze offer similar descriptions of the relationships between individuals, world, and God. Where their thinking overlaps, they both affirm an ontology that grounds the epistemological validity of aesthetic experience in the continuous material relationship between individual and world. Their departures from each other also provide an insight into the ways that divergent interpretations of aesthetic experience can fit into a shared ontological framework. This is important because while there are some environmental thinkers who utilize aesthetics as an ethical resource, there is no standard framework that allows for aesthetic disagreement.[4]

[4] For examples of aesthetics and subjectivity in environmental ethics, see Eugene C. Hargrove, *Foundations of Environmental Ethics* (Englewood Cliffs, NJ: Prentice Hall, 1989); Holmes Rolston III, *Conserving Natural Value* (New York: Columbia University Press, 1994).

Hart and Deleuze show how we can have aesthetic disagreements—divergent interpretations of sensory experience—within a shared framework. Science and law, for example, are frameworks within which truth and justice can be debated—the terms for debate are established and there exist recognizable protocols for mediating disagreements within those spheres. In the way that I compare disagreement between Hart and Deleuze, I hope to show how a framework might be developed for the mediation and use of aesthetics in environmental ethics.

Over the course of this chapter, I argue that the materialist ontology developed between Hart and Deleuze is compelling enough to cast doubt on the reasons aesthetic experience has not been fully integrated into environmental ethics. The idea that aesthetic experiences are 'subjective' is not taken to mean that the interior preferences implied therein are random attractions measured against an 'objective' exterior, and further, the same ontology rejects approaches to aesthetics—and aesthetic appreciation—which insist on any kind of qualifying distance or disinterest as a definitive condition for aesthetic experience. Instead, aesthetic experience and its attendant descriptors—experiences of ugliness, beauty, and so on—are understood precisely as active and entangled relationships between self and world, and the aesthetic motivation is no longer presumed to be free from all sense of self-interest. After sketching out the general framework of this shared ontology, I proceed to analyze Hart's critique of Deleuze to demonstrate the type of discourse that an aesthetic approach to environmental ethics can accommodate and support while accommodating divergent viewpoints, interpretations, or values.

The reason I have decided to evaluate a *disagreement* between two aesthetic thinkers is to demonstrate that the nature of their disagreement does not disqualify aesthetic perception from supporting and informing environmental ethics; the framework through which they describe aesthetic experiences as relationships between self and world share deep similarities even while their conclusions and interpretations of aesthetic experience diverge significantly. Hart and Deleuze both present an ontology of aesthetic experience that rejects a hard dualistic dichotomy between self and world, and thus objectifies aesthetic experience as a part of the world as much as a part of the self. The rejection of a hard dichotomy between self and world grants far more epistemological weight to aesthetic experience than if aesthetic experience is conceptualized as limited to the interior life of an individual.

We already know that the insights and perceptions generated through aesthetic experiences often evade empirical frameworks and expressions. While facts about the world are critical to the development of environmental ethics, framing environmental ethics as *primarily* a response to facts allows the validity of environmental ethics as a discipline to rest—at least to a significant degree—on whether various facts can be confirmed, and whether that confirmation is sufficient to modify human behaviors. This causes distinct problems for the public engagement goals of environmental ethics.[5] For example, if a 'climate skeptic' does not, cannot, or refuses to believe (or otherwise act upon the notion) that climate science generates proof that humans are driving up global temperatures, they are disposed to reject a fact-based environmental ethics because of their perception that such an ethics is based upon rejectable facts. And conversely, 'climate activists' are required by the same paradigms to present all their concerns and claims to a certain standard of evidence—usually empirical.[6]

However, not all concerns are best supported with empirical evidence. Often problems in environmental ethics are framed as 'threat X to object Y is real, and since object Y is valued, threat X must be opposed.' Yet while the status of identifying threats is for the most part uncontroversial in its

[5] For an extended examination of this problem and its related social and cultural dynamics, see Mike Hulme, *Why We Disagree About Climate Change* (Cambridge: Cambridge University Press, 2009).

[6] By pointing out this troubling dynamic I do not offer a new observation; the failures of climate communication in bringing about behavior change have been repeatedly described. For example: Susanne C. Moser and Laura Dilling provided an apt summary of this problem in their analysis of climate communication a decade ago: 'The science and scientists have been central to climate change communication… however, knowledge is not sufficient to motivate action; to the contrary, as the focus of much debate it has been used to delay action. Maybe even more problematically, science has been assumed to be of central interest to most audiences. However, the science of climate change is not nearly as interesting to most audiences as some might wish… when experts tell lay audiences about the technical aspects of climate change, typically in one-way communication, there is little room for dialogue, building a shared understanding of the problem and possible solutions.' Susanne C. Moser and Laura Dilling, "Communicating Climate Change: Closing the Science-Action Gap," in *The Oxford Handbook of Climate Change and Society*, eds. John S. Dryzek, Richard B. Norgaard, and David Schlosberg, 2nd ed. (Oxford: Oxford University Press, 2013), 165–166.

methodology, the methodology for identification of value is not.[7] Valuation emerges from subjectivity.[8] But insofar as value must be accounted for in environmental ethics, and inasmuch as environmental ethics justifies itself as a response to facts, the presentation of values is often shoehorned into the form of factual presentation and thus the adjudication of value disagreements can unfortunately mimic the methodologies used by factual discernment.[9] Furthermore, values of aesthetic origin in particular have a tendency to either be dismissed[10] or reframed as though they are something else.[11] There are thus two problems that an explicitly aesthetic

[7] Here I of course do not mean that climate science is uncontested; but rather that the basic structure of scientific enquiry is widely accepted even among those who profess skepticism about the danger posed by anthropogenic climate change. Most still accept the validity of a process whereby a hypothesis is tested, and the tests are repeated in the pursuit of consistent results, and where consistent results are eventually accepted as a demonstration of the validity of that hypothesis. Generally speaking, where climate science is denied, it is not *science itself* which is denied, but rather that catastrophic forecasts are speculation or exaggeration, that connections between warming and human activity are tenuous, that humans possess the ability to adapt to extreme climate change, and so on.

[8] Holmes Rolston III, "Value in Nature and the Nature of Value," in *Environmental Ethics: The Big Questions*, ed. David Keller (Wiley-Blackwell, 2010), 130–137.

[9] A very interesting example of this discrepancy between fact and value-based decision making is addressed in Katherine Reynolds' analysis of the legal challenges to the protected status of the Tongass National Forest in Alaska. Reynolds argues that because the courts do not have a viable framework for incorporating political influences (i.e. values) into their judicial process, they run the danger of making legal decisions as though they are assessing facts when the point of disagreement is actually value-based. Her concern is both that this dynamic makes the courts too powerful, because in treating value-disagreements as factual they are in effect imposing the values of the court upon the situation at hand, but also because a clarified legal process for adjudicating differences in value can more closely correspond and address the kinds of disputes the legal system is attempting to serve. Katherine Reynolds, "Alternative Reasoning: Why the Ninth Circuit Should Have Used NEPA in Setting Aside the Tongass Exemption," *Ecology Law Quarterly* 43, no. 2 (2016): 381–420.

[10] See Arnold Berleant's writing on the ongoing, though often understated, influence of Kantianism and the Cartesian dichotomy between self and world as justification for dismissing aesthetic experience on the basis of its subjective nature. Berleant, *The Aesthetics of Environment*; Arnold Berleant, *Aesthetics and Environment: Variations on a Theme* (London: Routledge, 2005).

[11] Holmes Rolston III, for example, makes the case that where 'intrinsic value' is used, the observer should actually be using the term '*extrinsic*,' as the very purpose of identifying 'intrinsic' value is to locate value outside of oneself or ones' own interests. So, to identify a value as 'intrinsic' is still a reference to self—that value would not be 'intrinsic' *for* the object being labeled as *having* 'intrinsic value.' Rolston's critique thus admits self-reference as an origin of valuation. Rolston III, "Value in Nature and the Nature of Value."

approach to environmental ethics can potentially address: first, it strikes closer to the heart of the environmental crisis in the sense that the environmental crisis is essentially a crisis of values which *results* in material threat. The environmental activist or ethicist is not stuck simply insisting that global warming is *real*, but can move directly to a critique of relationships between individuals and world (or 'environment') through a framework of aesthetics; the quality of experiences which are available to various creatures is a question which does not depend on climate or ecological science. In this sense, environmental ethics shifts from being a problem-solving paradigm toward a more explicit and self-conscious task of cultural critique. When the appeal of environmental ethics includes the consideration of sensory experience (i.e. aesthetics), the activist or ethicist is no less concerned with the possibilities of experience forecast by global climate projections, but can also make direct appeals for behavioral changes which do not necessarily need to reference climate models or ecological projections; the environmental ethicist employing an aesthetic approach can critique environments, and the impact of decision-making on environments, explicitly in terms of the *experiences* as well as the *lack of experiences* which both possibly and necessarily follow from the conditions facing critique.

Facts are, should, and will remain a critical part of this dynamic field—but the integration of aesthetic experience as an additional axis for ethical enquiry opens new possibilities for environmental engagement and allows longstanding claims and interests in environmental ethics to be expressed with greater depth specificity.[12] For a climate skeptic, the question of whether the world is warming, or whether that warming is dangerous, is not the only reason to engage ethically with 'the environment'; an aesthetic model invites the expression of their own environmental values and preferences as a legitimate basis for ethical testing and inquiry (even while we hope that environmental sciences continue to be utilized in the composition of environmental policy). If the preferences and values of any given creature are lent ethical weight from the outset, the possibility or prevention of valued experience forms at least one ethical baseline which doesn't necessarily depend on beliefs about the wider world or the

[12] For example, Eugene C. Hargrove tracks the trajectory of environmental activism/advocacy as originating initially in appeals to aesthetics, but later supplanted by appeals to justice and sustainability on the basis that they were less 'subjective.' In other words—dialogues shifted toward justice and sustainability for the sake of *efficacy* rather than a straightforward shift in the objects of interest/conservation. Hargrove, *Foundations of Environmental Ethics*.

believability of environmental science.[13] By the same token, environmental ethicists are no longer forced to advocate for the preservation or protection of a beloved place primarily or exclusively on the basis of its ecological importance or the recognition of its value within existing frameworks of justice. Of course, these frameworks *can* be used, but it is *also* possible to present compelling ethical arguments based on the potentials for aesthetic experience which follow from decisions made about our environments. The case for building ethics of the beautiful, the interesting, and other aesthetic superlatives is opened by adopting a stance toward aesthetic experience that presupposes such experiences are objectively meaningful despite their resistance to expression in the falsifiable form of 'facts.' The reason to integrate an aesthetic approach to environmental ethics is because the language(s) of aesthetics are able to capture values which cannot be accommodated in factual language. Yet, the following characterization of aesthetic experience as a material and continuous relation between subject and object, self and world, requires the acceptance of aesthetic descriptions as attempts to communicate something true about our relation to the world. We are not simply describing our own internal states, but supposing that our internal states are produced by the sensory relationship between ourselves and our environments. A description of our experience may not be as comfortably clear as factual language, but it is nevertheless a description of our environments as well.

The Materialist Ontologies of Hart and Deleuze

I have chosen to use David Bentley Hart and Gilles Deleuze as examples because they both offer accounts of the world in which perception consists in/as aggregates of experience and are produced in a material 'folding' inward of the world—the effect of which is our experience of consciousness. I have also chosen them because Hart draws extensively on Deleuze while offering his own Christian approach to a materialist aesthetics, and the overlaps as well as the departures between their thought are useful to the wider discussion I am developing around the use of aesthetics in environmental ethics.

[13] To be very clear, I am not writing as someone who is skeptical about climate science. I am making the specific point that one can make ethical claims about creaturely relationships to environments which do not rely on scientific claims for the strength of their moral impact.

Deleuze is introduced to the reader early on in Hart's *The Beauty of the Infinite*, and re-appears regularly throughout the work. He serves as one of Hart's primary foils in the 'Creation' section at the heart of the book, which is also where Hart characterizes human fulfillment as the ability to recognize the nature of God in the beauty of the world. I will start by explaining why Deleuze's project is so important to Hart and continue by probing the nature of their disagreement. This will provide a framework which incorporates aesthetic experience to the development of environmental ethics while also demonstrating how Hart's aesthetic epistemology operates in relation to Deleuze; his disagreement is not actually a rebuttal, but rather a theological adjustment to Deleuze's philosophical geometry.

The Deleuzean ideas with which Hart is mainly concerned appear in Deleuze's *The Fold*, a series of reflections on Leibniz which develop a perspectival geometry of experience. For Deleuze, ontological distinctions between entities or objects are accounted for by the 'curvature' of their world. As he puts it: 'change is mapped on the path of curvature, whereby curves are "the trace of the same line" in a continuous movement.'[14] This includes distinctions between objects and bodies, such that the 'change' between a body and the world is not an ontological interruption, but a 'curve.' In their guide to *The Fold*, Niamh McDonnell and Sjoerd van Tuinen suggest that the reason Deleuze was attracted to Leibniz is that 'a play in the creation of the world' animates Leibniz' philosophy insofar as he challenges the logic of the limit and its ontology.[15] Where the ontology of limit tries to identify a boundary where something finally becomes *not*—such as where our *interest* stops and an aesthetic qualification like *disinterest* could be established—an account of distinction on the basis of curvature leaves intact the ontological continuity between self and 'not'-self. Self is no longer divided by something absolutely distinctive or removed (*limited*, as it were), but identified as a perspective.[16] An

[14] Niamh McDonnell and Sjoerd van Tuinen, "Introduction," in *Deleuze and the Fold: A Critical Reader*, eds. Niamh McDonnell and Sjoerd van Tuinen (Basingstoke: Palgrave Macmillan, 2010), 5.

[15] McDonnell and Tuinen, 8.

[16] Which is not to say that any geometric point *is* a self because of the perspective it represents; a self has to possess the conditions to have a perspective (a brain or other sensory apparatus) and is then defined by that perspective (where 'perspective' is both a vantage point *and* the existence of some kind of sensory apparatus). What I am calling a 'sensory apparatus' is not necessarily limited to the sense organs of animals, either—the point is just that 'perspective' entails more than a location.

illustration which demonstrates this point is the distinction between a mountain and a hill; Everest and Kilimanjaro clearly and easily pass the definitional thresholds for a 'mountain' in most cases, while progressively shorter or less-famous mounds approach those definitional thresholds to the point where distinctions become murky: the difference between a 'hill' and a 'mountain' are at some point a matter of judgment, as is the specific point where a rise in elevation takes the name of an eventual peak; the peak is what carries and defines the name, not the first gentle rise from the nearest sea or valley. Thus while exemplars tend to be uncontroversial, that lack of controversy cannot be expected to run all the way down a list of candidates or examples. As a matter of perspective, Everest is a mountain to all, but as the mounds descend variance in their appraisal begins to reveal more about the perspective of their appraisers.

In a similar way, distinctions between persons and environment are clear enough to be named, and relatively uncontroversial. I carry my own name, and I am not suggesting that someone will be stumped if I ask them to differentiate 'me' from the air I breathe, the food I eat, or indeed my sensory experiences. Yet, as with mountains, the boundary between what is the person and what is not is not a *sharp* distinction; it is a certain ambiguous point along a curve, known by eventual difference rather than specific transition. This is not about being able to tell a foot from the pavement; it is rather that we are produced from such a combination of inputs (food, air, senses) that while we know ourselves and other subjects distinctly, the points where those inputs change from one thing to another—exterior to interior—is known by the *difference* it produces, rather than a specific transition where one thing clearly becomes (*only*) another.

Something similar occurs as we identify discrete objects which are composed of multiple constituent parts, or when the qualities of objects are essentially *summarized* by our use of aesthetic language. Our sensory environments contain *too many* objects to possibly know individually, and so we know them together in collections or impressions. As Deleuze puts it; 'all consciousness is a matter of threshold.'[17] In order to be conscious *of something*, in other words, an individual must have concept of difference that matches with a way of identifying some object as such. Our perception of a leaf as 'green' might serve as another example: that a leaf is recognized as 'green' requires both the differentiation of 'green' from other colors, but also that the combination and summation of sensory

[17] Gilles Deleuze, *The Fold*, trans. Tom Conley (London: Athlone Press, 2006), 88–89.

inputs received as the relation between the leaf and an eye passes the definitional threshold for 'greenness' by which that color is identified by the individual. On close inspection, a green leaf is nearly always (actually) mottled: blue, brown, yellow, red, orange flecks and veins become apparent the closer the eye moves in, yet, to move in this way is also to change the parameters of threshold. Closening the gaze obscures or ignores other inputs in order to focus on the colors of a leaf within a narrower threshold, and indeed to move in further still would simply repeat the process with respect to new observations. Consciousness of a 'green' leaf, therefore, is defined by a certain relation of distance that is characterized by a definitional threshold for 'greenness,' which in turn is defined as a learned definition as well as the material relation between the object and the organ which senses it. The leaf is not 'green' because it simply *is*; it is green *when* the eye of one who recognizes 'green' is in relation to it at a certain distance, and with other preconditions (such as being taught to label a frequency of light 'green'). When we call something 'green' we are reporting something *personal*; we are reporting a sensation we recognize—both the sensory proximity to 'greenness' and our history of learning to call it as such. If we explicitly describe an object as 'green' we implicitly reveal information about ourselves as well as the object in focus.

For Hart, whose work in aesthetic epistemology is likewise heavily influenced by and presented as a geometry of perspective, engaging with Deleuze amounts to collaboration in a branch in scholarship. Hart would agree with Deleuze that the 'perspectival geometry' emerging from Leibniz represented a profound re-imagining of the relationship between self and other, and an important rebuttal of atomistic alternatives. However, Hart's contention is that while Deleuze is right to identify an important genesis of thought in Leibniz and agrees that Deleuze (via Leibniz) offers an important corrective, Hart thinks that Deleuze moves Leibniz in the wrong direction. Or at least, Deleuze's expanded account of Leibniz, inasmuch as it supports his account of world-as-surface through the language of 'fold,' fails to adequately account for the *contents* of experience which a perspectival geometry endeavors to describe and frame. Hart thus takes issue specifically with Deleuze's treatment of the transcendent and the immanent—or rather, how Deleuze is able to distinguish one from the other if limited to the one:

> Deleuze ... suggests that the philosopher must henceforth put aside the desire to grasp the whole of things from above and instead become a crea-

> ture of the surface, a "tick or louse," never allowing his or her thought to stray from the planes of immanence; but Deleuze has quite rigid notions of where immanence, in general, comes to an end. In truth, though, thought that confines itself to an entirely "horizontal" axis of intelligibility could never have at its disposal means by which to determine where the immanent leaves off and the transcendent begins.[18]

Hart proposes that Deleuze has conceptualized the limit of the immanent without appreciating the necessity of a transcendent reference to differentiate between immanent and transcendent. In so doing, Hart is not taking issue with the *shape* of Deleuze's characterization of the relationship between self ('tick or louse') and the surface upon which it clings, but insisting that in order for such an analogy to be meaningful or comprehensible, an additional 'axis of intelligibility' is required. Hart is not disagreeing with Deleuze about the location of our vantages, or their relationship to the rest of the world: there is no escape or departure from the plane or surface. However, what is encountered from these vantages, which occurs to us and meets us on this surface, is for Hart *not only and always mere surface*. For Hart, the meaning of the surface is not limited to itself, but mediates something beyond itself. As creatures and thus features of the surface, we cannot ourselves go beyond the surface; but what *is* beyond the surface is mediated *through* the surface, *to* the surface, via its folds—which both form us materially and confer meaning to us sensorily.

As a proposal for theorizing the operations of consciousness, Deleuze is aiming at an account of our relation to the infinite that does not reach past the material while accounting for the irreducibility of an individual's perceptual encounter and interpretation. For individuals to make sense of their experience they have to do so at a level (threshold) higher than the essential or atomic. As I mentioned toward the beginning of this chapter—it is this sensory knowledge which defines my use of the 'aesthetic.'[19] While we can recite facts and secure various objects of knowledge within the bounds of language, this broader definition of aesthetics rightly

[18] David Bentley Hart, *The Beauty of the Infinite: The Aesthetics of Christian Truth* (Grand Rapids: Wm. B. Eerdmans Publishing, 2003), 12–13.

[19] This is not a unique claim, though 'aesthetics' is still most often treated as criticism of taste. See the introduction to Terry Eagleton's *The Ideology of the Aesthetic* for a detailed account of aesthetics' initial emergence as a broad philosophy of sensory knowledge and subsequent shrinkage to a philosophy of art appreciation and critique. Terry Eagleton, *The Ideology of the Aesthetic* (Oxford: Basil Blackwell Ltd., 1990).

describes our relation to various sensory experiences just as it does to works of art. In either our appreciation of a piece of art or our sensory contact with any other object (which includes environments-as-such), our sense of that object is conditioned both by our immediate apprehension as well as our sensory and epistemological context: the lighting in the room, the (other) thoughts in our head, our values and priorities, and so on. Along the same lines, claims and values of environments that emerge in and as combinations, which cannot be interrogated as their constituent parts in a way that preserves their meaning, are vulnerable whenever the interests of environmental ethics seemingly require reductive analysis in a bid to secure their justification. Accepting an aesthetic approach to environmental ethics requires patience and generosity to engage with values which cannot be articulated quickly, easily, or clearly. Its priority is not expedience, neither is it even necessarily *clarity*, but rather the recognition of those values' potential importance and the willingness to proceed along an assumption of importance even while clarity is, perhaps, still developing.

The cognitive operations that result in the perception of the greenness of a leaf also extend to other perceptions about environments that are directly relevant to ethics, especially aesthetic responses to environments belonging to the slippery domains of beauty or meaning. This includes the kinds of perceptions about environments that factor into evaluations of their value, meaning, purpose, and an individual's sense of self via those interpretive relations to material environment. If it were to be accepted that both (a) beauty and meaning (or even just enjoyment) are necessary to human flourishing, and (b) experiences of beauty or enjoyment are produced by material conditions which cause an individual observer to cross a perceptual threshold, then environmental ethics must accordingly develop responses to the language of aesthetic experience and the claims it produces.

Thresholds for beauty and meaning as products of material conditions require evaluation according to the distance and type of interaction of relevant observers. This applies to environmental initiatives such as the preservation of natural/wild/green space even in cases where it does not have a clearly defined ecological function. The *experience* green space provides to subjects (human or nonhuman) may be more valuable than its *function* as air filtration, and perhaps it is *most* valuable as a space where birdsong is heard, and flowers and trees are seen and explored (which is to say—*most valuable as the experiences it provides*). None of this needs to be taken as competition to the natural rights of nonhuman creatures to their own habitat, or indeed the full spectrum of ecological benefits provided by greenbelts, nature reserves, and so on—aesthetic experiences are an *additional and independent* area of concern which remains relevant even when creaturely rights or ecological benefits are not obviously or directly at issue. Thresholds also come into play here when thinking about *how much* of something is required in an environment for the attainment of an aesthetic experience: how wide does a green belt need to be for birdsong to be heard over the sound of traffic? What are the material conditions necessary for individuals to access and experience beauty and meaning?

Theological Stakes in Environmental Aesthetics

In Hart's *Beauty of the Infinite*, the geometry of consciousness is approached similarly to Deleuze, but Hart's interest in affectation and persuasion is presented specifically in relation to the origins and persistence of Christian faith. Hart's account of faith acquisition follows an analogous form to the recognition of a green leaf as discussed above. An individual must have both an idea of the difference between what it would be to believe or not, and from there it is a matter of their perspective: if a person is of a certain relation to the objects of perception (scriptures, tradition, or—especially important for environmental ethics—the world itself), Hart argues that the truth of Christianity effectively *becomes apparent*. The experience of conversion is as a gestalt shift; what was before seen in one way is now seen in another. A threshold is passed such that the truth of Christianity is recognized, internalized, and confirmed. Hart characterizes his book as a work of aesthetic epistemology, ultimately taking the position that experiences of beauty are tantamount to divine epiphany which can cause or support one's conversion to faith. Like Deleuze, Hart's treatment of the collation and amalgamation of experience presents the clarifying and productive operations of cognition as an adaptation which overcomes the impossibility of perceiving every detail in one's sensory environment. The implication of their mutual agreement is that objects of perception which are known at the level of our consciousness, at the threshold defining the limits of our own thoughts, are no less true than the details of those experiences. The fact that we can look closer at a leaf to find a completely different picture does not undermine the distinctiveness of the leaf *as a* leaf, as a coherent object of perception. We do not say, 'This is not a leaf after all, but a structure which contains photosynthetic systems!' even though a deepened understanding of a leaf involves knowledge of component parts which are always coming and going and a structure which becomes and unbecomes itself over time.

Hart's Christian ontology rests on the presumption that the transcendent and the immanent comingle. There is not just a surface which is self-contained while the term 'transcendent' refers to whatever falls outside the boundary or limit of surface; the surface, for Hart, is more like a richly textured and constantly moving fabric of glass, through which light shines, refracts, and changes according to the conditions which mediate its appearance to our vantages. Consider as contrast Deleuze's conception of God and God's relationship to creation: 'Clearly many areas are found in the understanding of God,' he writes in *The Fold*; 'we can state that the relations surge up in a region that no longer involves God himself, but the possibility of creation.'[20] Deleuze's conceptualization of God is reciprocal to the creaturely experience of being. Offered in the context of a study on Leibniz, 'God' is something quite near the opposite of a monad—that interior space which supports the existence of the world in and as its existence.[21] This leaves God as a pure abstraction for Deleuze; an abstraction which is meaningful only in terms of pure mathematics, and in a distinct sense irrelevant to the interior operations of being as experienced by observers and described in the geometry of the monad.

When God is limited to a principle, as in Deleuze, Hart's contention is that there is no reason to be interested in God beyond whatever appeal is contained in abstract thought. Yet Deleuze is right to recognize in Leibniz a philosophy with the potential to 'advocate the possibility of infinity to be thought within the restricted limits of our habitat.'[22] It is the hard and abstract distinction of the immanent from the transcendent—the inside from the outside—which Hart cannot allow. And this carries interesting implications for approaching environmental ethics through Christian aesthetics, for it insists that aesthetic relations *in* the world, in the immanent, carry and convey the transcendent to those vantages situated and disposed to receive them. Both Deleuze and Hart put forward ontologies of experience (or 'perspectival geometry') to describe an aesthetic epistemology, but Hart distinguishes himself from Deleuze in his insistence that the

[20] Deleuze, *The Fold*, 46.

[21] Deleuze formulates the relationship of God to a monad as 'God, whose formula is $\infty/1$, has as its reciprocal the monad, $1/\infty$,' Deleuze, *The* Fold, 49.

[22] Tom Conley, "Foreword," in *The Fold*, by Gilles Deleuze, trans. Tom Conley (London: Athlone Press, 2006), xvi.

importance of aesthetic experience of the immanent is in its potential to glimpse the transcendent. Thus aesthetic experiences—which cover all experiences of environment—always include the potential for relation to God.

What is most interesting about the potential contribution of reading Deleuze against Hart to inform environmental ethics is how similarly their ontologies are *shaped*, despite holding certain antithetical claims. Both treat knowledge of the world as emerging from the infinitesimal, on a gradient of intelligibility, which results in apprehension by the senses as the attainment of an epistemological threshold defined by the capacities of an individual. And furthermore, the basis of all understanding is conceived as a relation of the world to itself, where selves are produced in and *as* a peculiar sort of 'fold.' The key difference is thus not the dynamics of this process, but rather the ontological classification of what lies beyond the threshold of explicit or factual understanding. Neither Deleuze nor Hart actually have demonstrable access past this threshold, of course: they are limited (as is everyone) to projections of their understanding past the threshold of unambiguous language. Yet where Deleuze projects 'obscurity' past the threshold of 'clarity,' Hart insists that the wordless intensity of beauty directs not to obscurity but rather to God.[23] The rapturous nature of that wordless (or in fact word-defying) affectation is, essentially, its own proof that what is beyond our cognitive thresholds is not merely obscure to us but intimates the coherence of God. The differences between Deleuze and Hart are thus quite limited, though they signify major departures where relevant to the interpretation of faith experience. However, the overlaps between their thinking allow them to disagree about the interpretive limits of experience while allowing them to agree on the basic geometry of experience. Since they rely on the same materialist framework to justify their aesthetic epistemologies, they mutually corroborate the relevance of aesthetic experience to environmental ethics. Effectively, their respective frameworks both support claims about the importance of aesthetic experience in terms of continuity with the world; to conceive of experience as the consciousness produced by 'folds' is to make explicit the contribution of environments to any sense of self. An environment is the other face in a fold, with the 'curvature' at the bottom of the fold being

[23] Deleuze, *The Fold*, 90; Hart, *The Beauty of the Infinite: The Aesthetics of Christian Truth*, 275–288.

that transitional mode where we find metabolic processes such as respiration, digestion, and aesthetic experience.

Hart and Deleuze do not have to agree on the ability of individuals to glimpse the transcendent for the relevance of their shared positions to be evident to the purposes of environmental ethics: where they differ is precisely their respective interpretations of the meaning of experience—but *not* the way that meaning *arrives.* The arrival of meaning is produced in the material relations between environment and sensory organs: the eyes, the nose, the ears, and so on. Whether a given individual interprets their experiences or affectations as a glimpse of the transcendent (Hart) or the progressive culmination of disparate parts to an interpretive center (Deleuze), both refuse to allow the importance of sense relations between bodies and world to merely escape into 'subjectivity.' They are always the product of real material relationships between self and world. However interpreted, the relationships between bodies and world are real—and however interpreted, the relationships between bodies and world *must be interpreted.* For both, to be a conscious participant in the world is to experience its foldedness in ways which are defined aesthetically. Whether beautiful or ugly, meaningful or dull, oppressive or enlightening—aesthetic language is the framework by which individuals describe their experience in terms of the thresholds necessary to make the world intelligible. No one can perceive everything all at once—only in amalgamation. Because everyone has a different vantage, no amalgamations are the same.[24] Yet, as perceptual encounters emergent from material relations between self and world, all experiences fall directly into the purview of environmental ethics—which, of course, is the ethics of influencing the way the world materially is.

Conclusion

The ontologies of both Hart and Deleuze aim to buttress the epistemological validity of aesthetic experience on the same basis: by denying an ontological gap between self and world. They differ primarily in the *interpretation* of aesthetic experience, which is, and will continue to be, the first hurdle to incorporating references to aesthetic experience with environmental ethics. Their order of operations, however, is key. For both, interpretation follows ontology. Specifically, the interpretation of experiences is

[24] Which is not just a physical relative position, but also personal history.

secondary to their agreement that aesthetic experiences form the basis of identity. Indeed, for Hart, experiences also mediate relationship to God—a departure from Deleuze that fits easily into the category of 'interpretation' (of aesthetic experience). When aesthetic experiences are understood as the basis of identity, of a sense of self, then the significance of aesthetic experience exceeds simple taste, preference, pleasure; the availability or quality of aesthetic experiences in a given environment becomes an issue of justice, and the interpretation of aesthetic experience an urgent epistemological question.[25] Truth and justice are already well-integrated into the conversation of environmental ethics, and a materialist ontology in the vein of Hart or Deleuze offers a point of purchase for the integration of aesthetic experience to that project as well. Aesthetic commentary and critiques are not entirely absent from environmental ethics but *have* been marginalized on ontological grounds; rejecting an ontological gap between self and world not only validates the aesthetic category of critique, but explains its persistent presence despite the difficulty of, or resistance to, its use.

My final point highlights an ongoing challenge: no part of my argument has suggested that the adoption of a continuous or materialist ontology makes the mediation of aesthetic experiences *easier*—it just raises the stakes. Such a challenge is also, at the same time, straightforward: the difference between something difficult but not important, and something difficult but *very* important. If the ontologies of Hart and Deleuze are compelling, it is not because they can be read as interpreting the world the same way from divergent viewpoints: it is that despite offering contradictory interpretations of the limits of aesthetic perception at the climaxes of their respective works, the reception of meaning for the self arrives in the same way—in aesthetic experience, in the necessarily combinatorial mode of reception that produces meaning out of infinite sensory relationship to the material world.

This also brings us back to the question of disinterestedness to which I alluded toward the beginning of this essay. The above 'folded' ontology allows us to interact with other debates in environmental thought by

[25] The reader should note my avoidance of explicitly anthropocentric terms, though I have not included any extended discussion of the specifically non-human. These ontologies operate the same for any experiencing entity, though the composing of ethical systems remains a predominantly human enterprise.

showing how Hart and Deleuze could articulate interesting pushback against the requirement for disinterestedness; the differences between the two would have to do with their respective epistemological ranges for identifying creaturely interests. For example, Hart would likely argue that a requirement for human fulfillment is that humans are able to in some way recognize God—and moreover, that our experiences of beauty are the instances of that recognition. Deleuze does not share this position, but all the same, the fact that for both Hart and Deleuze aesthetic experiences are self-making means that within this framework the only way for aesthetic experience to be 'disinterested' is if an individual has no self-interest *at all.* Our aesthetic responses to natural environments cannot be truly disentangled from our awareness of ecological collapse, and neither can we avoid the fact that the contents of our minds—which is to say, the ingredients of who we find ourselves to be—are constantly derived from our sensory environments. We cannot help but be interested in our aesthetic experiences to whatever degree we are interested in ourselves. And if the self is continuous with the world, then we *cannot be* interested *only* in an object of perception, without any connection to the self. While the ontological approach described above presents a compelling basis for ethical inquiry as a general or religiously neutral framework, it again gains an additional and interesting valence when crossed with the common reference points for Christian ethics: if, for example, we are to love others as ourselves, then what experiences do we think should be available to others? We understand the stakes of this question in relation to ourselves, yet it is our own interpretation of our environment, our own aesthetics, which directs our ethical response.

Bibliography

Berleant, Arnold. 1992. *The Aesthetics of Environment*. Philadelphia: Temple University Press.

———. 2005. *Aesthetics and Environment: Variations on a Theme*. London: Routledge.

Brady, Emily. 2003. *Aesthetics of the Natural Environment*. Tuscaloosa: University of Alabama Press.

Conley, Tom. 2006. Foreword. In *The Fold*, by Gilles Deleuze, trans. Tom Conley. London: Athlone Press.

Deleuze, Gilles. 2006. *The Fold*, trans. Tom Conley. London: Athlone Press.

Eagleton, Terry. 1990. *The Ideology of the Aesthetic*. Oxford: Basil Blackwell Ltd.

Hargrove, Eugene C. 1989. *Foundations of Environmental Ethics*. Englewood Cliffs, NJ: Prentice Hall.

Hart, David Bentley. 2003. *The Beauty of the Infinite: The Aesthetics of Christian Truth*. Grand Rapids: Wm. B. Eerdmans Publishing.

Hulme, Mike. 2009. *Why We Disagree About Climate Change*. Cambridge: Cambridge University Press.

Kant, Immanuel. 2000. In *Critique of the Power of Judgement*, ed. Paul Guyer. Cambridge University Press.trans. Paul Guyer and Eric Matthews.

McDonnell, Niamh, and Sjoerd van Tuinen. 2010. Introduction. In *Deleuze and the Fold: A Critical Reader*, ed. Niamh McDonnell and Sjoerd van Tuinen. Basingstoke: Palgrave Macmillan.

Moser, Susanne C., and Laura Dilling. 2013. Communicating Climate Change: Closing the Science-Action Gap. In *The Oxford Handbook of Climate Change and Society*, ed. John S. Dryzek, Richard B. Norgaard, and David Schlosberg, 2nd ed. Oxford: Oxford University Press.

Reynolds, Katherine. 2016. Alternative Reasoning: Why the Ninth Circuit Should Have Used NEPA in Setting Aside the Tongass Exemption. *Ecology Law Quarterly* 43 (2): 381–420.

Rolston, Holmes I.I.I. 1994. *Conserving Natural Value*. New York: Columbia University Press.

———. 2010. Value in Nature and the Nature of Value. In *Environmental Ethics: The Big Questions*, ed. David Keller. Hoboken: Wiley-Blackwell.

CHAPTER 12

Zen Master Thich Nhat Hanh's Ecological Imagination

Victor Thasiah

Vietnamese Zen Master Thich Nhat Hanh's (1926–2022, hereafter TNH) imagination of the natural world offers a vision of environmental stewardship that contributes to a cultural transition from what we call ecological alienation to ecological alignment.[1] While the drivers of higher average temperatures, rising sea levels, extreme weather, ecosystem degradation, biodiversity loss, and mass extinction—threatening public health, water supply, food security, decent livelihoods, economic opportunities, domestic politics, and international relations—are by now familiar, ecological alienation as a summary term refers to the general, fundamental disconnect between corporate, state, and consumer behavior on the one hand, and the direct or indirect negative environmental consequences on the other. Simply put, this fallout is not shifting policies and practices to the extent that it should. In contrast, ecological alignment refers to a general,

[1] Thich Nhat Hanh, *Zen and the Art of Saving the Planet* (San Francisco: HarperOne, 2021).

V. Thasiah (✉)
California Lutheran University, Thousand Oaks, CA, USA
e-mail: vthasiah@callutheran.edu

T. J. Hastings, K.-W. Sæther (eds.), *Views of Nature and Dualism*,
https://doi.org/10.1007/978-3-031-42902-6_12

fundamental "reconnect," which, as we will see, involves conscious participation and partnership in nature-based and culture-based efforts toward making life and mutual flourishing possible for ecosystems, communities, and ourselves.

Meanwhile, informed by such ecological alignment, environmental stewardship in the United States that is responsive to ecosystem degradation and climate system destabilization is nowhere near as effective as it could be. Despite decades of efforts, we still lack the diversity of perspectives necessary to fully comprehend and justly solve environmental problems, even as the country rapidly becomes more racially and ethnically diverse. Moreover, those most often impacted (historically and today, and disproportionately so) by these problems—such as Indigenous communities, communities of color, and low-income communities—continue to be significantly underrepresented in environmental organizations, especially in leadership. While the current and significant involvement of youth and young adults in climate justice activism as a form of environmental stewardship is at once inspiring and insightful, and prodding and productive, we now need everyone involved, as these young leaders have insisted.

Our focus on one important, autobiographical account of TNH's imagination of the natural world from his early memoirs provides a timely vision for environmental stewardship understood as ecological alignment and a potential antidote to ecological alienation for many, given the mass appeal of the Zen master's writings. It can also, perhaps, support Asian Americans—projected to be the nation's largest immigrant group, surpassing Hispanics, by mid-century—in their environmental stewardship with another culturally accessible resource.[2] The unprecedented, global ecological and humanitarian challenges we face call not only for an expanded imagination of the natural world, but also for expanded participation in its protection. Especially welcome are accounts like TNH's that directly counter colonial and capitalist forms of exploitation and extraction that continue to endanger us today. After interpreting what ecological alienation means and the promise of accounts like TNH's, we critically examine the Zen master's imagination of the natural world and conclude with a proposal to view environmental stewardship as ecological

[2] Abby Budiman and Neil G. Ruiz, "Key facts about Asian Americans, a diverse and growing population," Pew Research Center, 29 April 2021, https://www.pewresearch.org/fact-tank/2021/04/29/key-facts-about-asian-americans/

alignment—i.e., as conscious participation and partnership in nature-based and culture-based efforts toward making life and mutual flourishing possible.

Ecological alienation, according to Buddhist philosopher David Loy, is a delusional sense of duality between ourselves and the world we are "in," leading to the destruction of the environment.[3] Favianna Rodriguez, a cultural strategist and the founder of the Center for Cultural Power, comments on the middle factor between alienation and destruction, namely, domination.[4] What we need, according to Rodriguez, are new, or old, stories that imagine the overturning of this domination and overcoming of this alienation, not to mention ones that imagine present and future flourishing out of past and present destruction. More specifically, James Miller, a scholar of Daoism, views our ecological alienation as a general, fundamental disconnect from the rest of the natural world. He observes,

> The source of … economic and ecological unsustainability is the inability of modern neoliberal ideology and its attendant cultural forms to conceptualize and operationalize a way of being in the world that inscribes human prosperity within the prosperity of planetary life. Rather, we have come to conceptualize human prosperity in a way that is alienated from the ecological systems that make such prosperity possible. As a result, the modes by which we pursue human prosperity serve only to diminish its long-term viability by destroying the ultimate foundation for prosperity: the capacity of the natural order to produce of its own accord the creative vitality that can support the flourishing and well-being of all species.[5]

Miller calls the natural order's various, wondrous, generative capacities—supportive of life and the mutual flourishing of ecosystems, communities, and ourselves—the "subjectivities" of nature. Ecological alienation is thus a lack of due recognition of, even an obliviousness to, these subjectivities, the distributed and differentiated agencies of the natural world—from microbes to mountains—that make up the substance and conditions of our own subjectivity. "Modern human culture, by denying nature's

[3] David Loy, *Nonduality: In Buddhism and Beyond* (Somerville, Massachusetts: Wisdom Publications, 1988), 324.

[4] Favianna Rodriguez, "Harnessing Cultural Power," in *All We Can Save*, eds. Ayana Elizabeth Johnson and Katharine K. Wilkinson (London: One World, 2020), 121.

[5] James Miller, *China's Green Religion: Daoism and the Quest for a Sustainable Future* (New York: Columbia University Press, 2017), xiv.

subjectivity and arrogating subjectivity and agency to itself alone," Miller likewise concludes, "has sowed the seeds of its own destruction."[6]

Finally, recent work on what we might call ecological intersectionality indirectly considers this alienation from the subjectivities of nature as it seeks to both comprehensively and critically examine environmental fall-out across social categories such as race, class, and gender, and ecological categories such as other-than-human organisms, landscapes, and geographies. This ecological justice perspective, according to critical environmental justice studies scholar David Pellow, "engages more deeply with the uneven relationships of human beings to the broader nonhuman world and embraces a more respectful and egalitarian relationship between the two."[7] Otherwise put, it "destabilizes the notion of the human as a biological category at the pinnacle of a human/nature hierarchy and, instead, embraces it as a political category that engages the broader ecological community."[8] Regarding the kind of ecological alignment with the subjectivities of nature that we have in mind: respectful, yes; engaging the broader ecological community, yes; while not necessarily egalitarian in the common sense of the word.

This work of seeking to analyze the often simultaneous and complementary ways (based on ideological frameworks that devalue peoples and places together) that communities and ecosystems are negatively impacted by and respond to environmental injustice includes centering not only other-than-human species, but also the physical environment.[9] In this context, Pellow's critical environmental justice studies (CEJ) further connects with Miller's subjectivities of nature.

> With respect to speciesism, CEJ extends beyond the category of the human to include the more-than-human world (from nonhuman animals and ecosystems to the built environment) as subjects of oppression and as agents of social change. Worms, viruses, ants, water, rocks, mountains, fish, elephants, krill, air/wind, and trees are just some of the infinite nonhuman beings and things that are affected by environmental injustice but that also exert their

[6] Miller, *China's Green Religion: Daoism and the Quest for a Sustainable Future*, xiv.

[7] David Pellow, *What Is Critical Environmental Justice?* (Cambridge, United Kingdom: Polity Press, 2018), 153.

[8] Pellow, *What Is Critical Environmental Justice?* 153.

[9] Pellow, *What Is Critical Environmental Justice?* 2.

> own influence on the character of those conflicts in particular and on the course and trajectory of human society and ecology more generally.[10]

Though not normally registered by us, environmental injustice impacts natural communities (nonhuman organisms and physical environments) as it impacts human communities. Pellow also includes the built environment as both subject of oppression and agent of social change; one thinks of dilapidated public housing and dangerous work sites. We will soon take up TNH's imagination of the natural world and social change, but before we do so, how can we counter this ecological alienation, domination, and destruction?

Rodriguez's cultural strategy for climate action is one based primarily on stories as opposed to an economic strategy. While many rightly center on reducing carbon emissions, Rodriguez centers on increasing imaginative alternatives.

> The power of culture lies in the power of story. Stories change and activate people, and people have the power to change norms, cultural practices, and systems. Stories are like individual stars. For thousands of years, humans used the stars to tell stories, to help make sense of their lives, to orient them on the planet. Stories work in the same way. When many stars coalesce around similar themes, they form a narrative constellation that can disrupt business as usual.[11]

We need stories that expand our imagination of the natural world and change our moral sensibilities at the level of what we notice, perceive, and experience, at the level of what we deeply feel and care about. By moral sensibilities we mean the dispositions (or cumulatively, the temperament) connected to living life fully, sustainably, and meaningfully in terms of the values, means, and ends that guide us. "We need our storytellers," Rodriguez maintains, "to help us shift our mythology and imagine a future where together we thrive *with* nature. That is the power we must harness, if we are to find our way out of the climate crisis."[12]

Such moral sensibilities would be better attuned to the subjectivities of nature. Miller contends, "One reason for the world's current ecological

[10] Pellow, *What Is Critical Environmental Justice?* 19 f. Cf. Bruno Latour, *Down to Earth* (Cambridge, United Kingdom: Polity, 2018).

[11] Rodriguez, "Harnessing Cultural Power," 123, 125.

[12] Rodriguez, "Harnessing Cultural Power," 121.

crisis is that the shared cultural world of human beings is producing a deterioration in the objective biophysical reality that is studied by climate scientists. For a culture to be sustainable in the long run, the physical world and the cultural world must be brought into close alignment."[13] In other words, the shared cultural world of human beings should be better attuned to the shared natural world of nonhuman beings, animate and inanimate. Aligning himself with flora and fauna on the one hand, and the landscape and atmosphere on the other—the subjectivities of nature—TNH demonstrates what overcoming ecological alienation can look and feel like, and suggests a vision for conceiving of environmental stewardship as the collective alignment and shared work of humans and the rest of the natural world for the mutual flourishing of ecosystems, communities, and ourselves. In what follows, we will critically examine and interpret a story told by TNH—who was informed by a Zen Buddhism that uniquely combines a Daoist attunement to the natural world with a Confucian attunement to the social world—about his time spent living in a mountain forest in the Central Highlands of Vietnam, the territory of the Indigenous Montagnards. We will also show how this story functions as a resource for reimagining our relationship with the natural and social world, supporting a cultural transition from ecological alienation to ecological alignment.

In his late thirties, while living abroad and teaching at Princeton and Columbia universities in the United States in the early 1960s, TNH, who would soon become one of the most well-known Buddhists in the world, wrote his literary memoir titled *Fragrant Palm Leaves: Journals 1962–1966*, translated into English in 1998. With no critical biography currently completed—and just a few published collections of thematic, anecdotal, autobiographical material available—the memoir covers one of the most significant periods of his life. The journals begin with entries from the mid-to-late 1950s, including accounts of the nearly two years he spent living in a mountain forest, the focus of what follows. They conclude with his return from the United States to Vietnam, where he resumed his peace and social work, and started a mindfulness community for a select-few fellow Vietnamese Buddhist volunteers called the Order of Interbeing, the precursor to what is now called the International Plum Village Community of Engaged Buddhism near Bordeaux, France.[14] Beyond the scope of the memoir, the Zen master would later be exiled from Vietnam and spend

[13] Miller, *China's Green Religion: Daoism and the Quest for a Sustainable Future*, 112.
[14] See the website for the community here https://plumvillage.org

most of the rest of his life in residence at Plum Village. If not foregrounded, looming in the background of the journal entries is the Vietnam War.

Vietnam in the mid-to-late 1950s, and on through the early-to-mid 1970s, was a catastrophe for Vietnamese communities and environments caught up in the conflict. Divided by the seventeenth parallel, Ho Chi Minh's authoritarian, pro-communist efforts on behalf of the Democratic Republic of Vietnam and Ngo Dinh Diem's authoritarian, anti-communist efforts on behalf of the Republic of Vietnam mirrored each other in both audacity and brutality, reminiscent of earlier French colonial violence. If you were caught stepping out of line on either side, forced labor at a concentration camp was the best you could hope for. Many endured arbitrary arrests and torture and were either executed or driven to suicide.[15] Remaining "sideless"—refusing to either actively support or passively endorse either side, pro-communist or anti-communist, or any faction thereof, TNH worked on reforming Vietnamese Buddhism from the village up—informed by earlier twentieth-century Chinese Buddhist revivalist values, practices, and institutions of secular education and social work—toward a more humanistic Buddhism and humane nationalism.[16] In his view, Vietnam, by which he means the peasants living in rural villages in both North and South Vietnam, desperately needed what we might call Buddhist community organizers who were more broadly educated and socially engaged, who could facilitate peasant self-reliance and sustainability, and who were committed to ending the war.

From the Zen master's journal entries, however, we know that the Buddhist leadership was not on board with his agenda. Without the benefit of other available sources, the conflict seems to be about predictable, intergenerational tensions around religious conventions and innovations, between what we would call older traditionalists and younger progressives, or more crudely, conservatives and liberals. TNH's recalls, "We were accused of sowing seeds of dissent when we challenged anything traditional. We were considered rabble-rousers who only want to tear things down … We sowed these seeds against steep odds, and while waiting for them to take root, we endured false accusations, hatred, deception, and

[15] Christopher Gosha, *Vietnam: A New History* (New York: Basic Books, 2016), 292.

[16] Elise A. DeVido, "The Influence of Chinese Master Taixu on Buddhism in Vietnam," *Journal of Global Buddhism*, vol. 10 (2009): 413.

intolerance."[17] It seems unlikely that he could have made any headway under these conditions. His influence would have to be channeled exclusively through the occasional underground circulation of his early writings on engaged Buddhism, anti-war tracts, and poetry.

The Zen master decides to reevaluate both his practice of Vietnamese Buddhism and resistance to the Vietnam War with a few friends in a relatively remote mountain forest. The inclination for such a place resonates with early Ch'an Buddhist perspectives on the healing and direction derived from more intensively experiencing flora and fauna, and winds and clouds, and from more intentionally aligning oneself with the natural world.[18] In Miller's Daoist terms, it is a desire to immerse oneself in, and more consciously and attentively interrelate with, the subjectivities of nature. This inclination is consistent with TNH's conception of Vietnamese Zen Buddhism, which combines Daoism's sense of connection with the natural world with Confucianism's sense of connection with the social world.[19] As one would expect from such a violent context, the Zen master needs healing from trauma, or the loss of connection with one's self and one's world.[20] Though he does not use the language of trauma, his feelings and emotions at the time certainly suggest such a condition. In Rodriquez's cultural terms, TNH undergoes oppression from dominant forces driven by colonial and capitalist narratives. The latter explains,

> We feel forced to comply with the dehumanizing demands of society, and we bow our heads and obey. We eat, speak, think, and act according to society's dictates. We are not free to be ourselves … We become cogs in the system, merchandise, not human beings. Our individuality is undermined, yet we comply because we lack the courage to refuse society's demands. We, too, have become so accustomed to our way of life with its conveniences and comforts that we allow ourselves to be colonized.[21]

[17] Thich Nhat Hanh, *Fragrant Palm Leaves: Journals, 1962–1966* (New York: Riverhead Books, 1999), 51.

[18] David Hinton (ed. and trans.). *Mountain Home: The Wilderness Poetry of Ancient China* (New York: New Directions, 2005).

[19] Thich Nhat Hanh, *Vietnam: Lotus in a Sea of Fire* (New York: Hill and Wang, 1967), 11.

[20] Gabor Maté, *The Myth of Normal* (Penguin Random House, 2022). Cf. also Resmaa Menakem, *My Grandmother's Hands* (Las Vegas: Central Recovery Press, 2017).

[21] Nhat Hanh, *Fragrant Palm Leaves: Journals, 1962–1966*, 92.

Elsewhere he summarizes, we were "invaded, occupied, assaulted, and destroyed by the system."[22] Healing from the resulting anxiety and despair, according to his memoir, involves reconnecting with one's body, humanity, authenticity, agency, community, and environment.

Turning to the wild Dai Lao Forest, TNH builds a hermitage amid the highest regional mountains with his friends and the Indigenous Montagnards in 1957. Surveilled by the Diem regime, the emerging community critically assesses the political and religious climate and determines its next steps in reforming Buddhism and supporting self-reliance at the village level. TNH is aware that this move locates him in a history of retreating to wild places under adverse political conditions. In his journals, he comments, "In ancient times, some Vietnamese chose to live in the highlands among wild beasts. They preferred the danger of being eaten alive to living under an oppressive regime."[23] The Zen master retreats for relief and recovery and to develop a more relevant way of being in the world for when he returns and re-engages with his peace and social work. More specifically, the turn to the wild connects him with the lives of the classical Chinese poets in the "rivers-and-mountains" tradition.

Scholar and translator of Chinese poetry David Hinton has charted the influence of Daoism on the Ch'an Buddhism of the artist-intellectual class, typically civil servants, who produced classical Chinese nature poetry.[24] Often driven by political corruption and corresponding disillusionment to retreat to wild places, these artist-intellectuals sought the raw and bracing immediacy of experience connected to both living and wandering amid mountains, forests, and rivers. Hinton explains,

> it was a complex political and personal gesture. Politically, it represented a criticism of the government in power: a refusal even to associate with such a government; a model of authenticity and simplicity for those in government whose vanity and greed were so destructive; and, finally, a kind of solidarity with the government's victims among the common people. On the personal level, retirement represented a commitment to a more spiritually fulfilling life in which one inhabits that wilderness cosmology in the most immediate day-to-day way … And this personal fulfillment had, in turn, clear political

[22] Thich Nhat Hanh, *The Raft Is Not the Shore* (Ossining, New York: Orbis Books, 2001), 129.

[23] Nhat Hanh, *Fragrant Palm Leaves: Journals, 1962–1966*, 49.

[24] Hinton, *Mountain Home: The Wilderness Poetry of Ancient China*, 6.

dimensions—for wisdom cultivated in such a recluse life was considered essential to sage governing.[25]

Turning to the wild for the above reasons is not of course unique to China and Vietnam. Many people across time and space have withdrawn to wild places with resistance, and perhaps, revolution in mind.[26] It is not clear from Hinton's work so far how such a reclusive life contributed to wiser governance, beyond whatever inspiration it may have provided public servants reading recluse poetry. What might be scaled up from an alternative wilderness community—such as social, political, economic, and ecological norms for a larger society—remains an open question.

Located at the base of the highest mountain in the region, TNH called the hermitage Phoung Boi, the name of the palm leaves on which the Buddha's teachings were written in the ancient world.[27] The name connected their lives in the forest to their Buddhist heritage, while encouraging them to renew their practice in ways relevant to the future of Vietnamese Buddhism and nationalism. What later became widely known as "engaged Buddhism"—which comprises "breathing regimens and other practices for sustaining both present attention to one's body and the social and natural world on the one hand, and present resistance to, dismantling of, and transition from systemic dehumanization and environmental degradation on the other"—originated during this time.[28] The hermitage as an alternative, experimental community drew a steady stream of friends, writers, artists, and activists. Life there followed a loose routine, without much attention paid to any rule or discipline. In addition to study, writing, and lounging around, the Zen master regularly joined others in hiking, exploring, meditating, sipping tea, reciting poetry, and making music. He eventually recovered from the depression and despondency he felt at the outset. As he recalls, "Living in the mountain forest, our strides and gestures grew bold and strong."[29] This is where TNH intensifies his encounters with and experiences of the natural world.

[25] Hinton, *Mountain Home, The Wilderness Poetry of Ancient China*, 6.

[26] See Andreas Malm, "In Wildness Is the Liberation of the World: On Maroon Ecology and Partisan Nature," Historical Materialism 26:3 (2018): 3–37.

[27] Nhat Hanh, *Fragrant Palm Leaves: Journals, 1962–1966*, 1.

[28] Victor Thasiah, "Collapsing Space and Time: Thich Nhat Hahn's Ecological Humanism," *Journal of Buddhist Ethics*, vol. 29 (2022): 33 f.

[29] Nhat Hanh, *Fragrant Palm Leaves: Journals, 1962–1966*, 28.

Having experienced alienating forms of domination and subjugation—by the Diem regime and Buddhist leadership (sharply critical of and threatened by his reform efforts, and maybe even more so by his increasing popularity)—TNH's recounts another form of domination that enchants and enlivens him.

> At night, the sacred forest declares its absolute authority … Sitting in the study at Phoung Boi, I heard the many eerie cries coming from the forest. By eight o'clock it was already night, and the forest's dominance was restored. The whole universe sank into a profound silence that, at the same time, vibrated with life. I could almost hear the majestic steps of the mountain god as he leaped between the towering trees … Some nights I stood gazing at the forest for hours. Just fifty meters away, the omnipotent forest pulled at me, with an irresistible force. It was wild and invigorating. I imagined seeing the shadowy form of a Montagnard tribesman from thousands of years ago, and I could feel the ancient tribesman in myself awakening. I felt the urge to leave civilization behind, throw away my bookish knowledge, tear off my clothes, and enter the forest naked. To do what? I didn't know. But I would enter the forest's depths. Even if wild animals devoured me, I knew I would feel no pain, terror, or regret. I might even enjoy being devoured. I stood at the window for a long time, struggling with the call of the forest and the moon.[30]

TNH perceives the mountain as at once imperial and imposing. The command of the forest suspends, relativizes, and repositions the commands of both Ho Chi Minh's Democratic Republic of Vietnam and Ngo Dinh Diem's Republic of Vietnam. The eerie, ambient noises of perhaps the gibbons, not to mention the tigers, as much as deep, surrounding silence, establish the natural world's authority, and its dominance is particularly demonstrated at night, as human awareness is heightened. Nature's pulsating immediacy transfixes and spellbinds the Zen master. The forest's omnipotence exerts an irresistible and awakening force, as opposed to the seemingly irresistible and deadening force exerted by the political regimes in conflict for control of Vietnam.

As he himself becomes Indigenous to place, TNH can feel the spirit of the Indigenous Montagnards grow within him, those who have long sought to maintain their relative independence against Vietnamese

[30] Nhat Hanh, *Fragrant Palm Leaves: Journals, 1962–1966*, 22 f.

majority domination and expectations of assimilation.[31] In the Dai Lao Forest at the time, the Montagnards resist Diem's Viet centralization policies, threatening their political autonomy and cultural continuity. While the Zen master concedes that his admiration for the Montagnards is influenced by a romanticism, it is important to consider the ecological perspective such an identification with indigeneity provides. It offers TNH critical detachment from dominant social, political, economic, and environmental worldviews, frameworks, institutions, and practices ("civilization") on the one hand, and dominant cultural forms ("bookish knowledge" and "clothes") on the other. He wants to "enter the forest naked" without such knowledge and corresponding predispositions, in order to "enter the forest's depths."

To enter the forest depths is an immersive, intensification of exposure to the subjectivities of nature, a submitting of oneself to their communications and agencies. While nature, like society, certainly poses risks of being eaten alive, the Zen master expresses that he would feel no pain, terror, or regret while subject to the natural world, feelings more readily associated with one's existence in both North and South Vietnam's political worlds. He is engrossed by the communications of the forest and moon, by what he refers to as the call of the cosmos, related to what Hinton designates as the "wilderness cosmology" of the Ch'an Buddhist, classical Chinese nature poets. This call of the cosmos attracts TNH to the subjectivities of nature and an alignment with its agencies. From the perspective of trauma studies, the call also reconnects one to one's body, and perhaps expands one's range of feeling. TNH explains,

> I heard the call from the heart of the cosmos. I wanted to turn into an areca tree or become a branch bending in the wind. I wanted to be a bird testing the strength of its wings against the wind. I wanted to run outside in the rain and scream, dance, whirl around, laugh, and cry ... My spirit was absorbed by the storm's majesty. I became one with the storm's powerful music, and I felt wonderful![32]

[31] On becoming indigenous to place, see Robin Wall Kimmerer, *Braiding Sweetgrass: Indigenous Wisdom, Scientific Knowledge and the Teachings of Plants* (Minneapolis: Milkweed Press, 2015). On the Indigenous Montagnards, see Oscar Salemink, *The Ethnography of Vietnam's Central Highlanders: A Historical Contextualization, 1850–1990* (Honolulu: University of Hawai'i Press, 2003).

[32] Nhat Hanh, *Fragrant Palm Leaves: Journals, 1962–1966*, 30.

The communication of the natural world from both living organisms and physical environments draws the Zen master into ecological alignment, even a oneness, as it draws him into an alignment, even a oneness, with himself. "When I hear [the call of the cosmos] now," he observes, "I pause, and, with all my body, with every atom of my being, every vein, gland, and nerve, I listen with awe and passion."[33] TNH recovers his humanity in such an alignment and can now continue in the resistance and reformation work.

It is difficult to read his story today and not think about increasing instances of extreme weather and storm activity, heatwaves and hurricanes, and droughts and flooding due to climate destabilization. Unlike in the Zen master's narrative above, such "absolute authority" and "calls of the cosmos" do result in pain, terror, and regret. Would a deeper alignment with such realities lead to further resentment and despondency? Or would it be beneficial in attempting to manage and minimize such pain, terror, and regret in the future? While this ecological alignment perspective connects with an ecological justice perspective in terms of calling for a more respectful and communal relationship from the human side toward living organisms and physical environments, it prioritizes humans aligning with the natural world and the latter's creative vitality, rather than the natural world aligning with colonial and capitalist human interests, such as those involving domination, exploitation, and extraction. These interests, for all of the suffering they can possibly alleviate, clearly continue to devastate both natural and social worlds.

Determining what this ecological alignment looks like in detail is complex and would require the very secular education and social work TNH's reforms called for at the time. Ecological alignment is more of a vision and aesthetic, than a policy or program. Using a koan, TNH seems to say as much.

> For someone who has seen the nature of things, knowledge gives rise to action. For those who have truly seen, there is no philosophy of action needed. There is no knowledge, attainment, or object of attainment. Life is lived just as the wind blows, clouds drift, and flowers bloom ... If asked a philosophical question, you might answer with a poem, or ask, "Have you had your breakfast? Then please wash your bowl." Or point to a mountain forest.[34]

[33] Nhat Hanh, *Fragrant Palm Leaves: Journals, 1962–1966*, 30 f.

[34] Nhat Hanh, *Fragrant Palm Leaves: Journals, 1962–1966*, 105.

Immersion in the subjectivities of the natural world is the first step. Seeing and feeling the nature of things—as the wind blows, clouds drift, and flowers bloom—and acting accordingly, without the Zen master making explicit any normative parameters, is apparently the way, the way to address the alienation driving environmental degradation and climate destabilization. The way to fold in our flourishing with the broader flourishing of the natural world, rather than pursuing the former at the expense of the latter. And, the way to shape culture—our values, behaviors, and worldviews—so that we can thrive with nature.

TNH's view can be compared to what Glen Coulthard and Leanne Betasamosake Simpson call "grounded normativity."[35] Grounded in a deep and intimate reciprocity with the natural world—the subjectivities of nature—this normativity is an ethical framework that emerges from Indigenous place-based practices and associated forms of knowledge. "Grounded normativity," they explain, "teaches us how to live our lives in relation to other people and nonhuman life forms in a profoundly nonauthoritarian, nondominating, nonexploitive manner … Our relationship to the land itself generates the processes, practices, and knowledges that inform our political systems, and through which we practice solidarity."[36] This is the kind of ecological alignment that this chapter proposes, in contrast to the ecological alienation we have considered. Nature's authority is the primary consideration, and political authority, however important, is secondary and derivative.

Even in poor and unhealthy environmental conditions, such ecological alignment—an affection for and attachment to the natural world and the subjectivities of nature—is the basis of our well-being and aliveness as we, as the Earth ourselves, take care of the Earth's capacity to take care of us. In his magnum opus on environmental stewardship, *Zen and the Art of Saving the Planet*, TNH acknowledges the challenges of this perspective:

> some of us feel tired of the Earth and find it hard to love the Earth. We may resent, blame, or reproach the Earth for bringing us into a life of such suffering. We may wish we'd never been reborn or wish to be born somewhere else. But, with deep looking, it is possible to overcome all suffering and resentment and see the true nature of the Earth and of ourselves.[37]

[35] Glen Coulthard and Leanne Betasamosake Simpson, "Grounded Normativity / Place-Based Solidarity," *American Quarterly*, Johns Hopkins University Press, Vol. 68, no. 2, (June 2016): 249–255.

[36] Coulthard and Simpson, "Grounded Normativity / Place-Based Solidarity," 254.

[37] Nhat Hanh, *Zen and the Art of Saving the Planet*, 227.

One way to see the true nature of the Earth and of ourselves, according to the Zen master, is by taking refuge in Mother Earth, what we have called an immersion in the subjectivities of nature. He explains what this means by commenting on the so-called practice of non-practice,

> We need to learn how to take refuge in Mother Earth—it is the best way to heal and nourish ourselves. We can do it if we know how to allow the Earth to *be*, within us and around us—just being aware that *we are the Earth*. And we don't have to do much. In fact, we don't have to do anything at all. It's like when we were in our mother's womb. We did not have to breathe, we did not have to eat, because our mother breathed for us and ate for us. We did not have to worry about anything. You can do the same now when you sit. Allow Mother Earth to sit for you. When you breathe, allow the Earth to breathe for you, when you walk, allow the Earth to walk for you. Don't make any effort. Allow her to do it. She knows how to do it. Don't try to do anything. Don't try to fight in order to sit. Don't try to breathe in and breathe out. Don't even try to be peaceful. Allow the Earth to do everything for you. Allow the air to enter our lungs and to flow out of our lungs. We don't need to make any effort to breathe in or breathe out. Just allow nature, allow the Earth to breathe in and out for us. And we just sit there, enjoying the breathing in and the breathing out. There is the breathing but there is no 'you' who is breathing in or breathing out. We don't need a 'you' or an 'I' in order to breathe in and out. The breathing in and the breathing out happen by themselves.[38]

This practice of non-practice recapitulates TNH's recovery in the mountain forest a half-century earlier. He took refuge in the Dai Lao Forest in the Central Highlands of Vietnam, and allowed the Earth to be around and within him, the subjectivities of nature. As he identified with the Earth, the Earth healed and nourished him. Going forward, the call of the cosmos was a call back into this state, which he would later refer to as interbeing. This is what ecological alignment-based environmental stewardship looks and feels like: conscious participation and partnership in nature-based and culture-based efforts toward making life and mutual (ecosystems, communities, and ourselves) flourishing possible.

[38] Nhat Hanh, *Zen and the Art of Saving the Planet*, 227.

Bibliography

Budiman, Abby and Neil G. Ruiz. 2021. *Key facts about Asian Americans, a diverse and growing population.* Pew Research Center, April 29, 2021. https://www.pewresearch.org/fact-tank/2021/04/29/key-facts-about-asian-americans/

Coulthard, Glen, and Leanne Betasamosake Simpson. 2016. Grounded Normativity / Place-Based Solidarity. *American Quarterly* 68 (2): 249–255.

DeVido, Elise A. 2009. The Influence of Chinese Master Taixu on Buddhism in Vietnam. *Journal of Global Buddhism* 10: 413–458.

Gosha, Christopher. 2016. *Vietnam: A New History.* New York: Basic Books.

Guha, Ramachandra. 2020. *Environmentalism: A Global History.* Essex: Pearson.

Hinton, David (ed. and trans.). 2005. *Mountain Home: The Wilderness Poetry of Ancient China.* New York: New Directions.

Khong, Sister Chan. 2007. *Learning True Love: Practicing Buddhism in a Time of War.* Berkeley: Parallax Press.

Kimmerer, Robin Wall. 2015. *Braiding Sweetgrass: Indigenous Wisdom, Scientific Knowledge and the Teachings of Plants.* Minneapolis: Milkweed Press.

King, Robert. 2001. *Thomas Merton and Thich Nhat Hanh: Engaged Spirituality in an Age of Globalization.* New York: Continuum.

Latour, Bruno. 2018. *Down to Earth: Politics in the New Climatic Regime.* Cambridge: Polity.

Loy, David. 1988. Nonduality. In *Buddhism and Beyond.* Somerville, Massachusetts: Wisdom Publications.

Malm, Andreas. 2018. In Wildness Is the Liberation of the World: On Maroon Ecology and Partisan Nature. *Historical Materialism* 26 (3): 3–37.

Maté, Gabor. 2022. *The Myth of Normal.* New York: Penguin Random House.

Menakem, Resmaa. 2017. *My Grandmother's Hands.* Las Vegas: Central Recovery Press.

Merchant, Carolyn. 2013. *Reinventing Eden: The Fate of Nature in Western Culture.* 2nd ed. Milton Park: Routledge.

Miller, James. 2017. *China's Green Religion: Daoism and the Quest for a Sustainable Future.* New York: Columbia University Press.

Nhat Hanh, Thich. 1999a. *Call Me By My True Names: The Collected Poems of Thich Nhat Hanh.* Berkeley: Parallax Press.

———. 1999b. *Fragrant Palm Leaves: Journals, 1962-1966.* New York: Riverhead Books.

———. 2008. History of Engaged Buddhism: A Dharma Talk by Thich Nhat Hanh, Hanoi, Vietnam, May 6-7, 2008. *Human Architecture: Journal of the Sociology of Self-Knowledge* 6 (3): 29–36.

———. 1999c. *The Miracle of Mindfulness: An Introduction to the Practice of Meditation.* Boston: Beacon Press.

———. 1993. On Simplicity. In *Love in Action: Writings on Nonviolent Social Change*, 113–118. Berkeley: Parallax Press.
———. 2001. *The Raft Is Not the Shore*. Ossining, New York: Orbis Books.
———. 1967. *Vietnam: Lotus in a Sea of Fire*. New York: Hill and Wang.
———. 2021. *Zen and the Art of Saving the Planet*. San Francisco: HarperOne.
Nixon, Rob. 2013. *Slow Violence and the Environmentalism of the Poor*. Cambridge: Harvard University Press.
Pellow, David. 2018. *What Is Critical Environmental Justice?* Cambridge: Polity Press.
Rodriguez, Favianna. 2020. Harnessing Cultural Power. In *All We Can Save: Truth, Courage, and Solutions for the Climate Crisis*, ed. Ayana Elizabeth Johnson and Katharine K. Wilkinson, 121–127. London: One World.
Salemink, Oscar. 2003. *The Ethnography of Vietnam's Central Highlanders: A Historical Contextualization, 1850-1990*. Honolulu: University of Hawai'i Press.
Slingerland, Edward. 2003. *Effortless Action: Wu-wei as Conceptual Metaphor and Spiritual Ideal in Early China*. Oxford: Oxford University Press.
Strain, Charles. 2014. *The Prophet and the Bodhisattva: Daniel Berrigan, Thich Nhat Hanh, and the Ethics of Peace and Justice*. Eugene, Oregon: Wipf & Stock.
Thasiah, Victor. 2022. Collapsing Space and Time: Thich Nhat Hahn's Ecological Humanism. *Journal of Buddhist Ethics* 29: 21–46.
Thu, Nguyen Tai (ed.). 2008. *The History of Buddhism in Vietnam*. The Council for Research in Values and Philosophy.

CHAPTER 13

Lines of Distinction and Circles of Connection: Toward a Holistic Epistemology

Thomas John Hastings

Introduction: The Ubiquity and Axiology of Lines and Circles

Before touching on our need to move beyond binary thinking, I want to begin with a meditation on the line and the circle, figures, symbols, and metaphors that are employed in disparate ways. First, consider the humble line, which in mathematics is defined as "a continuous extent (whether straight or curved) of length without breadth or thickness; the limit of a surface; the trace of a moving point"[1] (OED). Where I live in the so-called global north, the line appears in every sphere of life. Presidents and prime ministers draw "lines in the sand," corporations obsess over the "bottom

[1] "line, n.2", *OED Online*, September 2022, Oxford University Press. https://www.oed.com/view/Entry/108603?rskey=xh17Tk&result=2 (accessed October 27, 2022).

T. J. Hastings (✉)
International Bulletin of Mission Research, Princeton Theological Seminary, Princeton, NJ, USA
e-mail: thomas.hastings@ptsem.edu

T. J. Hastings, K.-W. Sæther (eds.), *Views of Nature and Dualism*,
https://doi.org/10.1007/978-3-031-42902-6_13

line," culture warriors on the right and left mark off lines of separation between "us and them," schools and religious communities line their respective constituents up in rows of desks or pews, computer programmers make mathematical distinctions between "zero for off" and "one for on," artists use "linear perspective" to create the illusion of three dimensions in two-dimensional space, social scientists distinguish "insider or emic" from "comparative or etic" perspectives when analyzing societies and cultures, natural scientists differentiate "matter and non-matter," "organic and inorganic," "living and non-living," philosophers differentiate "truth from belief," theologians distinguish "human and divine," and so on.

Physical, mental, and metaphorical lines are found everywhere and perhaps nowhere as commonly as in their homeland in mathematics, the discipline the Greeks believed to be the purest and highest expression of reason. While Greek belief in the rational and spiritual purity of mathematics arose out of religious and philosophical traditions, it is no longer possible to claim that mathematics is a purely abstract, virtual, or disembodied field, inasmuch as it operates within and belongs to the attentive, "conscious" interactions of the cerebral cortex with the world.[2] Even before neuroscientists had new tools to explore attention and hypothesize about consciousness, child psychologists such as Piaget observed that prelinguistic children show evidence of "embodied mathematics" when, for example, they play with sugar or salt on a table, dividing it into neat piles of roughly equivalent sizes.[3] In light of all this, it is reasonable to claim that, for *Homo sapiens*,[4] making distinctions, commonly symbolized by the line, is both a biological and a cultural adaptation.

It is worth remembering that the line has also been employed to mark time, as in a "time line," or "progress" and "development" over time depicted as a line with an arrow attached to one end. In modern Western societies, as the linear metaphor of "time as an arrow" directed into the future became dominant and was fused with notions of difference,

[2] Michael S. A. Graziano, *Rethinking Consciousness: A Scientific Theory of Subjective Experience* (New York: Norton, 2019).

[3] Catherine McCluskey, Joanne Mulligan, Penelope Van Bergen, "Noticing mathematical pattern and structure embodied in young children's play," in *Making waves, opening spaces: Proceedings of the 41st conference of the Mathematics Education Research Group of Australasia* (Adelaide: MERGA 2018).

[4] While other species clearly demonstrate arithmetical ability, it seems the jury is still out on whether or not they can do mathematics. See Jamie I. D. Campbell, *The Handbook of Mathematical Cognition* (New York and Hove: Psychology Press, 2005).

progress, hierarchy, and superiority,[5] we witnessed the appearance of a range of anthropocentric "stage" theories, constructed almost exclusively by European men, with compelling explanatory power in sociobiology, anthropology, sociology, psychology, and economics.[6] Think of taxonomies that placed the male *Homo sapiens* at the top as an example of how the metaphorical "line of progress or development" distinguished "primitive" or "inferior" from "advanced" or "superior" forms of life.

At least in the modern West, the line stands both for first-order perceptions or intuitions of duality or distinction as well and as a convenient device in constructing second-order axiologies of difference, progress, hierarchy, and superiority.[7]

For comparison and contrast, we turn now to the circle, another ubiquitous symbol, which like the line, also hails from the homeland of mathematics. Ironically, the circle is actually a curved line. Per the *OED*, the circle is "A perfectly round plane figure. In *Geometry* defined as *a plane figure bounded by a single curved line*, called the circumference, which is

[5] Even the so-called "Mars symbol," which signifies "male," is a line with an arrow protruding from a circle.

[6] Of course, post-modern, post-critical, and post-colonial theorists have attempted to deconstruct stark distinctions, binaries, and hierarchies of all kinds (racial, gender, sexual, economic, political, etc.), but quotidian experience and the current world situation (Spring, 2023) give more than ample evidence of their enduring presence and power.

[7] Also, given the social, cultural, and political power of Christianity in the West for so long, it is not surprising that stage theories bear a striking family resemblance to the inherited classical spiritual stages of progress or development (awakening, purgation, illumination, unification).

everywhere equally distant from a point within, called the K. But often applied to the circumference alone, without the included space."[8] In contrast to the line that splits, sorts, analyzes, divides, or marks discontinuity and hierarchical notions of progress, the curved line of the circle draws together, encloses, synthesizes, and unites, leaning toward continuity or familial resemblance. Think of a family circle, a circle of friends, circles of trust, circles of influence, political, social, literary, and academic circles, and so on. If the line points to difference or otherness, the circle embraces resemblance or sameness. If we were to reimagine the above taxonomy of species as a circle instead of a line, it would resemble the branches of a single tree, all deriving from a common root.[9]

In Japan, where I lived for more than two decades, and also in other non-Western societies, a higher social value is placed on the circle than the line, as it represents the ideal of continuity, harmony, peace, and wholeness. In East Asia, as with the more familiar Taoist philosophical image of the *yin-yang* (陰陽) relationality uniting all things in heaven, earth, and

[8] "circle, n.", *OED Online*, September 2022, Oxford University Press. https://www.oed.com/view/Entry/33187?rskey=ZymSMY&result=1&isAdvanced=false (accessed October 29, 2022). Italics added.

[9] "We share roughly 90 percent of our DNA with mice, dogs, cattle, and elephants. Coming closer to home, the DNA of human beings and chimpanzees is 98 to 99 percent identical. The differences between us that we (and presumably the chimps) regard as significant depend on only 1 or 2 percent of our DNA." James Franklin Crow, "Unequal by Nature: A Geneticist's Perspective on Human Differences" (Deadalus: Journal of the American Academy of Arts and Sciences, Winter 2002), 81. Of course, some would argue that this 1–2% difference in DNA and neurological capacity represents an infinitely qualitative distinction.

humanity (天地人), the preference for the circle is also seen in the Zen brush drawing of the 円相 (*ensō*) below. We will return to this open-ended image of the *ensō* below.

At this point in our discussion, we will conclude that, like the line, the circle is also a biocultural adaptation of *Homo sapiens*, symbolizing first-order perceptions or intuitions of *relationality*, as well as being a useful cultural device for constructing second-order, holistic notions of relationality, constancy, unity, and equality.

If we think of the relation between lines and circles as a continuum rather than as an antinomy, lines tend to be associated with individuality, autonomy, hierarchy, and discontinuity, while circles tend to be associated with community, belonging, equality, and continuity. Though we may have a personal or cultural bias toward one side of this continuum over the other, given our daily encounters with both symbols in the digital era, perhaps it is wisest to remain agnostic on the question of whether the line is better than the circle or vice versa. Though it would take us beyond the scope of this chapter into an exploration of current and competing theories of consciousness,[10] it seems safest to hold to the modest claim that the drawing of lines and circles is hard-wired into our brains and guided by particular sociocultural norms and conditions.

Of course, while lines and circles are often employed in neutral ways, the use of either figure may raise serious ethical questions. Here we have

[10] See Patrick House, *Nineteen Ways of Looking at Consciousness* (New York: St. Martin's Press, 2022).

in mind the ever-present, often unconscious, and difficult-to-acknowledge *habitus*[11] that guides the exercise of power in specific sociocultural contexts; that is, who is drawing a particular line or circle, and why is it being drawn in this particular way? Or who or what is included and, who or what is being excluded in this process and why? Both lines that divide and circles that enclose may exclude or marginalize other ideas, people, nonhuman species, and entire ecosystems. Such exclusions run the gamut from the more benign to the more malignant.

In reference to this volume's theme of "Views of Nature," we are now witnessing the frightening consequences of what scholar of Japanese religion and philosophy James Heisig has called our ***ad naturam*** molestation and silencing of nature's voice. In his brilliant essay, *In Praise of Civility*, Heisig says,

> From the time we learned to control fire and began to forge tools, humankind have been caught up in an argument with the forces of nature. Civilization has progressed not only by using nature's laws to its own advantage, but also by taking the law into its own hands. As we grow more conscious of how the rewards of winning have knocked nature out of the conversation, so, too, have we come to question the logic of insult at work in the background. The inflation of our self-image as masters of everything below us on the evolutionary hierarchy has arrived at a threshold that threatens civilization itself.[12]

In drawing our lines and circles, we have far too often excluded the shared, lived environment upon which we all depend for survival. Of course, this kind of homicidal and suicidal erasure has many precedents in history.

Here we want to introduce one example from American history that illustrates the deep moral ambiguity attending the drawing of lines and circles. When the Puritans settled the rolling terrain of New England in the seventeenth century, they encountered rugged fields full of massive granite deposits. They had to dig up, split, and remove countless heavy boulders to clear the land and make it suitable for English farming methods. They repurposed those field stones to build walls that redefined New

[11] In Pierre Bourdieu's terms, *habitus* is "a subjective but not individual system of internalised structures, schemes of perception, conception, and action common to all members of the same group or class." Pierre Bourdieu, *Outline of a Theory of Practice* (Cambridge: Cambridge University Press, 1977), 86.

[12] James W. Heisig, *In Praise of Civility* (Eugene: Resource Publication, 2021), 96–97.

England's rolling hills and open fields as private properties. These stone walls eventually became a symbol of the fiercely independent spirit of New England society and the Puritans' descendants. Any student of historical geography would identify the New England stone wall as a "New World" adaptation of the hedgerow and stone walls that demarcate properties in England, Ireland, Scotland, and Wales.

Employing the English legal theory of *vacuum domicilium* as a justification for this appropriation of native lands, Puritan settlers judged these lands to be vacant since they lacked the familiar fixed property markers or permanent dwellings of the "Old Country." Where native populations had once erected sacred cairns from the granite, the Puritans now super-imposed the meandering lines of their stone walls on these lands while pushing out the Algonquin peoples that included the Abenaki, Micmac, Pennacook, Pequot, Mohegan, Nauset, Narragansett, Nipmuc, Woronoco, and Wampanoag. Tragically, this massive land grab took place in the midst of a smallpox epidemic, which, according to one estimate, reduced the native population from around 90,000 to around 10,000 between 1600 and 1674.[13]

Against the background of this ignominious history, the celebrated poet Robert Frost, himself a descendant of an early Massachusetts settler, wrote a meditation on those walls called "Mending Wall,"[14] which expresses the generic human ambivalence about walls. On the one hand, the narrator begins, "Something there is that doesn't love a wall." Harsh New England winters wreaked havoc on these inherited property lines, so mending the walls was a necessary but wearisome Spring ritual for the descendants of the Puritans. After repairing the wall and reconfirming their respective boundaries, the more practical neighbor in Frost's poem concludes, "Good fences make good neighbors."

Frost's mediation touches on the sense of security we feel when we are part of the "inner circle." How we draw lines and circles around our individual selves or group of belonging contributes significantly to our sense of identity. Still, it is easy to forget that every wall that protects also excludes. Walls signal distinctions between "insiders" and "outsiders," "us" and "them," "natives" and "foreigners." Notably, the indigenous

[13] John Peacock, "Principles and Effects of Puritan of Appropriation of Indian Land and Labor," *Ethnohistory*, Vol. 31, No. 1 (Winter), 39–44 (Durham: Duke University Press, 1984).

[14] To see the whole poem, see https://www.poetryfoundation.org/poems/44266/mending-wall

peoples who lived on these lands for generations are absent in the Frost poem. Indeed, Frost legitimates, albeit unwittingly, his ancestors' land grab that redrew the spatial boundaries and redefined time in the so-called "New World." Native New Englanders over 60 years old, like the present author, recall being taught in school sentimental legends about the convivial relations between the first Puritans and the native peoples, legends centered around a harvest festival now glossed with the quasi-religious name of Thanksgiving.[15] Not surprisingly, later generations of formerly enslaved Blacks from the Southern US and Roman Catholic and Jewish immigrants from Europe also found themselves outside the well-established walls of New England society. They had to carve out their own social lines and circles to survive as "acceptable outsiders."

To repeat, while you or I may have a personal or cultural preference for the circle over the line, or vice versa, we must face the moral and ethical consequences of how we draw our lines and circles. In light of the subject of this volume, perhaps the most pressing question for us today is, "How might we rethink lines and circles in ways that move beyond binary logics and exclusionary ethics and press our perceptions and ideas of difference and relationality into the service of life-giving and life-sustaining ends for all inhabitants of this tiny, threatened planet?"

Educational Systems, Epistemologies, and the Exile of Theology

Next, I want to touch briefly on academia, another sphere of action redefined by how we draw lines and circles. It is clear that Western modernity, especially in its long-dominant colonizing, technologizing, and capitalizing expressions, has preferred intellect, reason, and analysis over instinct, affect, and synthesis. At the risk of oversimplification, it is safe to say that Science, Technology, Engineering, and Mathematics (STEM fields) are metaphorically more "linear" and associated with intellect, reason, and analysis. At the same time, the arts and humanities are metaphorically

[15] Growing up in central Massachusetts, we inherited some vague sense of nostalgic reverence for the pre-colonial inhabitants of our suburban habitat. We romanticized those ancient inhabitants roaming the hills as ghost-like playmates we knew only by the generic label "Indians." We were "Pow Wow the Indian Boy," the Saturday morning cartoon character who "loved all the animals in the woods," treading softly through pine forests to avoid disturbing our non-human friends. We were archeologists excavating a craggy granite outcrop we called "Indian Cliffs," where one friend had dug up an exquisite flint arrowhead.

more "circular" and associated with instinct, affect, and synthesis. In part, these distinctions are a consequence of the hyper-specialization of the academic curriculum that came about as a response to the explosion of knowledge and the transition from the medieval university, with its "unified" curriculum (*trivium and quadrivium*) as the basis for training professionals in medicine, mathematics/natural philosophy/science, law, and religion, to the modern research university, which developed increasingly specialized curricula in a wide variety of subdisciplines in the humanities, natural sciences, and social sciences.

With the ascendency of modern science—again a relatively recent phenomenon—intellect, reason, and analysis have been valued in higher education over instinct, affect, and synthesis as better and more accurate epistemological strategies. Along the way, there have been a few in the West who questioned such a view, such as French philosopher Henri Bergson, whose notion of the *elan vital* attempted to reorder the relationship of intellect and instinct. While acknowledging the positive role of intellect, Bergson hypothesizes an organic evolution of consciousness that gives instinct an advantage over intellect.

> We see that the intellect, so skillful in dealing with the inert, is awkward the moment it touches the living. Whether it wants to treat the life of the body or the life of the mind, it proceeds with the rigor, the stiffness and the brutality of an instrument not designed for such use ... Instinct, on the contrary, is molded on the very form of life. While intelligence treats everything mechanically, instinct proceeds, so to speak, organically. If the consciousness that slumbers in it should awake, if it were wound up into knowledge instead of being wound off into action, if we could ask and it could reply, it would give up to us the most intimate secrets of life.[16]

While Bergson's view has been contested by leading Western philosophers, most notably Bertrand Russell,[17] we will turn below to Kagawa Toyohiko who was influenced by Bergson's notion, albeit from his very different location in Japan.[18]

Returning to the claim that modern higher education in the West, and in its Westernizing expressions far beyond the geographical West, opted

[16] Henri Bergson, *Creative Evolution* (New York: Holt and Company, 1911), 165.

[17] Bertrand Russell, *The Philosophy of Bergson. With a Reply by Mr. H. Wildon Carr* (London: Macmillan and Co., Ltd., 1914).

[18] Bergson and other "vitalist" thinkers may have grasped certain aspects of recent neuroscience-based theories long before the advent of brain imaging technologies.

for intellect over instinct, we offer the cautionary story of Aum Shinrikyo, an apocalyptic Japanese religious group led by the late Asahara Shoko who was executed in 2018 for his role in the 1995 gas attacks in Tokyo.[19] On the morning of March 20, 1995, five of Asahara's disciples boarded Tokyo subways at his direction and released the deadly nerve agent sarin, killing 12 and injuring more than 1000 people. Besides the subway attack, Aum followers used poisonous chemicals to hurt other people on several occasions. At least 11 of Asahara's highest-raking followers—effectively his "cabinet members"—were educated in science and engineering at Japan's elite universities, and many of the group's devotees were also graduates from these leading universities. As a society that places such a high value on education, the Japanese public was shocked when they realized that several leaders behind the deadly sarin gas attack were graduates of Japan's best universities. Some scholars tried to explain the bizarre religious justifications that led to Aum's deadly attacks. Still, there is little doubt that the nation's confidence in its educational system was dealt a severe blow. Of course, we should be cautious not to overgeneralize from a single case, but the Aum incident raises serious questions about the adequacy of education centered purely on STEM subjects.

While STEM subjects were ascending, humanities such as religious studies, philosophy, and theology were becoming increasingly specialized and distant from sites of embodied practice. Take contemporary New Testament scholarship within the theological curriculum as one example of this hyper-specialization. There you will find synoptic specialists on Matthew, Mark, or Luke, Johannine specialists, Pauline and sub-Pauline specialists, Pastoral Epistle specialists, Apocalyptic specialists, and more. As a consequence of this specialization, theological studies were divorced from concrete sites of professional practice where intellect and intuition were once presumed to come together. Such was the rationale for placing practical or "applied subjects" at the tail end of a clerical education centered on "theoretical subjects" in bible, history, and dogmatics.[20] The following image of isolated circles divided by clear lines reflects the current reality in academic theology and other humanities.

[19] For a thorough religious studies treatment of this incident, see Ian Reader, *Religious Violence in Contemporary Japan: The Case of Aum Shinrikyo* (London: Routledge, 2003).

[20] This pattern continues in many divinity schools and seminary around the world today.

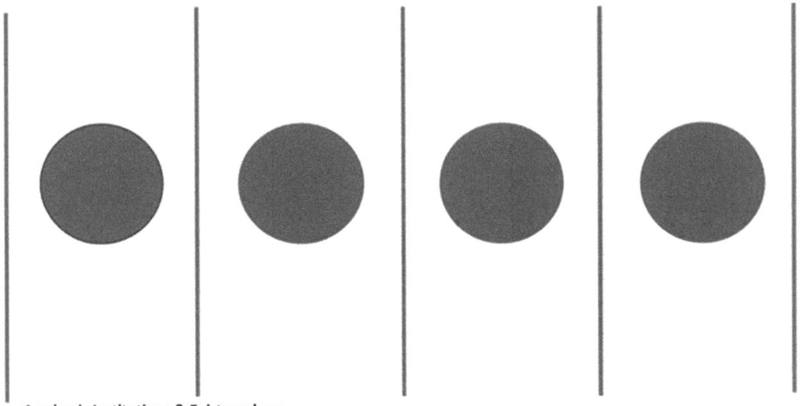

As a consequence of these lines, scholars in biblical, historical, systematic, and practical fields speak different languages because, with some notable exceptions, the primary audience for theological scholarship today is other theological specialists within one's particular academic guild. In contrast, the primary audience for religious professionals is a community of faith, where clergy and laypeople perform various leadership roles together as preachers, teachers, counselors, worship leaders, elders, deacons, and so on. Sharp lines have been drawn around biblical, historical, systematic, and practical fields, perhaps except for practical theology, which must account empirically for contexts of lived religion while also considering normative claims. Unsurprisingly, specialization has opened a deep chasm between the theological academy and religious communities.[21] As a former professor in a leading Japanese Protestant seminary and associate director at a once renowned theological think tank, I grew weary of theologians apologizing for their narrow research agendas or inability to offer practical guidance to leaders of religious communities. While many recognize this problem, we are still awaiting a new vision for integrating study and practice within a particular religious tradition.

[21] In my view, the deep chasm in the United States between the Protestant theological academy and lived religion in communities of faith is a major reason that 80% of white evangelicals in the United States voted to elect a reality TV star as president in 2016. I have described this problem in terms of hardened epistemological divisions between religious subjectivists, activists, and objectivists in my recent book for educational ministers, pastors, and practical theologians, entitled *Worshiping, Witnessing, and Wondering: Christian Wisdom for Participating in the Mission of God* (Cascade, 2022).

While theology and other humanities continue to strive for more integrative paradigms, the shared anthropogenic threats to our planet—environmental destruction, economic inequality, pandemic, war, unregulated capitalism, and artificial intelligence—are front and center in the daily news. Responding to these threats, some in academia are beginning to push back against the consequences of hyper-specialization. Today, as more and more funding is being allocated to STEM fields and less and less to humanities, some are trying to open the doors and windows of disciplinary citadels through participation in multidisciplinary, cross-disciplinary, interdisciplinary, and transdisciplinary research circles.

Transdisciplinary circles convene scholars from a range of fields in response to one of the threats mentioned above.[22] In spite of these noble efforts, academia still languishes under the privileging of intellect, analysis, and reason over intuition, affect, and synthesis. And not surprisingly, theologians and religious practitioners are *de jure* excluded from such efforts. Indeed, when we cannot even renegotiate the lines dividing our own subdisciplines, why should we expect to be invited to speak with colleagues in other fields?

Softening the Line and Expanding the Open Circle

I will conclude by briefly introducing two majority world exemplars of what I call a "holistic epistemology." First, we will consider the case of Kagawa Toyohiko (1888–1960), the Japanese evangelist, social reformer, artist, scientific mystic, and prolific writer, and then we will look at a painting by Sri Lankan artist Nalini Jayasuriya (1927–2014) and compare and contrast it with an iconic image from the history of Western art. Like the open-ended *ensō*, I want to suggest that the "holistic epistemology" of non-Western thinkers and artists like Kagawa and Nalini may hint at a path beyond the harsh binaries of the line and the exclusionary ethics of the closed circle, which have been so characteristic of human and interspecies relations during the Anthropocene.

[22] For a philosophical approach to transdisciplinarity, which is close to what I have in mind, see Basarab Nicolescu, *Manifesto of Transdisciplinarity*, translated by Karen-Claire Voos (New York: SUNY Press, 2002).

First, we must situate Kagawa's life from 1888–1960 within Meiji, Taisho, and Showa Japan. Kagawa was a second-generation Japanese convert to Protestant Christianity, a faith to which he had been drawn as a junior high school student in Tokushima after meeting American Presbyterian missionaries. Like many first-generation Japanese Protestant converts and church leaders from former samurai clans disenfranchised in 1876 by the Meiji oligarchy, Kagawa was also born into a former samurai family. However, first- and second-generation Japanese Protestant leaders came of age within starkly different social, cultural, and political circumstances. After American gunboats had forced Japan's reopening to commerce with the West (1853–), the 250-year-old Tokugawa *Bakufu* (Shogunate) collapsed, and the new Meiji government was established in 1868. During an era of dramatic change, the nation and the people were groping to develop a new identity. Having been reintroduced by missionaries from the West, Christianity offered one possible path forward for ambitious young men from former samurai families. By contrast, during the formative years of second-generation Protestant leaders like Kagawa, Japan was emerging as East Asia's unrivaled military and industrial leader, defeating China in 1894–1895 and Russia in 1904–1905. There was a growing sense of national pride and self-confidence that Japan could compete on an equal footing with the dominant Western imperial and colonial powers.

As a lonely child, Kagawa found solace and purpose in exploring the rivers, fields, and mountains of Tokushima on the island of Shikoku. As a sensitive teenager, he was drawn to the parables of Jesus that spoke to him of a deep intimacy between the human and natural worlds common in agrarian societies.[23] Just as Jesus had taken his parables from the natural environment of Palestine, Kagawa situated his own sense of the nature-human-divine relationality in his own *Sitz im Leben*.

> All creation is mine. I live by penetrating every created thing. In the kitchen I am one with the spirit of the fire, one with the spirit of the water, one with the spirit of the blazing range. All things charm me, and I am fused with everything. I can dwell in the soot in the chimney or find a peaceful place with the flea under the tatami mat. Set free, I can fly upward to the Great Bear Constellation, frolic from star to star, or hide myself under the mirror on my lover's dresser. As long as I love the whole creation I can travel about it with the utmost freedom. Mount Fuji and the Japan Alps are wrinkles on my brow, the Atlantic and Pacific Oceans are my robes. All creation is mine, the earth is one part of my body, and I hold the solar system in the palm of my hand and scatter millions of stars across the heavens. When God gave me Christ, all creation was thrown in as a gift.[24]

Anxious about the rapid pace of Japan's industrialization and militarization, Kagawa embraced the pacifism and agrarian romanticism of Tolstoy and others, as evidenced by his refusal to participate in the military training exercises at Tokushima Junior High School and by his study of biology.[25] Kagawa felt called to practice the redemptive love of God he saw in Jesus Christ. Further, having absorbed the writings of Christian socialists and cooperative theorists and practitioners, Kagawa made a dramatic decision to live out his adopted faith by making his home from 1909–1923 in Shinkawa Fukiai, Kobe's worst slum. Except for two years of study at Princeton Theological Seminary and Princeton University, he and his wife Haru engaged their neighbors in evangelism, childhood education, settlement work, medical care, and microfinancing. They became leaders of a cooperative movement seeking to improve conditions for industrial laborers, consumers, and farmers. While Japanese church leaders were flummoxed by this anomalous flurry of social reform activities in and beyond

[23] Thomas John Hastings, *Seeing All Things Whole: The Scientific Mysticism and Art of Kagawa Toyohiko (1888–1960)* (Eugene: Pickwick, 2015), 17–47.

[24] Kagawa Toyohiko, *A Few Words in the Dark: Selected Meditations*, translated and annotated by Thomas John Hastings (Princeton: Bridges to Peace Publications, 2015), 102.

[25] Kagawa, *A Few Words in the Dark*, 32.

the slums, the Kagawa story attracted young readers when he published a best-selling semi-autobiographical novel in 1920 and Kagawa became a household name. While he abandoned his pacifism during the late 1930s,[26] after WWII he was nominated twice for the Nobel Prize in Literature and four times for the Noble Peace Prize.

Kagawa developed a lifelong passion for studying the latest scientific findings because of his love and affinity for nature. He argued that faith should welcome science's emerging insights into nature's physical, chemical, and biological properties and processes. Fascinated by what he called "revelatory" scientific breakthroughs, he devoured the latest physics, chemistry, and biology books. Nurtured in a mostly Buddhist ethos, he wove together Taoist, Confucian, and Christian ideas with Bergson's vitalism and Borden Parker Bowne's personalism. He described the evolutionary endowment of consciousness as the key to unifying science, religion, and philosophy.[27] For Kagawa, who called himself a "scientific mystic," consciousness is a window that opens up the mysteries of nature, and nature is a window that opens up the mystery of ultimacy or God.

> … I believe that nothing has done so much to lay bare the world of mystery as modern science. The reason science has lost its sense of mystery is because it is disconnected from life. Once we realize life inspires science, it will be clear that mechanism, law, and reason are all supporters of life and thus windows opening into the world of mystery. I am a scientific mystic. The more scientific I am, the more deeply I feel I am breaking into the divine realm. Especially in the field of biology do I feel as though I am participating in a direct interview with God. Through life, I discover purpose even in the sphere of mechanism. Science is the mystery of mysteries. It is the divine revelation of revelations.[28]

Kagawa wrote these words in 1926, ten years after graduating from Princeton Theological Seminary and completing several advanced biology, mathematics, psychology, and philosophy courses at Princeton University.[29]

[26] Bo Tao, *Imperial Pacifism: Kagawa Toyohiko and Christianity in the Asia Pacific War*, Yale University Dissertation, 2019.

[27] Kagawa, *Seeing All Things Whole*, 48–105.

[28] Kagawa, *A Few Words in the Dark*, 67.

[29] Kagawa, *Seeing All Things Whole*, 124–132.

Before the 1925 Scopes Trial, a flash point for the so-called Fundamentalist-Modernist split that transformed American Protestantism and sent ripple effects around the world through the agency of missionaries and global dissemination of literature, Protestant views on natural selection and other recent scientific theories were far from uniform, running the gamut from enthusiastic support to outright rejection, with varying degrees of moderation in between. Some saw in natural selection the potential for a bold modern synthesis of biblical faith and reason, while others countered positivist assertions that modern science would solve all of humanity's problems. One way to retrieve a more nuanced picture of this range of views is to explore how majority world Christians like Kagawa viewed nature and how they saw the relationship between their adopted faith, inherited religious and philosophical traditions, and modern science.

This positive approach to Japan's religious and philosophical traditions and modern science caused some to wonder about Kagawa's view of God. Responding to certain Japanese church leaders and theologians who were worried that Kagawa might be a pantheist, he wrote,

> Someone may ask this question: Are God and the universe one? And are God and human beings one? Pantheism takes that stand. But I am not a pantheist. I am an advocate of the Holy Spirit. No, beyond that I am one who rejoices in the Spirit-filled life. Is the child living in the womb identical to the mother? Although conceived in the mother, the child is a different person from the mother. The mother transcends the child. Still, the child is living in the mother. And the child comes from the mother. In like manner, the absolute God transcends human beings while embracing human beings, and human beings are created by God. We can think of the relation of God and the universe in the same way. The material world is not itself God. But God transcends it, dwells in it, and through it manifests Godself. I wonder if it is not most appropriate to think of the material world as the garment of God.[30]

Like French Jesuit Teilhard de Chardin, to whom he has been compared but whose work he did not know, Kagawa pursued a positive

[30] Toyohiko Kagawa, *Meditations on the Holy Spirit*, translated by Charles A. Logan (Nashville: Cokesbury, 1939), 345.

rapprochement between Christian faith and natural science.[31] Unlike Teilhard, however, who was a trained paleontologist, Kagawa's calling was not as a scientist but as a Christian evangelist. He chastised Western missionaries for failing to consider Japanese holistic views of nature.

> So closely do the Japanese feel themselves akin to nature and her ways that their thought of God takes on cosmic dimensions. Therefore, to bring home to the Japanese Christ's revelation of God as Father we must teach, as Ostwald does, that matter also has definite direction. We must show, as Driesch does, that there is harmony among organic bodies. We must stress the fact, as J. H. Fabre does, that God's purpose is built even into animal intuition. If we fail to make these things convincingly clear the Japanese will not believe in a God of love. Even the urbanized Japanese are never weaned from nature. A religion, therefore, which fails to interpret nature will not win their allegiance. No matter how much they are taught regarding human love, unless shown that there is love in nature they will not find faith possible.[32]

Kagawa's willingness to embrace seemingly contradictory epistemological positions anticipates the "transdisciplinary viewpoint" advocated by Nicolescu and others. "The transdisciplinary viewpoint allows us to consider a multidimensional Reality, structured by multiple levels replacing the single-level, one-dimensional reality of classical <Western> thought."[33]

As a way of concluding this chapter, I will introduce a painting by Nalini Jayasuriya (1927–2014), the late Sri Lankan artist.[34] Like Kagawa, Nalini identified with a minority Christian community in a mostly Buddhist society and had a positive appreciation for Sri Lankan philosophical, artistic, and religious traditions. Here is how she described what she calls her "multidimensional Reality." "I come from a land of rich, ancient, and diverse cultures and traditions. While I carry the enriching influences of

[31] Hideshi Kishi, "The Religious Aspects of Cosmic Consciousness: A Comparison of Pierre Teilhard de Chardin and Toyohiko Kagawa" *Christian Century* 87 (1970), 1533–1536.

[32] Kagawa Toyohiko, *Christ and Japan*, translated by William Axling (New York: Friendship, 1934), 40–41.

[33] Nicolescu, *Manifesto of Transdisciplinarity*, 49.

[34] Nalini was OMSC's first artist in residence from 2001–2003 (see https://omsc.ptsem.edu/artist-jayasuriya/)

both East and West, I express myself through an Asian and Christian consciousness with respect for all confessions of religious faith."[35]

The Nalini painting, entitled "Birds Sing, St. Francis Dances," offers a playful, subversive, and contemporary counterpoint to Giotto's famous fresco "St. Francis Preaching to the Birds."

[35] Nalini Marcia Jayasuriya, *A Time of My Singing: Witness of a Life* (New Haven: OMSC Publications, 2004), 8.

Giotto painted his iconic image around 1299, about 70 years after the canonization of Francis. Accompanied in the forest by one of his devoted disciples, Giotto's golden-haloed Francis bends down, bestowing a blessing on the birds as he preaches. For their part, the eager birds line up and attentively face the saint as if gathering in pews under a huge tree that doubles as a roof in this arcadian cathedral. Like the legendary saint, Giotto is not without humor, as he depicts some tardy birds swooping down to join the gathered congregation. Though the colors are muted, Giotto's fresco is a warm, beautiful, and touching scene depicting the artist's reverence for the great saint of Assisi, whose influence was spreading rapidly across Christendom.

Nalini knew Giotto's painting well, yet she invites us to imagine the great saint from a different epistemological, aesthetic, and spiritual perspective. The title, "Birds Sing, St. Francis Dances," captures the playful subject-object reversal she seeks to evoke. Here, it is neither the word nor the blessing of the haloed male saint that charms the birds; rather, the colorful circling birds set the androgynous saint's feet to dancing.[36] In contrast to the muted fresco of Giotto, Nalini chose bright pastels to illustrate the sacred vibrancy of creation, eliding borders between natural and human worlds and merging them in a single plane. This whirling dance wraps the birds and the saint in a shared halo. Is it the birds who reflect the saint's multi-colored robe, or is the robe a reflection of the multi-colored birds? She invites us to imagine the latter possibility. Indeed, suppose we were to draw lines between the evolutionary age of modern birds and modern homo sapiens. In that case, we should be humbled by our relative youth, having appeared just 160,000 years ago compared with our feathered friends who showed up around 60 million years ago. Coming from a Sri Lankan, mostly Buddhist context where ancestors and elders are revered, Nalini's painting is a fitting invitation for us to awaken to the deep interdependence and relationality of human and non-human species. As we seek new paths that challenge the anthropogenic threats to our shared habitat, does this not point to the epistemological, aesthetic, and spiritual metanoia that is called for?

Bibliography

Bergson, Henri. 1911. *Creative Evolution*. New York: Holt and Company.

Tao, Bo, and Imperial Pacifism. 2019. *Kagawa Toyohiko and Christianity in the Asia Pacific War*. Yale University Dissertation.

Bourdieu, Pierre. 1977. *Outline of a Theory of Practice*. Cambridge: Cambridge University Press.

Campbell, Jamie I.D. 2005. *The Handbook of Mathematical Cognition*. New York and Hove: Psychology Press.

Crow, James Franklin. 2002. Unequal by Nature: A Geneticist's Perspective on Human Differences. *Deadalus: Journal of the American Academy of Arts and Sciences*.

Graziano, Michael S.A. 2019. *Rethinking Consciousness: A Scientific Theory of Subjective Experience*. New York: Norton.

Hastings, Thomas John. 2015. *Seeing All Things Whole: The Scientific Mysticism and Art of Kagawa Toyohiko (1888–1960)*. Eugene: Pickwick.

[36] Nalini typically depicts male figures from Scripture or church history as women expressing her hope to break open sacred stories to include those who have far too often been excluded.

———. 2022. *Worshiping, Witnessing, and Wondering: Christian Wisdom for Participating in the Mission of God*. Cascade.

Heisig, James W. 2021. *In Praise of Civility*. Eugene: Resource Publication.

House, Patrick. 2022. *Nineteen Ways of Looking at Consciousness*. New York: St. Martin's Press.

Kagawa Toyohiko. 1934. *Christ and Japan*. Translated by William Axling. New York: Friendship.

Kagawa, Toyohiko. 1939. *Meditations on the Holy Spirit*. Translated by Charles A. Logan. Nashville: Cokesbury.

———. 2015. *A Few Words in the Dark: Selected Meditations*. Translated and annotated by Thomas John Hastings. Princeton: Bridges to Peace Publications.

Kishi, Hideshi. 1970. The Religious Aspects of Cosmic Consciousness: A Comparison of Pierre Teilhard de Chardin and Toyohiko Kagawa. *Christian Century* 87: 1533–1536.

Jayasuriya, Nalini Marcia. 2004. *A Time of My Singing: Witness of a Life*. New Haven: OMSC Publications.

McCluskey, Catherine, Joanne Mulligan, and Penelope Van Bergen. 2018. Noticing mathematical pattern and structure embodied in young children's play. In *Making waves, opening spaces: Proceedings of the 41st conference of the Mathematics Education Research Group of Australasia*. Adelaide: MERGA.

Nicolescu, Basarab. 2002. *Manifesto of Transdisciplinarity*, translated by Karen-Claire Voos. New York: SUNY Press.

Peacock, John. 1984. Principles and Effects of Puritan of Appropriation of Indian Land and Labor. *Ethnohistory* 31 (1): 39–44.

OED Online, September 2022, Oxford University Press. https://www.oed.com/view/Entry/108603?rskey=xh17Tk&result=2. Accessed 27 Oct 2022.

Reader, Ian. 2003. *Religious Violence in Contemporary Japan: The Case of Aum Shinrikyo*. London: Routledge.

Russell, Bertrand. 1914. *The Philosophy of Bergson. With a Reply by Mr. H. Wildon Carr*. London: Macmillan and Co., Ltd.

Index[1]

[1] Note: Page numbers followed by 'n' refer to notes.

T. J. Hastings, K.-W. Sæther (eds.), *Views of Nature and Dualism*,
https://doi.org/10.1007/978-3-031-42902-6

Q

R

S

SPRINGER NATURE

GPSR Compliance

The European Union's (EU) General Product Safety Regulation (GPSR) is a set of rules that requires consumer products to be safe and our obligations to ensure this.

If you have any concerns about our products, you can contact us on ProductSafety@springernature.com

In case Publisher is established outside the EU, the EU authorized representative is:

Springer Nature Customer Service Center GmbH
Europaplatz 3
69115 Heidelberg, Germany

Kolibri 360 GmbH
Ferdinand-Jühlke-Straße 7
99095 Erfurt, Deutschland
produktsicherheit@kolibri360.de